MODERN CHINESE FOODWAYS

Food, Health, and the Environment

Robert Gottlieb, Henry R. Luce Professor of Urban and Environmental Policy, Occidental College

Nevin Cohen, associate professor, City University of New York (CUNY) Graduate School of Public Health

Keith Douglass Warner, *Agroecology in Action: Extending Alternative Agriculture through Social Networks*

Christopher M. Bacon, V. Ernesto Méndez, Stephen R. Gliessman, David Goodman, and Jonathan A. Fox, eds., *Confronting the Coffee Crisis: Fair Trade, Sustainable Livelihoods and Ecosystems in Mexico and Central America*

Thomas A. Lyson, G. W. Stevenson, and Rick Welsh, eds., *Food and the Mid-Level Farm: Renewing an Agriculture of the Middle*

Jennifer Clapp and Doris Fuchs, eds., *Corporate Power in Global Agrifood Governance*

Robert Gottlieb and Anupama Joshi, *Food Justice*

Jill Lindsey Harrison, *Pesticide Drift and the Pursuit of Environmental Justice*

Alison Alkon and Julian Agyeman, eds., *Cultivating Food Justice: Race, Class, and Sustainability*

Abby Kinchy, *Seeds, Science, and Struggle: The Global Politics of Transgenic Crops*

Vaclav Smil and Kazuhiko Kobayashi, *Japan's Dietary Transition and Its Impacts*

Sally K. Fairfax, Louise Nelson Dyble, Greig Tor Guthey, Lauren Gwin, Monica Moore, and Jennifer Sokolove, *California Cuisine and Just Food*

Brian K. Obach, *Organic Struggle: The Movement for Sustainable Agriculture in the U.S.*

Andrew Fisher, *Big Hunger: The Unholy Alliance between Corporate America and Anti-Hunger Groups*

Julian Agyeman, Caitlin Matthews, and Hannah Sobel, eds., *Food Trucks, Cultural Identity, and Social Justice: From Loncheras to Lobsta Love*

Sheldon Krimsky, *GMOs Decoded: A Skeptic's View of Genetically Modified Foods*

Rebecca de Souza, *Feeding the Other: Whiteness, Privilege, and Neoliberal Stigma in Food Pantries*

Bill Winders and Elizabeth Ransom, eds., *Global Meat: The Social and Environmental Consequences of the Expanding Meat Industry*

Laura-Anne Minkoff Zern, *The New American Farmer: Immigration, Race, and the Struggle for Sustainability*

Julian Agyeman and Sydney Giacalone, eds., *The Immigrant-Food Nexus: Food Systems, Immigration Policy, and Immigrant Foodways in North America*

Benjamin R. Cohen, Michael S. Kideckel, and Anna Zeide, eds., *Acquired Tastes: Stories about the Origins of Modern Food*

Karine E. Peschard, *Seed Activism: Patent Politics and Litigation in the Global South*

Jennifer Gaddis and Sarah Robert, eds., *Transforming School Food Politics around the World*

Mariana Chilton, *The Painful Truth about Hunger in America: Why We Must Unlearn Everything We Think We Know—and Start Again*

Jia-Chen Fu, Michelle T. King, and Jakob A. Klein, eds., *Modern Chinese Foodways*

MODERN CHINESE FOODWAYS

EDITED BY JIA-CHEN FU, MICHELLE T. KING, AND JAKOB A. KLEIN

The MIT Press
Cambridge, Massachusetts
London, England

© 2025 Massachusetts Institute of Technology

This work is subject to a Creative Commons CC-BY-NC-ND license.

This license applies only to the work in full and not to any components included with permission. Subject to such license, all rights are reserved. No part of this book may be used to train artificial intelligence systems without permission in writing from the MIT Press.

The MIT Press would like to thank the anonymous peer reviewers who provided comments on drafts of this book. The generous work of academic experts is essential for establishing the authority and quality of our publications. We acknowledge with gratitude the contributions of these otherwise uncredited readers.

This book was set in Stone Serif and Stone Sans by Westchester Publishing Services. Printed and bound in the United States of America.

Library of Congress Cataloging-in-Publication Data

Names: Fu, Jia-Chen, editor. | King, Michelle Tien, editor. | Klein, Jakob, editor.
Title: Modern Chinese foodways / edited by Jia-Chen Fu, Michelle T. King, and Jakob A. Klein.
Description: Cambridge, Massachusetts ; London, England : The MIT Press, [2025] | Series: Food, health, and the environment | Includes bibliographical references and index.
Identifiers: LCCN 2024017185 (print) | LCCN 2024017186 (ebook) | ISBN 9780262551311 (paperback) | ISBN 9780262381642 (pdf) | ISBN 9780262381659 (epub)
Subjects: LCSH: Food—China—History. | Food habits—China—History. | Food industry and trade—China—History. | Food consumption—China—History. | Cooking, Chinese—History | Cultural fusion—China.
Classification: LCC TX360.C6 M63 2025 (print) | LCC TX360.C6 (ebook) | DDC 641.300951—dc23/eng/20240611
LC record available at https://lccn.loc.gov/2024017185
LC ebook record available at https://lccn.loc.gov/2024017186

10 9 8 7 6 5 4 3 2 1

CONTENTS

SERIES FOREWORD   IX
FOREWORD   XI
E. N. Anderson
ACKNOWLEDGMENTS   XVII

    INTRODUCTION   1
    Jia-Chen Fu, Michelle T. King, and Jakob A. Klein

I   CREATING VALUE: MARKET STRUCTURES AND COMMODITIES

1   LORD MILLET IN ALIBABA'S CAVE: THE RESURRECTION OF AN ICONIC FOOD   25
    Francesca Bray

2   BECOMING AN EVERYDAY FOOD: SOY SAUCE IN MODERN CHINA (CA. 1800–1930)   49
    Angela Ki Che Leung

3   BEEF IN CHINA: A HISTORY IN EIGHT DISHES   69
    Thomas DuBois

II   SYSTEMATIZING EXPERTISE: FOOD, SCIENCE, AND TECHNOLOGY

4   ABSORBING VITAMINS: HOW A NUTRITIONAL PARADIGM WAS REINVENTED IN REPUBLICAN CHINA   89
    Hilary A. Smith

5   TASTE 100 HERBS: MATERIAL SCARCITY AND LOCAL PLANT KNOWLEDGE IN THE MAO-ERA CAMPAIGN FOR NATIVE PESTICIDES   109
Sigrid Schmalzer

6   FOOD DELIVERY, THE PLATFORM ECONOMY, AND DIGITAL CULTURE IN CHINA: THE HUMAN-NONHUMAN ENTANGLEMENT OF URBAN CHINESE FOODWAYS   129
Fan Yang

III   CONSTRUCTING CULINARY IDENTITIES: GENDER, NATION, AND ETHNICITY

7   DOMESTIC COOKBOOKS AND THE EMERGENCE OF FEMALE CULINARY AUTHORITY IN TWENTIETH-CENTURY CHINA   151
Michelle T. King

8   TAIWANESE CUISINE AND NATIONHOOD IN THE TWENTIETH CENTURY   169
Yujen Chen

9   GETTING SMASHED: DRINKING AND ETHNIC SPACE CONSTRUCTION IN A HUBEI TOURIST SPOT   191
Xu Wu

10   GASTROGRAPHISM IN CONTEMPORARY CHINA: FROM A LITERARY PASTIME TO A PROFESSIONAL ACTIVITY   209
Françoise Sabban

IV   "CHINESENESS" IN MOTION: MIGRATION AND MOBILITIES

11   JAPANESE CUISINE IN CHINESE FOODWAYS   229
James Farrer and Chuanfei Wang

12   CHIFAS: HOW CHINESE FOOD BECAME A PERUVIAN NATIONAL TREASURE   253
Lok Siu

13   PIGS FROM THE ANCESTORS: CANTONESE ANCESTRAL RITES, LONG-TERM CHANGE, AND THE FAMILY REVOLUTION   271
James L. Watson

AFTERWORD: CHINESE FOOD FUTURES   291
Jia-Chen Fu, Michelle T. King, and Jakob A. Klein

RECOMMENDED WORKS   305
CONTRIBUTORS   323
INDEX   329

SERIES FOREWORD

Modern Chinese Foodways is the twenty-third book in the Food, Health, and the Environment series. The series explores the global and local dimensions of food systems and the issues of access, social, environmental and food justice, and community well-being. Books in the series focus on how and where food is grown, manufactured, distributed, sold, and consumed. They address questions of power and control, social movements and organizing strategies, and the health, environmental, social and economic factors embedded in food-system choices and outcomes. As this book demonstrates, the focus is not only on food security and well-being but also on economic, political, and cultural factors and regional, state, national, and international policy decisions. Food, Health, and the Environment books therefore provide a window into the public debates, alternative and existing discourses, and multidisciplinary perspectives that have made food systems and their connections to health and the environment critically important subjects of study and for social and policy change.

Robert Gottlieb, Occidental College
Nevin Cohen, City University of New York (CUNY)
Graduate School of Public Health

FOREWORD

E. N. ANDERSON

I got into studying Chinese food by a route very common among ethnographers, but rarely acknowledged: an ethnographer winds up studying what the people being interviewed love to talk about. I worked in Hong Kong, because at that time—the 1960s and 1970s—only Taiwan and Hong Kong were open to foreign researchers, and my advisor, Jack Potter, did his research in the latter. I found that when I interviewed Hong Kong people about kinship, or medicine, or economics, or anything else, we invariably wound up talking about food, and it was food that inspired the most enthusiastic and detailed discussions. So Marja Anderson (who did research along with me) and I published a brief article on Cantonese food, which was discovered by the eminent Yale University anthropologist K. C. Chang, who had been one of my teachers. He invited us to do a chapter for his edited volume, *Food in Chinese Culture: Anthropological and Historical Perspectives* (1977). By the time it came out, I had spent three years in Chinese communities and published one or two other articles on food. Another of my teachers, Ed Schafer (a Berkeley Sinologist), was involved in the project, writing about the Tang Dynasty. All this led me to further research in that area. At that time, there was virtually no social-science writing about Chinese food. Historians had worked on it, including the other authors in the Chang volume, and there was some nutritional and medical research, but social scientists had concentrated on kinship, village organization, and local politics (all of which I also studied).

The Chang volume was part of a wider trend toward taking food consumption seriously. This trend began, unsurprisingly, with the French. The

Annales school of historians was particularly important in making food production and consumption visible to historians and social scientists. Most American social scientists at the time regarded food consumption as a lowly topic suitable only for cookbook writers, but the brilliant work of the *Annales* historians like Lucien Febvre, Emmanuel Le Roy Ladurie, Marc Bloch, and Fernand Braudel was having its effect. The *zeitgeist* changed rapidly. The Chang volume was a critical part of bringing that change to China studies. After the Chang work, studies of Chinese food increased rapidly. Ethnographers began to realize that food was vitally important in Chinese society. James and Rubie Watson, who like me had studied under Potter and done research in Hong Kong, were particularly important in putting it on the anthropological agenda.

A new generation of anthropologists and historians, many of them ethnically Chinese, grew up without the anti-foodways bias. Researchers today are better trained, much more diverse in gender and ethnicity, and more specialized in topic. The opening of China led to a rush to study both familiar and remote, often romantic, areas, so we now know about the food of Yunnan Province, Gansu Province, and other areas once closed. However, current restrictions on research and travel are returning us to the bad old days. One hopes that current tensions between China and the West, especially the United States, will not last.

Topics of research have steadily broadened, from simple documentation of foods and eating patterns to serious studies of the place of foodways in society. The studies in the present book are a varied sampling of current paths of research. Studies of Chinese food in diaspora seem to be particularly exciting at the moment.[1] Research on China's contacts with the rest of the world, especially west and central Asia in early times and then expanding to cover the planet, have also been especially productive and revealing.[2]

Can one define Chinese food? I have always been persuaded by Elisabeth Rozin's "flavor principle."[3] Chinese food is easily recognized by the combination of soy sauce, fresh ginger, garlic, and, often, other fermented bean products such as fermented black beans and *doubanjiang* (豆瓣酱 bean paste). Yet the task of classifying China's local cuisines remains a daunting challenge for today's scholars. Carolyn Phillips recognizes fully thirty-five in her enormous and highly useful cookbook.[4] Hu Shiu-Ying's encyclopedic *Food Plants of China* (2005) lists about 1,700 plants eaten somewhere in the vast country, a figure comparable to the total number of herbal remedies in Li Shizhen's

Bencao gangmu (本草纲目 *Compendium of Materia Medica*).[5] Many more have turned up since.

It seems to me that by far the most important thing to do today is to document more foodways from more regions. There is an astonishing variety of foodways of which we had absolutely no knowledge back in the 1970s. Chinese foodways are more diverse than we ever dreamed. Coupling all this with historical research is now seriously needed. There are also more focused questions to be researched on specific techniques, for example, and the histories of cleverly preserving food by means of natural plants and fungi. Pickling, fermenting, brewing, autodigestion in salt (as in making fish sauces), heavy herb use, and even rose oil (powerfully antibiotic), all made Chinese food perhaps the leader in the world at varied techniques for food conservation and fermentation. One recalls Francesca Bray's great studies on the *Qimin yaoshu* (齊民要術 Essential techniques for the common people) and the late H. T. Huang's magisterial work in the same series.[6] Bray's major relevant question was found in her *Agriculture* volume in the Needham *Science and Civilisation in China* series; Huang's was found in the *Fermentations and Food Science* volume in the same series, which constituted his life work.

A lifetime obsession with good food has not only taken me around the world; it has taught me a great deal about how culture works. The most important message is that people do not usually do what they do because of grand abstract ideologies. They are usually more interested in getting through the day. Food and drink are immediate necessities. Humans abstract the grand principles and great ideas from daily interactive practice.[7] Moreover, people can adapt to anything, make almost anything edible, and change quickly and smoothly when they need to. This leads to the age-old question, easy to ask but surprisingly difficult to answer, of why people eat what they eat. Why do people of the Yangtze delta like their foods sweeter than Cantonese people do? Why did yogurt and cheese persist in a few spots in the south but die out in most of southern China after the medieval period?

Cultural conservatism is astonishing, but so is the speed with which people can abandon a staple food or adopt a new one. The worldwide triumph of white and sweet potatoes, for example, was far from immediate but still quite fast. They became so vitally important that one cannot imagine Irish or German or Chinese life without them. Another particularly interesting example is the chile pepper.[8] They were brought to the Old World by Portuguese and Spanish voyagers. By 1600 they were adopted in almost every area tropical or subtropical enough to grow them easily. Then a process of developing less fiery and more quick-growing varieties allowed them to move north into most of Europe as well as northeast Asia. In China, they slotted easily into a cuisine already spicy with black pepper and its relatives (*Piper* spp.), as well as brown peppers of various kinds (*Zanthoxylum* spp.) and spicy-hot herbs. This cuisine flourished most in Hunan, which thus became China's chile center, spreading them to Sichuan in the seventeenth century (if not earlier).

When we consider the variety of topics and approaches covered in this volume, the unifying theme bringing them together is the worldwide perspective. The chapters integrate the old East Asian world-system—the orbit influenced by Chinese culture—and now the modern world-system in its fully globalized form. Globalization may be breaking down, as pandemics, wars, supply chain collapses, and other density-dependent phenomena attack, yet the world remains united in one vast exchange network, with goods, cultural creations, money, and social forms all flowing in vast patterns.

What can one conclude from the wonderful chapters here, and from my stray observations? First, the myth of a changeless, closed, culturally isolated China could not be more wrong. Second, China has broken out on the world stage in the last 150 years, revolutionizing food from the Arctic to the Antarctic; there are Chinese restaurants on the north coast of Alaska, and Chinese food has been served in the Antarctic research station. Third, more interesting from the perspective of food studies, we must conclude that cultural borrowing is universal and continual. Cultures are not closed steel-walled spheres. They are vague, loose, shifting entities that move like amoebas, stretching out and incorporating foods and other good things. They adapt, but culture is not simply an adaptive mechanism, as both the good and the bad can be adopted. Nonetheless, on the whole, borrowing is a good idea, enriching cultures and making them more flexible and resilient. This volume brings together a particularly wide and striking diversity of approaches, regions, traditions, and ideas. It is exactly what the world needs to understand better the food and foodways of one-fifth of humanity.

NOTES

1. See, for example, Jenny Banh and Haiming Liu, eds., *American Chinese Restaurants: Society, Culture and Consumption* (New York: Routledge, 2020) and Sidney C. H. Cheung, ed., *Rethinking Asian Food Heritage* (Taipei, Taiwan: Foundation of Chinese Dietary Culture, 2014).

2. Paul D. Buell, E. N. Anderson, Montserrat de Pablo, and Moldir Oskenbay, *Crossroads of Cuisine: The Eurasian Heartland, the Silk Roads, and Foods* (Leiden: Brill, 2020).

3. Elisabeth Rozin, *Ethnic Cuisine: The Flavor-Principle Cookbook* (Lexington: The Stephen Greene Press, 1983).

4. Carolyn Phillips, *All Under Heaven: Recipes from the 35 Cuisines of China* (San Francisco: McSweeny's; Berkeley: Ten Speed Press, 2016).

5. Hu Shiu-Ying, *Food Plants of China* (Hong Kong: Chinese University of Hong Kong Press, 2005).

6. Francesca Bray, *Agriculture*, in *Science and Civilisation in China*, ed. Joseph Needham, vol. 6, *Biology and Biological Technology*, pt. 2 (Cambridge: Cambridge University Press, 1984); H. T. Huang, *Fermentations and Food Science*, Science and Civilisation in China, ed. Joseph Needham, vol. 6: *Biology and Biological Technology*, pt. 5 (Cambridge: Cambridge University Press, 2000).

7. Pierre Bourdieu, *Outline of a Theory of Practice*, trans. Richard Nice (New York: Cambridge University Press, 1977).

8. Brian R. Dott, *The Chile Pepper in China: A Cultural Biography* (New York: Columbia University Press, 2020).

ACKNOWLEDGMENTS

The chapters in this volume were developed over the course of two conferences. The first conference, "Modern Chinese Foodways," was held online from April 22 to 24, 2021 (https://tarheels.live/modernchinesefoodways/; see also the Modern Chinese Foodways YouTube channel, https://www.youtube.com/channel/UC3CQiV0V9DflZlSlCe9Xdpw). The second conference, "Chinese Food Futures," was held at Emory University and online from April 21 to 23, 2022 (https://chinesefoodfutures.wordpress.com/).

The editors would like to thank all of the participants in both of these conferences for their robust contributions, despite the challenges of the coronavirus pandemic. We would like to give special thanks to Francesca Bray for her keynote speech (included in revised form in this volume) at the first conference, and the participants of the roundtable, "Fifty Years of Chinese Food Studies," which included Francesca Bray, Françoise Sabban, Eugene Anderson, James Watson, and David Wu. Likewise, we would like to give special thanks to Lucas Sin, Fuchsia Dunlop, and Xiaowei Wang for their participation in the "Chinese Food Futures" roundtable at the second conference. Jeffrey Pilcher provided marvelously helpful comments on all of the papers and the structure of the volume itself, as did three anonymous peer reviewers for the MIT Press. Our thanks also go to Mukta Das and Chenjia Xu for their preconference interviews, Xiaowei Wu for the beautiful design of both conference websites, Cynthia Peng and Lynn Zhang for logistical support, Gabe Moss for the map design, and Donny Santacaterina and Jim Sojourner for bibliographic assistance.

The editors would like to acknowledge the support of the National Endowment for Humanities for its generous Collaborative Research Grant (2020–2023) and the Chiang Ching-kuo Foundation for its Conference Planning Grant (2020–2021), which made both conferences and the drafting of the editorial portions of this volume possible. The editors would also like to thank the Department of Russian and East Asian Languages & Cultures at Emory University, the Confucius Institute at Emory in Atlanta, Emory College's Hightower Speaker Funds, and the Center for Faculty Development and Excellence at Emory University, as well as the Institute for the Arts and Humanities at the University of North Carolina at Chapel Hill for additional financial support.

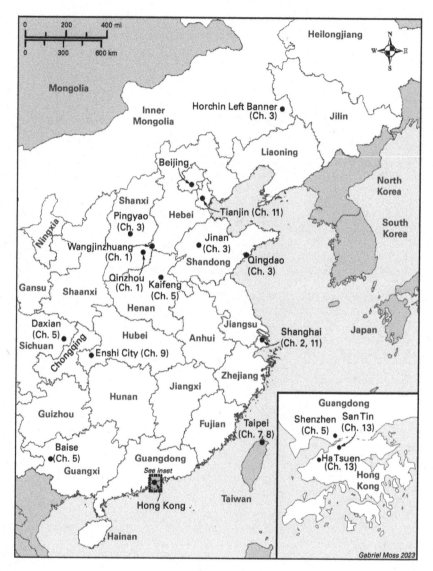

FIGURE 0.1
Map by Gabe Moss.

INTRODUCTION

JIA-CHEN FU, MICHELLE T. KING, AND JAKOB A. KLEIN

In 2022, the United Nations Educational, Scientific, and Cultural Organization (UNESCO) inscribed "Traditional tea processing techniques and associated social practices in China" onto its Representative List of Intangible Cultural Heritage of Humanity. This was the country's first food and drink-related item selected for the honor. The promotional video accompanying the bid to UNESCO, produced by the Ministry of Culture and Tourism of the People's Republic of China (PRC), opens with footage of lush, green, beautifully terraced mountainside tea farms in Yunnan Province, where female tea leaf pickers happily labor in matching colorful ethnic costumes and iconic bamboo cone hats. The video's depiction of "traditional manual techniques passed down for generations" according to "natural conditions and local customs" is equally intensive on a human scale, as the tea is dried, agitated, heated, crushed, and roasted—all actions performed by hand in small batches, with woven baskets and equally preindustrial brick ovens.[1] After watching the video, anyone who has never been to China might be excused for imagining that tea drinkers there imbibe only while wearing colorful ethnic costumes and sitting in courtyard enclosures set among traditional peaked roof buildings, or while contemplating nature in moist, green forests, next to unsullied waterways.

But for anyone who has visited China in the last four decades, these idyllic scenes of traditional tea culture are virtually unrecognizable—except as carefully curated visions of "Chinese culinary tradition" for domestic and international tourists. As of 2022, China produces 3.35 million metric tons of tea annually and is the world's leading exporter of tea, accounting for almost

45 percent of global production—numbers and scales that could never be achieved by painstaking artisanal manual labor alone.² Other recent videos produced by China's state news agency, Xinhua, proudly depict a very different vision of China's modern tea industry, boasting of technological innovation at every stage of production and processing. One soil scientist, who analyzes phosphorus and nitrogen compounds in the soil of tea farms in the Wuyishan region of Fujian Province, explains that for biologists, botanists, and agronomists like her, "Our job is to bring science and technology to the fields."³ Meanwhile, robots, computers, and automation typify the industrial processing of one company, Enjoy Manor, in Yingde, Guangdong Province, where few humans are to be found on the factory floor and driverless forklifts soundlessly move enormous pallets of tea packaged in metal and plastic. The news video boasts that through automation, the company "has already been able to reduce the numbers of workers from dozens to less than five," helping to make the industry "more efficient and profitable."⁴ Experiments are now underway to deploy AI sensors to replace human judgement of the quality of the tea, which has traditionally been based on sight, touch, sound, smell, and taste.

The explicit aim of this volume is neither to valorize the artisanal imaginary of nostalgic agricultural Chinese pasts (particularly as performed in a supposedly ethnically harmonious nation-state) nor to celebrate the technological prowess of China's hypermodern industrial food production, but rather to see both impulses as part and parcel of the same complex and contradictory bundle of attitudes that characterize modern Chinese foodways. Critically, what we recognize as Chinese food today is *not* simply the result of five thousand continuous and glorious years of culinary history, as is commonly touted on dozens of food websites and in countless popular and academic books, particularly those produced in the PRC.⁵ Instead, we argue that Chinese food as we know it today is very much the result of concurrent processes of commodification, scientization, identity formation, and migration taking place across the late nineteenth to the late twentieth centuries. In the first instance, then, our title *is* our argument: Chinese foodways *are* very much modern.

But beyond this, we want to argue that the story of modern Chinese foodways is central, not ancillary, to understanding the phenomenon of culinary modernity writ large. Culinary modernity is not merely something that first took place elsewhere (primarily in Europe and America), only to arrive late to

Chinese shores. Rather, we argue that the Chinese story, having compressed the world's most wrenching political, economic, and social changes (culinary and otherwise) into one dramatic century for more than one billion people, is crucial to understanding culinary modernity as a phenomenon in itself. China now looks ahead as the leading contemporary bellwether of possible dystopian futures of environmental degradation, food insecurity, and techno-oppression, while also presenting advanced technological solutions to these problems. Learning from China's culinary past, present, and future—and understanding the intricate ways in which those temporal states can be entangled and experienced simultaneously—can shed light onto other examples of culinary modernity taking place around the world.

This edited volume on modern Chinese foodways draws together the research of scholars working in fields such as history, anthropology, sociology, food studies, ethnic studies, and media and communication studies—an interdisciplinary range of expertise demonstrating the scope and reach of the field today. We have taken great pains to ensure international participation, with individuals representing institutions in mainland China, Taiwan, and Hong Kong, as well as the United States, the United Kingdom, France, and Japan. This robust participation of international contributors underscores our editorial conviction that rigorous analysis of modern Chinese foodways must necessarily be a collective scholarly endeavor with global perspectives, since the geographic and temporal scope encompassed (to say nothing of their political, social, and economic dimensions) defies mastery by any single individual scholar.

One major motivation for putting together this volume is that despite a great number of excellent monographs and articles in recent years about different aspects of Chinese food history and anthropology, no scholars have attempted to make any kind of broad collective statement about the state of the field since the seminal 1977 publication of K. C. Chang's edited volume, *Food in Chinese Culture: Anthropological and Historical Perspectives* (Yale University Press).[6] Chang and his fellow contributors made a huge impact on framing Chinese food studies with their essays on the foodways of early, middle, and late imperial China, dynasty by dynasty. To be sure, the two last chapters of the volume did cover "modern China," north and south, but the bulk of the book remained firmly rooted in China's dynastic past. This structure suggested a sense of coherence and continuity in Chinese foodways from ancient times to the present, and Chang himself concluded that "continuity vastly outweighs change in this aspect of Chinese history."[7] This is a monolithic claim that calls out for closer examination today. More importantly, however, whatever "modern China" may have looked like in 1977 bears only passing resemblance to what it looks like now. Since the late 1970s, the economic reforms ushered in by Deng Xiaoping and continued by each subsequent leader of the PRC have radically transformed the country from a collectivist agrarian party-state into one of the dominant global economic and political superpowers of the twenty-first century. It is high time, then, for scholarly reassessment.

What, then, are some of the distinctive characteristics of modern Chinese foodways? A few initial observations are in order. First, insofar as "Chinese" foodways often refers to the People's Republic of China, or geopolitical China (which, as we will explain later in this introduction, is only one way to define "Chinese" foodways), two of the most obvious and concrete aspects shaping its food practices are the country's size and scale, of both its land and population. As the second-most populous country in the world and the fourth largest in territorial size, China's vast range of geographic terrains and distinct cultures has produced a culinary diversity virtually unmatched by any country in the world, with perhaps the exception of India (which is roughly one-third of China's landmass). Outside of the culinary sphere, other scholars have argued that China as a region is more comparable to all of Europe than to a single country.[8] This diversity and complexity becomes even more apparent once we include Chinese diasporas in our purview of Chinese foodways.

Another factor that shapes our approach to Chinese foodways is the extent to which both modern Chinese and non-Chinese observers have come to associate the words "China" and "Chinese" with food and eating, in both celebratory and anxious ways. For instance, some patriotic writers of the early twentieth century described the cultural supremacy of Chinese foodways and emphasized the major role thinking and talking about food took in Chinese life. Writing in English for Western audiences, philosopher, essayist, and bon vivant Lin Yutang asserted in 1937 that the importance of food for Chinese people surpassed religion or learning. "How a Chinese spirit glows over a good feast!" he wrote. "How apt is he to cry out that life is beautiful when his stomach and his intestines are well filled! . . . The Chinese relies upon instinct and his instinct tells him that when the stomach is right, everything is right."[9] (Let us set aside, for the moment, who Lin meant by the term "Chinese" and his insistence on a default male subject.) It is not the heart or the brain, Lin argues, but the stomach that centers the Chinese experience of life.

Such celebrations stand in stark contrast to the fears and anxieties surrounding China's food production and consumption. In the early twentieth century, China was described by some Western observers as "the land of famine," driving racialized fears of a mass exodus of Chinese to Western countries.[10] Meanwhile, national leaders, such as Sun Yat-sen, worried about the poverty and food scarcity facing its then four hundred million people, who by and large still led agrarian lives on the land. Food security in the sense of national food self-sufficiency became a central concern of the Nationalist government[11] and remains a central policy concern, as reflected in President Xi Jinping's recent statement that "a country must strengthen its agriculture before making itself a great power, and only a robust agriculture can make a country strong."[12]

For many outside observers, China's 1.4 billion consumers are once again seen as an essential part of the global food supply conundrum, now because of how much and what they eat, not how little. The demands of China's allegedly insatiable middle-class consumers are blamed for everything from global overfishing to the overuse of pesticides to increased daily demands for previously unaffordable meats such as pork.[13] (China now accounts for half of the world's pork consumption.[14]) Scientists have recently confirmed that statistical evidence points to one section of a wet market in Wuhan selling live farmed "wild" animals as the likeliest point of origin for the zoonotic spillover of the SARS-CoV-2 coronavirus,[15] while the emergence and spread of avian flu (H5N1) has been associated with the rapid development and concentration of industrial poultry production in southern China.[16] Anxieties around the environmental and public health impacts of China's food system and the quality and safety of its foods are by no means limited to outside observers; they are keenly felt within the country as well and are profoundly shaping everyday foodways across China.[17] To speak of modern Chinese foodways, then, requires holding several impossible things in mind at once—perhaps not specifically six before breakfast time but several nonetheless: size and scale generative of immense local diversity yet coupled with strong, synthesizing pressures to unify and homogenize; pronounced joy and celebration in rooting identity in food yet always constrained by the fear of contamination, excess, and depravation; and the allure of the traditional matched by a dedication to a technologically driven future. That three words—"modern," "Chinese," and "foodways"—can carry within them such complexities and contradictions is exactly the premise of this book.

To map out the other salient features of Chinese culinary modernity, we turn now to a closer examination of each word in our title. We begin with the most tangible term, "foodways," then grapple with the most difficult concept to grasp, that of the "modern," before tackling the many shades and contours of "Chinese" identity. Within the space where these three words intersect to form "modern Chinese foodways," we have highlighted four distinct themes around which we have organized our volume, all of which span the late nineteenth, twentieth, and early twenty-first centuries: (I) "Creating Value: Market Structures and Commodities," (II) "Systematizing Expertise: Food, Science, and Technology," (III) "Constructing Culinary Identities: Gender, Ethnicity, and Nation," and (IV) "Chinese in Motion: Migration and Mobilities." Within each part, individual chapters focus on different moments of Chinese history, allowing readers to gain a better sense of both the continuities and discontinuities of specific thematic patterns across traditional temporal divides.

WHAT DO WE MEAN BY "FOODWAYS"?

Scholars such as Francesca Bray and Huang Hsing-tsung (H. T. Huang), writing for Joseph Needham's ground-breaking series *Science and Civilization in China* (1954–present), have done a masterful job of tracing the material history of Chinese food, in terms of agricultural production and fermentation processing.[18] But the centrality of food to China and modern Chinese history—and by extension the value of Chinese food studies—is not limited to the history of rice, the history of the soybean, the history of soy sauce, or other individual commodities such as sugar. Building on previous work by Bray and Huang, as well as more recent Chinese "food biographies,"[19] in this volume we emphasize that, in order to understand the profound and complex roles of food in shaping Chinese lives, past and present, we need to study not only how foods, condiments, and drinks are produced and processed but also how they are distributed, exchanged, cooked, served, eaten, and disposed of, as well as valued, discussed, and classified.

The term "food system"[20] encourages scholars to look at this whole gamut of processes from production to consumption, and it draws attention to the economic, environmental, and political dimensions. But it does so often at the expense of the cultural and the social. We find the term "foodways" useful for our purposes in that it encourages investigation into the cultural dimensions of cooking, eating, and sharing food, and how these are embedded in food systems. We also find it useful because it suggests that the culinary practices of a group of people, community, or place are not random but in some sense patterned and distinctive.[21] In contrast to some early uses of the term, however, we do not see "real" foodways as necessarily being grounded in a preindustrial past and localized agroecosystems, but rather we argue that distinctively Chinese foodways can be urban, industrial, and geographically mobile.[22] Moreover, while some have criticized the foodways concept for encouraging an overly bounded view of culture and "an essentializing gaze on the food-related practices of marginalized communities,"[23] our articulation of "foodways" with "Chinese" is intended instead to encourage investigation into the complexities, diversities, and ambiguities of "China" and "Chineseness" while also insisting that there *is* a value to exploring what may be distinctively Chinese—if not necessarily unique—about certain practices, experiences, and discourses of food.

In this volume, Chinese foodways encompass a myriad of intimately connected activities in all of their cultural, political, economic, social, environmental, and technological dimensions, including agriculture and crop management, industrialization and commodification, marketing and advertising, technology and delivery systems, scientific evaluation and government regulation of food, and all manner of shopping, preparation, processing, cooking, consumption, and ritual practices, as well as the consequences of food waste and environmental pollution. Our approach to "Chinese foodways" also includes discourses that celebrate or critique these foodways. There are of course other topics that might have been included in our volume but were not due to limitations of time and space. Rather than aim for comprehensive inclusion, we have instead attempted to identify the major thematic elements of Chinese culinary modernity, in the hopes that other scholars will continue to join us in building a more robust analytical framework.

WHAT DO WE MEAN BY "MODERN"?

This is perhaps the most difficult of the three terms to pin down, for a variety of reasons. Yet we can say with some confidence that one of the most notable aspects of Chinese culinary modernity involves contradictory extremes of expansion and compression of history and time. While it is essential to remain skeptical of triumphal claims of five thousand years of continuous Chinese culinary traditions, it is also necessary to acknowledge that the culinary past (or some imagined version of it) weighs heavier on China's present than perhaps any other place in the world, if only because there has been such a long, documented written tradition from which to choose. The video from the PRC Ministry of Culture and Tourism for its successful UNESCO traditional tea processing bid, for example, made sure to reference the eighth-century *Classic of Tea*, by Lu Yu, emphasizing the age of the text by showing it in the form of calligraphic script and woodblock print.[24]

To speak of Chinese culinary modernity could, of course, refer to discrete moments of historical transformation: (1) the encounter with the West (and Western cuisines) and imperialism in the nineteenth century, (2) the transformation of society and food supply through industrialization and technological innovation after 1978, and (3) globalization in the 2000s. Within this schema, however, China occupies an uncomfortable and contradictory position as being "too little, too late" or "too much, too fast" when compared with Euro-American experiences. Culinary modernity becomes too easily conflated with whether or not, or to what degree, the Chinese accepted or adopted Western *qua* modern diets of bread, beef, and milk,[25] supermarkets and processed foods, or industrial agriculture and food production from CAFOs (concentrated animal feeding operations) to McDonald's. These definitions would seem to require a persistent evaluation of China's place within a linear unfolding of modernization, marked as industrialized agriculture, commercial and industrial standardization, and concentration of state and capitalist forces making mass foods available more quickly to more people. But while these three moments of transformation were indeed significant, they can obscure other equally telling moments that differentiate China's experiences from Euro-American antecedents, such as Republican-era political and scientific experimentations or radical state socialism in the Maoist era, which are often regarded as setbacks or failures but are in fact crucial to understanding Chinese culinary modernity.[26]

Chinese culinary modernity cannot be reduced to a simple, linear story of greater industrialization or Westernization, and, perhaps for this reason, this has meant that food studies scholars outside of China have largely ignored its experiences, instead promoting, whether intentionally or not, the idea of Europe and the United States (and by extension their cuisines) as the epicenters of modern culinary innovation. Edited volumes purporting to cover global food histories include China, India, and the Middle East as part of their coverage of the ancient or medieval world but invariably exclude or minimize the non-West when it comes to modern culinary developments in the nineteenth and twentieth centuries.[27] The pattern of culinary development as experienced in the United States and Europe, focused on the postwar industrial development of processed foods, stands in as representative of the "modern age," with only a passing nod to experiences in the rest of the world.[28] For example, rather than seeing the "restaurant" as a specific historical and linguistic phenomenon that blended eighteenth-century French ideas about digestion and nourishment with a specific physical site for culinary consumption, the predominant impulse has been to equate one form of hospitality and public dining as *the* definitive one by which all others are judged, as if the markers of a modern restaurant must necessarily be its menus and professional waitstaff.[29] Our goal is not to replace a Eurocentric definition of culinary modernity with a Sinocentric one, but we do insist that grappling with Chinese culinary modernity can help provincialize Euro-American experiences by revealing both the inherent plurality of culinary modernity as it has emerged in different places at different times and the social and moral values they encode.[30]

What would happen if we began with China and Chinese experiences in our exploration of culinary modernity or made food central to our analysis of Chinese modernities? Without setting forth an exhaustive list and yet beginning with the assumption that culinary modernity was both multiple and patterned, Chinese culinary modernity expands our temporal sense of the modern; it forces us to carefully consider the social, economic, and cultural values encoded in material and technical systems without being predetermined by those same systems; it heightens our attentiveness to the reflexive nature of being both Chinese and modern; and it challenges our often unexamined assumption of capitalism as the driving force in culinary modernity. For example, social welfare has often been cast as a modern and national problem to which European governments in the late nineteenth and early twentieth centuries turned their attention and resources, but this was clearly not the case in China, where long-standing state concern for the welfare of the people, rural and urban, can be documented through the institutionalization of famine relief ranging from physical granaries to market-directed policies by the mid-seventeenth century.[31] In this, we see continuities between imperial and postimperial Chinese states, which took food security and subsistence to be a cornerstone of good governance. (Whether or not such political regimes succeeded in this regard is a separate matter.) At the very least, we can assert with confidence that food was, and continues to be, politically important, and the implications of this position can be demonstrated not only through specific policies but also in and through the technologies, technical systems, and forms of knowledge organizing food production, provision, and consumption.[32]

Two significant themes we have identified as constituent of the "modern" are commodification and scientization. Part I, "Creating Value: Market Structures and Commodities," tackles large-scale historical processes of agricultural industrialization and commodification from the nineteenth and twentieth centuries to the present by exploring the many different ways in which state and nonstate agents have engaged in value formation for specific food commodities. Few will disagree that some of the most striking transformations of Chinese society have occurred since the beginning of the reform era in 1978: vast urbanization, the creation of a property-owning middle-class population, and the rapid industrialization of agriculture. Yet while the scale and magnitude of such post-1978 changes, especially the rapid industrialization of agriculture and the creation of a national food system, cannot be denied, the roots of these changes reach back into the past in surprising and complicated ways. How foods such as millet, soy sauce, and beef became integrated into an emerging national economy, and how such foods passed into and out of being everyday foods draws our attention to the material, social, and discursive practices surrounding production, trade, and consumption as objects circulate in and out of commodity status.

In chapter 1, Francesca Bray explores two different strategies and ongoing choices adopted by farming communities in contemporary China in their efforts to reintroduce and rebrand an ancient Chinese staple, millet. By situating her discussion within a longer history of millet as food and crop, Bray draws out the tensions and selectivity of today's invented food traditions. In contrast to millet, which served historically as both the material and symbolic foundation of Chinese civilization, soy sauce was a much later entry into the canon of Chinese "traditional" foods. In chapter 2, Angela Leung demonstrates its transformation from an obscure condiment for the elite into an everyday food with the cultural power to symbolize China itself. In chapter 3, Thomas DuBois highlights the coproduction of commodity and cuisine in his examination of six centuries of beef in China as eight iconic dishes. Each dish illuminates a different aspect of a modernizing foodscape, ranging from preindustrial livestock chains and later economic liberalization to political ideology and franchising.

Part II, "Systematizing Expertise: Food, Science, and Technology," focuses on the ways in which individual and corporate actors throughout the twentieth and twenty-first centuries have sought to codify and normalize scientific knowledge, practices, and skills, thereby shaping Chinese ideas about dietary health, foodstuffs, and patterns of culinary production and consumption. Modern science was often cast as both integral to and a necessary precondition for the makeup of a modern people and its food and foodways, but given China's own robust native knowledge systems, what counted as "science" was a matter of ongoing debate and messy negotiation: Which sources of expertise should one draw on? Who could embody such expertise? How should such expertise be gathered and disseminated? And what ends—behavioral, economic, political—should science and technology serve? Rather than approaching science and technology as simply the transfer of predetermined ideas and practices, the authors in this section underscore the self-consciously dynamic ways in which Chinese actors have insisted that science and technology be subject to human agency even as they become entangled in human-nonhuman assemblages that seem to exceed human control.

In chapter 4, Hilary Smith explores the world of Republican-period dietary supplementation and shows how scientists, Chinese medical practitioners, and advertisers alike seized upon the idea of vitamins as a source of vitality, thereby articulating a scientific way of eating that could resonate in multiple registers and incorporate preexisting discourses of deficiency. Sigrid Schmalzer (chapter 5) also draws attention to the ways in which different knowledge forms and practices were integral to Mao-era attempts to rapidly industrialize agriculture. During the Great Leap Forward (1958–1961), Chinese agricultural scientists, state agents, and rural people sought to systematize and incorporate native knowledge of wild and cultivated plants as part of a broader developmental approach to address the problem of scarcity and accelerate agricultural production. Their purposeful integration of native and scientific knowledge underscores a common goal in search of nourishment and healing, even in circumstances of terrible destruction. From the attempt to create a different accounting of human-nonhuman entanglements during the Mao period to our contemporary world built by information and communication technologies (ICTs), in chapter 6, Fan Yang interrogates the gendered ways in which the platform economy and algorithmically driven food delivery systems have both re-entrenched and reordered domestic and public spaces. ICTs have changed not only Chinese food practices, but also conceptions of human agency, desire, and waste. Vitamins, native pesticides, and the food delivery industry in China's rapidly growing platform economy each flag the complex ways in which science and technology have been shaping and reshaping food and eating in modern China.

WHAT DO WE MEAN BY "CHINESE"?

The term "Chinese" buries within it several ideas that exist in tension with one another. On the one hand, it can and does refer to the PRC nation-state, the Republic of China that preceded it, and even the long history of dynastic empires that came before the twentieth century. Using the same English term to describe all of these time periods, however, reinforces a sense of monolithic continuity that does not necessarily bear up under historical scrutiny. What is often translated as a single English term, "Chinese," may have in fact been represented by different Mandarin terms at different historical moments, including *Han*, *Tang*, *Zhonghua*, and *Zhongguo* (漢, 唐, 中華, 中國). Moreover, to limit the term "Chinese" to a geopolitical or geographical designation alone is highly problematic, as it suggests a tacit support of the present-day boundaries of the PRC and the political claims of the CCP over contested territories. Instead, although we acknowledge the common usage of "Chinese" in geopolitical terms, recognize the real power of the nation-state over the lives of people living under its rule, and observe the nationalist pride of many of its citizens, we want to underscore in this volume an alternative vision of "Chinese" identity that does not adhere to the nation-state alone and instead escapes its hegemonic shadow. This expansiveness is essential in thinking through "Chinese" identity in particular because of the historical legacies of the multiethnic Qing Empire (1644–1911), the unresolved conclusion to the Chinese Civil War (1945–1949), and massive out-migration of Chinese around the world during the late nineteenth and early twentieth centuries.

The contours of Chinese identity, culinary and otherwise, are intimately connected with the modern period precisely because they arose with greatest urgency in the late nineteenth and early twentieth centuries, as the multiethnic Qing dynasty empire (1644–1911) was overthrown and transformed into the fledgling nation-state of the Republic of China. Before the end of the Qing dynasty, individuals most likely would have identified and distinguished themselves—and the cuisines they ate—regionally through their hometowns or home provinces (e.g., Yangzhou, Sichuan), or alternatively through ethnic (e.g., Han, Manchu, Mongol) or religious identities (e.g., Buddhist, Muslim).[33] There was no commonly used collective term to identify oneself or one's way of eating as "Chinese"; this was instead a national culinary identity that came into being largely around the end of the nineteenth century, partly in response to the pressure of Western imperialist incursions.

Ethnic identity emerged as a particularly salient issue at that moment because the Qing rulers had been Manchu, an ethnic group originating beyond the limits of the Great Wall in the area north of the Korean Peninsula, while the majority of subjects they dominated were ethnically Han Chinese. Sun Yat-sen and other national revolutionaries thus framed the overthrow of the Manchu Qing as a "return" to Han rule, as the rulers of the preceding Ming dynasty (1368–1644) had also been Han. But Qing conquests and relationships also brought other border territories with different ethnic populations into the imperial fold, such as Xinjiang, Tibet, and Mongolia. Neither Sun Yat-sen, Chiang Kai-shek, nor Mao Zedong had any desire to let those territories go, and they actively worked to retain them, recasting the Qing imperial project as a new, multiethnic Chinese nation under Han rule.

The political tensions of this Qing multiethnic inheritance still manifest today, as Uyghur Muslims inside and outside of Xinjiang, as well as Tibetans inside and outside of Tibet, have strained against Han rule. According to official ethnic classifications, Han Chinese comprise 92 percent of the PRC's population today, with the remaining 8 percent divided among fifty-five officially recognized ethnic minorities. This imbalance has always favored the majoritarian Han, with minority ethnic identities only supported insofar as they add "color" and "variety" to the Chinese nation-state (and in the case of the foodways, a range of new tastes), absent any real political demands. Though the PRC state may insist on "Chinese" as a geopolitical identity, connoting shared citizenship, encompassing the Han as well as ethnic minority groups, foodways and cuisine offer an ideal vehicle for critiquing these hegemonic claims. What should we call the foodways of ethnic minorities, such as Tibetans or Uyghurs, whose contested political pasts and futures do not necessarily align neatly with that of the PRC state?[34] Are their foodways indeed "Chinese," or is the term more synonymous with Han ethnic identity, connoting shared foodways and a shared culture that stretches beyond the borders of the PRC, to include diasporic overseas Chinese? At the same time, in some places people classed as "ethnic minorities" are quite willing to work within existing state frameworks of ethnicity to leverage recognition for themselves, be it in political terms or in terms of their livelihoods, capitalizing on ethnic tourism.[35]

The early twentieth century also saw the emergence of the "woman question," or the place of women in modern Chinese society, as reformers sought to break free of traditional Confucian norms. No longer would footbinding, arranged marriages, concubinage, female chastity, and a host of other traditional social practices mark the cloistered lives of Chinese women. Instead, women established new lives and possibilities for themselves outside the home as students, journalists, educators, nurses, and even political revolutionaries. Reckoning with gender identity is also crucial to the story of Chinese culinary modernity. It was not, of course, that women had no roles in food preparation or cooking in early or late imperial times; it was more that the Chinese men who wrote about food (and by extension an earlier generation of Chinese food scholars, such as K. C. Chang and his fellow contributors) paid very little attention to women's contributions or their presence in kitchens. The "Chinese" perspective embedded in Chang's *Food in Chinese Culture* is centered on the experience of Han Chinese men (whether the emperor or literati elites) geographically located within the urban centers of the Middle Kingdom.

Another fissure of "Chinese" identity arises from China's contested political history. The unresolved legacy of the Chinese Civil War (1945–1949), which saw the victory of the Communist Party on the mainland and the retreat of the Nationalist Party to Taiwan, has of course dramatic implications today for Taiwan's future and continued existence as a self-governing polity. Yet the Civil War is far from the only historical event of significance in Taiwan's modern past that shaped its identity and foodways. While the majority (95 percent) of Taiwan's population is also of Han Chinese descent from centuries past, the legacy of both Spanish and Dutch colonialism in the late seventeenth century, fifty years of Japanese colonial rule (1895–1945), as well as an Indigenous population of Austronesians have also influenced its foodways. We have felt it crucial to include a consideration of Taiwan within this volume, not to lend tacit support to PRC political claims over this contested territory but to probe the limits and boundaries of the concept of Chineseness and ultimately to suggest a complex diversity of potentially "Chinese" culinary identities.

Part III, "Constructing Culinary Identities: Gender, Nation, and Ethnicity," highlights the question of who has been allowed to speak for and about Chinese cuisine, shape its communal and ritual practices, and actively construct commensality through imagined communities of eaters. The twentieth century overturned the overwhelming imperial dominance of male, Han literati over Chinese food writing and the formulation of food practices by giving women, ethnic minorities, and newly minted national citizens distinctly new modes of expression and culinary authority. The construction of each of these modern culinary identities—gendered, ethnic, and national—is implicitly a boundary-making project, often articulating self and other in rigid dichotomies: household tasks are deemed suitable for either women or men, eating customs are described as either Han or of an ethnic minority, dishes are categorized as either Taiwanese or Chinese. Although these identities are often taken for granted as always-already existing or fixed, part of the power of the close examinations in these chapters is to explain how certain identities have come into being over time, or how they have been actively deployed.

In chapter 7, Michelle T. King traces the mid-century publication of a plethora of cookbooks by Chinese women, both in Taiwan and the United States, as a new mode of culinary writing for female audiences, distinct from imperial models privileging the hierarchical relationship between the epicurean tastes of the male gourmand literatus and the manual skills of his male chef. In chapter 8, Yujen Chen discusses the emergence of a distinctly Taiwanese national culinary identity, one that has been shaped by a range of colonial, geopolitical, and ethnic forces and now stands defiantly against implicit inclusion in a PRC version of geopolitical culinary order. In chapter 9, Xu Wu takes us to the present-day tourist hotspot of Enshi in Hubei Province, where Han tourists visit a Tujia ethnic minority theme park to drink local alcohol and afterward smash the bowls in which it was served. In so doing, visitors literally consume a commercial vision of transgressive minority ethnicity and enable certain spaces to be repackaged as ethnic spaces, while reifying normative Han behavior. Finally, in chapter 10, Françoise Sabban brings us up to the reform era after 1978 under Deng Xiaoping, when market reforms not only opened urban areas to a restaurant boom but also inspired a renewal of literary works dedicated to China's food, cuisine, culture, and history. The debates between writers as to their status as professional *meishijia* (美食家 gourmets) underscores the ubiquity of this new, reform-era culinary identity, situated at the crossroads of marketization, branding, epicureanism, and professionalization.

A similar process has shaped the contours of Chineseness from the outside, through the movement of Chinese migrants abroad. Although Chinese migrants had already begun traveling to various trading ports in Southeast Asia in the fifteenth century, the number of Chinese migrants going to Southeast Asia and other parts of the world expanded exponentially in the late nineteenth and early twentieth centuries, with over nineteen million going overseas.[36] And wherever Chinese migrants went, their foodways and favorite dishes followed. Today, one could readily assert that Chinese cuisine is the ultimate global cuisine, with a Chinese restaurant in practically every country in the world, including at the Chinese research station in Antarctica.[37] Yet at the same time, the movement has not only extended outward: diasporic Chinese (not least the wandering overseas Chinese identity of Sun Yat-sen himself) have likewise had a crucial role in shaping the modern Chinese nation, including its foodways, and in raising critical questions about what it means to be and to eat "Chinese."[38] Moreover, migrants from outside China are now settling in the country and changing its urban foodscapes.[39]

Part IV, "Chinese in Motion: Migration and Mobilities," emphasizes movement and circulation. All of these chapters force us to reconsider the usual, limited definitions of "Chinese" foodways, and how multiple varieties of Chinese food cultures and practices have come to be constituted through translocal and transnational mobilities. This has included internal migrations of Chinese within China as well as different waves of overseas migration. Each of these cases challenges standard definitions of "Chinese" cuisine, encouraging readers to rethink what is contained by or excluded from the boundaries of this term, moving beyond the borders of the PRC to a more fluid, capacious, and globalized understanding.

Chinese food is now an integral part of the foodscapes of numerous countries around the world, with hybridized creations appealing to the palates of local customers savoring their favorite Indian-Chinese, Mexican-Chinese, or Peruvian-Chinese dishes, as described by Lok Siu in chapter 12. Siu explains how over the course of the nineteenth and twentieth centuries, Chinese migrants and *chifas* (local Chinese restaurants and food) have become an integral part of Peru's national culinary identity. At the same time, foreign foods have been absorbed into Chinese cuisine, from the milk teas and curry puffs of Hong Kong to the taste for Japanese cuisine evident among Chinese urban consumers, as described by James Farrer and Chuanfei Wang in chapter 11. Farrer and Wang argue that the long-time popularity of Japanese restaurants in Shanghai and Tianjin throughout the twentieth century has cemented their status as an indisputable part of urban "Chinese" food culture. Similarly, overseas capital has remade Cantonese foodways, as James Watson describes in his experience of fifty years of fieldwork in a village in the New Territories of Hong Kong in chapter 13. Watson explores how ritual pork division practices of family lineages, which were traditionally exercises in asserting patriarchal authority and hierarchy, have been completely transformed by overseas migration into genial extended family reunions, often organized and managed by female lineage members.

As a "constant need but perishable good," food is a "highly condensed social fact" and a key vehicle for meaning-making and politics, on scales ranging from the household to the nation-state to international diplomacy and transnational networks.[40] Foodways are a site where hegemonic struggles are played out, where power is reproduced and contested, where boundaries of identity, place, and time may be both defined and blurred. As such, the study of food and foodways offers an important way of understanding experiences and contradictions of Chineseness in the modern world.

This volume offers a starting point for understanding these experiences and contradictions through a wide range of studies into key areas of modern Chinese foodways. Rather than attempting to be comprehensive, we aim to lay the groundwork for further exploration into Chinese culinary modernity. We emphasize the importance of *Chinese* culinary modernity, but not because we seek to make a claim for Chinese exceptionalism. On the contrary, this volume offers a crucial counterpoint to dominant approaches to culinary modernity, which center on Euro-American experiences and treat culinary modernities in non-Western societies as derivative of these. Chinese culinary experiences and practices may resonate with other non-Western culinary modernities in ways that Euro-American models do not and will provide new perspectives on Western foodways, too. As we expand on in the afterword, the authors writing in this volume insist on the increasingly central relevance of modern Chinese foodways for culinary modernity writ large and, in doing so, hope to enable a new generation of researchers to forge a truly global field of food studies.[41]

NOTES

1. "Traditional Tea Processing Techniques and Associated Social Practices in China," UNCESCO Intangible Cultural Heritage, accessed October 29, 2023, https://ich.unesco.org/en/RL/traditional-tea-processing-techniques-and-associated-social-practices-in-china-01884.

2. "Tea industry in China—Statistics & Facts," Statista, accessed October 29, 2023, https://www.statista.com/topics/4688/tea-industry-in-china/#topicOverview.

3. "GLOBALink / How Technology Boosts Tea Industry in China's Wuyishan," New China TV, YouTube, uploaded June 19, 2023, https://www.youtube.com/watch?v=FKxvCumFD80.

4. "Tea Production in South China City Uses AI for Improved Efficiency," CCTV Video News Agency, YouTube, uploaded September 30, 2021, https://www.youtube.com/watch?app=desktop&v=qhUn8TFHJkc.

5. For a recent example, see Chong Li Xiansheng, *Shishang wuqian nian: Zhongguo chuantong meishi biji* [Five thousand years of food and counting: Notes on China's traditional cuisine] (Nanjing: Jiangsu fenghuang kexue jishu chubanshe, 2022).

6. In the meantime, there have been several edited collections on overseas Chinese foodways outside of China. See David Y. H. Wu and Tan Chee-beng, eds., *Changing Chinese Foodways in Asia* (Hong Kong: The Chinese University Press, 2001); Sidney Cheung and David Y. H. Wu, eds., *The Globalization of Chinese Food* (Honolulu: University of Hawai'i Press, 2002); Tan Chee-Beng, ed., *Chinese Food and Foodways in Southeast Asia and Beyond* (Singapore: NUS Press, 2011).

7. K. C. Chang, "Introduction," in *Food in Chinese Culture: Historical and Anthropological Approaches*, ed. K. C. Chang, (New Haven, CT: Yale University Press, 1977), 20.

8. Kenneth Pomeranz and R. Bin Wong, "China and Europe: The New Units of Analysis," *China & Europe, 1500–2000 and Beyond: What Is "Modern"?*, Columbia University Asia for Educators website, http://afe.easia.columbia.edu/chinawh/web/s4/index.html.

9. Lin Yutang, *The Importance of Living* (New York: John Day, 1937), 46.

10. Walter H. Mallory, *China: Land of Famine* (New York: American Geographical Society, 1927).

11. Seung-Joon Lee, *Gourmets in the Land of Famine: The Culture and Politics of Rice in Modern Canton* (Palo Alto, CA: Stanford University Press, 2011).

12. As quoted in Daisuke Wakabayashi and Claire Fu, "China's Bid to Improve Food Production? Giant Towers of Pigs," *New York Times*, February 8, 2023, https://www.nytimes.com/2023/02/08/business/china-pork-farms.html.

13. Hongzhou Zhang and Genevieve Donnellon-May, "China's Efforts to Reel in Overfishing," *East Asia Forum*, August 3, 2022, https://www.eastasiaforum.org/2022/08/03/chinas-efforts-to-reel-in-overfishing; Nerissa Hannink, "Overuse of Fertilizers and Pesticides in China Linked to Farm Size," *Stanford Earth Matters*, June 17, 2018, https://earth.stanford.edu/news/overuse-fertilizers-and-pesticides-china-linked-farm-size.

14. Wakabayashi and Fu, "Pigs."

15. Michael Worobey et al., "The Huanan Seafood Wholesale Market in Wuhan Was the Early Epicenter of the COVID-19 Pandemic," *Science* 377, no. 6609 (July 26, 2022): 951–959, https://doi.org/10.1126/science.abp8715.

16. Mike Davis, *The Monster at Our Door: The Global Threat of Avian Flu* (New York: New Press, 2005).

17. Amy Hanser and Jialin Camille Li, "Opting Out? Gated Consumption, Infant Formula and China's Affluent Urban Consumers," *The China Journal*, no. 74 (2015): 110–128; Jakob A. Klein, "Buddhist Vegetarian Restaurants and the Changing Meanings of Meat in Urban China," *Ethnos* 82, no. 2 (2017): 252–276; Ingrid Fihl, "Risky Eating: Shanghai Families' Strategies to Acquire Safe Food in Everyday Life," *Journal of Current Chinese Affairs* 48, no. 3 (2019): 262–280.

18. Francesca Bray, *Agriculture*, in *Science and Civilisation in China*, ed. Joseph Needham, vol. 6 *Biology and Biological Technology*, pt. 2 (Cambridge: Cambridge University Press, 1984); Hsing-tsung Huang, *Fermentations and Food Science*, in *Science and Civilisation in China*, ed. Joseph Needham, vol. 6 *Biology and Biological Technology*, pt. 5 (Cambridge: Cambridge University Press, 2000).

19. Francesca Bray, *The Rice Economies: Technology and Development in Asian Societies* (Berkeley: University of California Press, 1994); Seung-Joon Lee, *Gourmets in the Land of Famine: The Culture and Politics of Rice in Modern Canton* (Stanford, CA: Stanford University Press, 2011); Jia-Chen Fu, *The Other Milk: Reinventing Soy in Republican China* (Seattle: University of Washington Press, 2018); Sucheta Mazumdar, *Sugar and Society in China: Peasants, Technology, and the World Market* (Cambridge, MA: Harvard University Asia Center, 1998); Brian R. Dott, *The Chile Pepper in China: A Cultural Biography* (New York: Columbia University Press, 2020).

20. Jack Goody, *Cooking, Cuisine and Class* (Cambridge: Cambridge University Press, 1982).

21. Richard Wilk, "From Wild Weeds to Artisanal Cheese," in *Fast Food/Slow Food: The Cultural Economy of the Global Food System*, ed. Richard Wilk (Lanham, MD: Altamira Press, 2006), 18; Charles Camp, *American Foodways: What, When, Why, and How We Eat in America* (Atlanta, GA: August House Publishing, 1989).

22. Timothy Charles Lloyd, "The Cincinnati Chili Culinary Complex," *Western Folklore* 40, no. 1 (1981): 28–40.

23. Ashanté M. Reese and Hanna Garth, "Black Food Matters: An Introduction," in *Black Food Matters: Racial Justice in the Wake of Food Justice*, ed. Hanna Garth and Ashanté M. Reese (Minneapolis: University of Minnesota Press, 2020), 16.

24. "Traditional Tea Processing."

25. Rachel Laudan, *Cuisine and Empire: Cooking in World History* (Berkeley: University of California Press, 2015).

26. See Schmalzer's chapter 5 in this volume. The broader role of state socialism in shaping the current global food system has been largely neglected apart from those working in post-socialist studies such as Yuson Jung, Jakob A. Klein, and Melissa L. Caldwell, eds., *Ethical Eating in the Postsocialist and Socialist World* (Berkeley: University of California Press, 2014) and Melissa L. Caldwell, ed., *Food & Everyday Life in the Postsocialist World* (Bloomington: Indiana University Press, 2009).

27. See for example Ken Albala, ed., *The Food History Reader: Primary Sources* (London: Bloomsbury Academic, 2014); Paul Freedman, ed., *Food: The History of Taste* (Berkeley: University of California Press, 2007).

28. Amy Bentley, ed., *A Cultural History of Food in the Modern Age* (London: Bloomsbury Academic, 2014). The series editors acknowledge that the contents deal almost exclusively with Euro-American examples, yet they still claim that the work stands in for "the Modern Age" writ large. The nod to the rest of the world comes in a single chapter at the end of each volume, while the other nine chapters focus on the Western world.

29. Priscilla Parkhurst Ferguson, *Accounting for Taste: The Triumph of French Cuisine* (Chicago: University of Chicago Press, 2004), and Rebecca Spang, *The Invention of the Restaurant: Paris and Modern Gastronomic Culture* (Cambridge, MA: Harvard University Press, 2000).

30. Jeffrey M. Pilcher, "Culinary Infrastructure: How Facilities and Technologies Create Value and Meaning around Food," *Global Food History* 2, no. 2 (2016): 105–131.

31. Lilian M. Li, *Fighting Famine in North China: State, Market, and Environmental Decline, 1690s–1990s* (Stanford, CA: Stanford University Press, 2007) and Pierre-Étienne Will, *Bureaucracy and Famine in Eighteenth Century China* (Stanford, CA: Stanford University Press, 1990).

32. Francesca Bray, *Technology, Gender, and History in Imperial China: Great Transformations Reconsidered* (Oxford: Routledge, 2013), and Mark Swislocki, "Nutritional Governmentality: Food and the Politics of Health in Late Imperial and Republican China," *Radical History Review* 2011, no. 110 (May 1, 2011): 9–35. https://doi.org/10.1215/01636545-2010-024.

33. Michelle T. King, "What Is 'Chinese' Food? Historicizing the Concept of Culinary Regionalism," in *Global Food History* 6.2 (Summer 2020), 89–109, https://doi.org/10.1080/20549547.2020.1736427.

34. M. Cristina Cesaro, "Consuming Identities: Food and Resistance among the Uyghur in Contemporary Xinjiang," *Inner Asia* 2, no. 2 (2000): 225–238.

35. Brendan Galipeau "Tibetan Wine Production, Taste of Place, and Regional Niche Identities in Shangri-la, China," in *Trans-Himalayan Borderlands: Livelihoods, Territorialities, Modernities,* ed. Dan Smyer Yü and Jean Michaud (Amsterdam: Amsterdam University Press, 2017), 207–228; Xu Wu, "The Farmhouse Joy (*nongjiale*) Movement in China's Ethnic Minority Villages," *The Asia Pacific Journal of Anthropology* 15, no. 2 (2014): 158–177.

36. Adam M. McKeown, *Melancholy Order: Asian Migration and the Globalization of Borders* (New York: Columbia University Press, 2008) and "Global Migration, 1846–1940," *Journal of World History* 15, no. 2 (2004): 155–189, http://www.jstor.org/stable/20068611.

37. Katie Thornton, "Researchers at This Base in Antarctica Eat Better Than You Do," *Atlas Obscura* (April 5, 2019): https://www.atlasobscura.com/articles/food-in-antarctica.

38. Ien Ang, *On Not Speaking Chinese: Living Between Asia and the West* (London: Routledge, 2001); Tu Wei-ming, ed., *The Living Tree: The Changing Meaning of Being Chinese Today* (Stanford, CA: Stanford University Press, 1994).

39. Mukta Das, "Making It in China: Informality, Belonging and South Asian Food in Guangzhou, Macau and Hong Kong," PhD diss., SOAS University of London (2019); Elaine Lynn-Ee Ho, "African Student Migrants in China: Negotiating the Global Geographies of Power through Gastronomic Practices and Culture," *Food, Culture & Society* 21, no. 1 (2018): 9–24.

40. Arjun Appadurai, "Gastro-politics in Hindu South Asia," *American Ethnologist* 8, no. 3 (1981): 494–511.

41. See, for example, Krishnendu Ray and Tulasi Srinivas, eds., *Curried Cultures: Globalization, Food, and South Asia* (Berkeley: University of California Press, 2012); James McCann, *Stirring the Pot: A History of African Cuisine* (Athens: Ohio University Press, 2019); Sami Zubaida and Richard Tapper, eds., *A Taste of Thyme: Culinary Cultures of the Middle East* (London: Tauris Parke Paperbacks, 2000); Katarzyna J. Cwiertka, *Modern Japanese Cuisine: Food, Power and National Identity* (London: Reaktion Books, 2006); and Hanna Garth, ed., *Food and Identity in the Caribbean* (London: Bloomsbury Academic, 2013).

1 CREATING VALUE: MARKET STRUCTURES AND COMMODITIES

1 LORD MILLET IN ALIBABA'S CAVE: THE RESURRECTION OF AN ICONIC FOOD

FRANCESCA BRAY

"This year's new millet from the Taihang terraces, grown with donkey manure." "Millet from the terraces of the Agricultural Heritage Site of Wangjinzhuang, traditional varieties grown with traditional techniques." "Yellow millet from Qinzhou in Shanxi, peasant-grown, in its bran." These are just a handful of the hundreds of advertisements on Taobao, Alibaba's online sales platform, purveying *xiaomi* (小米 millet grain) from the rugged hills of China's ancient loesslands to retailers, eco-consumers, health faddists, village homestays, and specialist restaurants across North China and beyond.[1]

Through the violent upheavals and titanic socialist development projects of the twentieth century, many remote rural regions of China remained desperately poor. Among them were Qinzhou in Shanxi and Wangjinzhuang in Hebei, both high in the Taihang mountains. Now the millets that used to be a mark of their destitution, and the tradition that used to be a mark of their backwardness, have become a source of wealth and respect.

In today's prosperous China, the taste of millet evokes both bitterness and sweetness for older people: a "coarse" food cheap enough for communal canteens and army rations, in Maoist times millet also featured in *yi ku fan* (忆苦饭 "remembering bitterness meals"), unappetizing items served before the New Year feast to remind people of pre-Revolutionary hardships. In contrast, younger people worried about cholesterol or weight-gain appreciate millet's complex carbs. Others seek out peasant-grown grains as "green" foods untainted by pesticides or pollution. Or as gourmets they may simply enjoy millet congee as an alternative to rice porridge. Eaten in the countryside, in a farm homestay, or village eatery, millet is an essential part of the *nongjia fan* (农家饭 peasant food) experience that prompts warm conversations, and a feeling of solidarity and trust, between locals and visitors.[2] Consumer enthusiasm for millet as a healthy, authentic, enjoyable food is just one strand in a complex braid of factors that today are rescuing an ancient Chinese staple—the bowl of millet—from near extinction.

The long history of millets is almost exactly the inverse of the history of soy sauce discussed in chapter 2 by Angela Leung, where, thanks to an escalation in supplies of the raw beans, an expensive delicacy became a ubiquitous and indispensable condiment. Millet, in contrast, began as the staple food on every table—not so much a commodity as the staff of life around which not only the foodways but also the whole production and administrative system of the Chinese empire were organized. But as other staple grains rose to prominence in the empire's diet, millet dwindled into insignificance, becoming a food associated with subsistence farming, poverty, and backwardness.

Today, however—as with the beef producers discussed by Thomas DuBois in chapter 3—millet producers in China have begun making their way up the value ladder, reestablishing millet as an authentic, locally rooted food that nourishes in many ways. In the process, not only has the millet grain itself become a valuable commodity, but so too have some of the impoverished farming landscapes where it is grown. Villages in the eroded hills and mountains around Beijing, Taiyuan, and other cities of the north have become popular destinations for weekend getaways and internal tourism. Thanks to new marketing and distribution technologies combined with burgeoning consumer demand for honest-to-goodness peasant-grown foods, the millet grown in the northern highlands has shifted from a marginal subsistence crop to a profitable commodity exported not only throughout China but also to Canada and the United States.

Meanwhile, agronomists and policymakers working in networks that span local, provincial, and national level, often collaborating with local farmer groups or NGOs such as Farmers' Seed Network China, have identified the hardy, drought-tolerant millets as sustainable crops whose breeding and cultivation should be expanded nationally and internationally to meet the challenges of climate change and increasing environmental stress.[3]

Northern Chinese millet farming systems are among those being scientifically plotted and fashioned into Nationally or Globally Important Agricultural Heritage Systems, icons of resilient tradition refashioned as models for a sustainable future, historically grounded but innovative systems for emulation at home and abroad.[4] As exports, China's crop research and its rapidly expanding portfolio of farming heritage sites both add to the nation's capital not through cash sales but as prestigious contributions to the global commons, in a soft-power project that positions China as a world leader in sustainability research and "ecological civilization."[5]

But in what sense is a half-kilo packet of Taihang millet sold on Taobao *traditional*, or a donkey plowing land for millet in a remote village a form of *heritage*? In contrast to such "heritagized" delicacies as cheese or walnuts, millets have a long, richly documented history as China's major food, the material and symbolic mainstay of the country's civilization.[6] Millets were the earliest domesticated crops in North China and the empire's iconic staples through much of its history—the food of rich and poor, the basis of the tax system, and a potent symbol of the Mandate of Heaven. So, in heritagizing millet, there is plenty of potential tradition to choose from, plenty of scope for strategic selection and framing.

The bags of Qinzhou and Wangjinzhuang millet on display on Taobao look almost identical. But while investigating the current resurrection of millet in China, I was intrigued by striking differences in how the two districts constructed a history and created value for their millet, Qinzhou focusing on the pedigree of its local variety and Wangjinzhuang emphasizing the traditional and sustainable system of farming, the landscape, that produced the grain. To put their choices in context, and to evaluate the potential of each strategy for reestablishing millet's longer-term prospects as part of modern Chinese foodways, I begin by sketching the long history of millet in China and some of its milestones as a food and crop.

THE MANORIAL MILLETSCAPE: AN ICON OF ABUNDANCE

Contrary to popular belief, millet, not rice, was the staple grain on which early Chinese civilization was founded. According to Chinese legend, it was Hou Ji (后稷 Lord Millet), a magical being conceived when his hitherto barren mother stepped in a footprint left in the soil by the supreme deity, who taught the ancient Chinese how to grow grain.[7] The grains in question were millets: foxtail millet (*Setaria italica*) and broomcorn millet (*Panicum miliaceum*), both shown in figure 1.1.[8] Broomcorn millet predominated in earliest times, but as the climate cooled, from around 5000 BCE, foxtail millet replaced broomcorn as the main staple in all but the harshest zones of North China.[9]

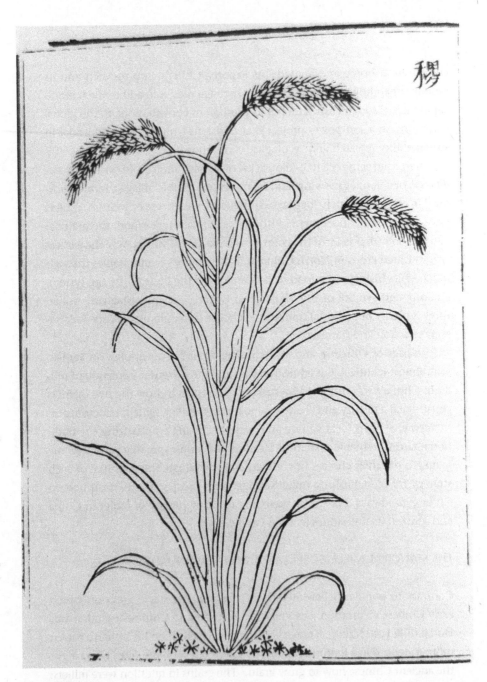

FIGURE 1.1
Chinese millets. Left, setaria (*Shoushi tongkao*, Compendium of granting the seasons, 1743, 23/5b); right, panicum (*Bencao gangmu* [*Compendium of Materia Medica*] of 1596, 2/24a in 1885 ed.). Note that both species are labeled with the same term, *ji* (稷 grain/millet).

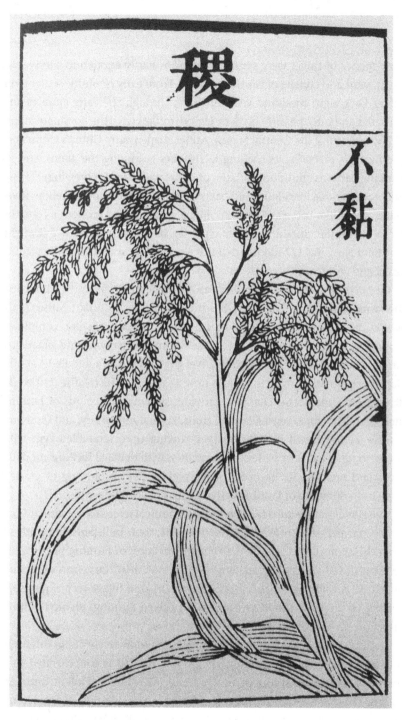

FIGURE 1.1 (continued)

Both species of millet were spring-sown crops hardy enough to survive the harsh, semi-arid climate of Northern China. From early Neolithic settlements dating back eight thousand years or more, through the early empires and dynasties and into the early modern period, millets were the dominant staple of *Zhongguo* (中国 the Central States). Millet shaped early China's cropscapes and farming practices, its cooking techniques (steaming the grain), and its drinking patterns (glutinous varieties of millet were used for brewing).[10] From the mountainous loess-lands of Shaanxi Province down to the Yellow River plains, the millet belt was China's political, cultural, and economic core, the site of its capital cities and its principal tax base. Only with the fall of the Northern Song in 1127 did rice definitively supplant millet as China's iconic cereal and cropscape.[11]

Our richest source for the millet-based diet of medieval North China and the farming system on which it depended is the agricultural treatise *Qimin yaoshu* (齊民要術 Essential techniques for the common people), completed around 540 CE.[12] The author, Jia Sixie, was an estate owner and practicing farmer who had served the Northern Wei government as a middle-level official. His estate was located in what is now Shandong. The treatise (ten books divided into ninety-two chapters) provides systematic accounts of farming methods, field crops, vegetables and fruits, animal husbandry, and food processing techniques. It includes recipes ranging from scrambled eggs with chives to meat jellies or barley-sugar candies, with detailed sections on dairy foods and brewing. As Jia notes in his preface, "From plowing to pickles, there is no domestic or farming activity that I have not described."[13]

The *Qimin yaoshu* quickly became the canonical reference for agricultural writers, farmers, and officials in North China, their indispensable guide to what they considered an unequaled sophistication of farming practice. Jia Sixie portrayed an iconic Northern farming system at the zenith of technical excellence and abundance, its intensive dryland tillage techniques supporting an enviably productive system of mixed farming, pivoted around the staple cereal, foxtail millet.

Northern China is not an easy place to farm productively. It has intensely cold winters and burning summers; the low rainfall is concentrated in a couple of months in spring or summer and often falls as violent thunderstorms. The *Qimin yaoshu* lays out the principles of an effective dryland farming system. Land was plowed and harrowed several times. Millet seed was sown in straight furrows using a drill, carefully spaced to reduce competition between plants for water and nourishment, and to facilitate weeding. The seed was covered and mulched by drawing a light bush-harrow across the furrows. Hoeing was repeated again and again, to get rid of weeds but above all to keep the surface-soil in light crumbs, a dust mulch to protect the crop's roots and conserve moisture.

If you got the millet right, the rest would follow. The *Qimin yaoshu* laid out a complex farming system that resembled a patterned brocade, interweaving multiple combinations and options. Crops were rotated; green manures such as mung beans or alfalfa were plowed in; precious manure from livestock fed on cereal straw, beans, or turnips, was divided between field crops, gardens, and orchards. But whatever the pattern of the weave, setaria millet was always the warp thread, the organizing principle. Millet was the staff of life.

Within the medieval foodscape reflected in the *Qimin yaoshu*, setaria millet was the *fan* (饭 staple), eaten either as steamed grains or as porridge. For relish, rich and poor alike added stews or soups of fresh or preserved vegetables, sauces, pickles, or *shi* (豉 fermented soybean paste), a cheap and tasty condiment that amplified the savor of expensive salt and also (as we now know) provided a valuable source of protein.[14] Rich and poor shared a diet based on combinations of staple grain and relish. But the diet of the poor was essentially vegetarian: food was boiled or steamed, cooking oil was a luxury, and instead of tasty millet the poorest families might have to make do with a bitter and indigestible pot of boiled soybeans.[15] Wealthy households, like Jia Sixie's, had much richer and more varied diets. Their manorial estates contained vegetable gardens and orchards, their fields rotated between different crops, and they often kept large flocks of animals, especially sheep and goats, for meat and milk. Rich households' stews, pickles, and sauces typically contained fresh, dried, or fermented meat or fish as well as vegetables. One of Jia's simplest recipes for soup called for: "Five pounds of calabash leaves, three pounds of mutton and two pounds of onions with half a pint of fine salt added to taste."[16] Homemade yogurts and cheeses, home-grown apricots, chestnuts, Chinese dates, mulberries, melons, and Sichuan peppers as well as home-processed pickles, beers, and soybean paste were all part of the manorial mixed-farm repertory, part for home use and part for sale.

The variety and richness of the northern food system as described in *Qimin yaoshu*, its supplies of fruits and vegetables, ales and meats, dairy and preserves, depended on the scale of operation and resources. Manorial farms could afford plentiful draft animals and equipment. They had enough land to intensify and diversify their production, enough labor to brew or garden at commercial scale, enough resources to weather a fall in prices or a bad harvest. Peasant farmers usually had no choice but to devote most of their land to millet if they were to feed their families and pay their taxes, and with fewer resources, they were less productive than large estates. Following devastating civil wars and invasions, from around 800 CE, the estates of the northern Chinese aristocracy dwindled, and as millet farming shifted to the peasant scale it became associated with hardship rather than abundance.

THE IMPERIAL MILLETSCAPE: AN ICON OF RESPONSIBLE RULE

China was an agrarian empire. Good farming was the foundation of the imperial world order, ever shadowed by the threat of shortages, hunger, and disorder. Bounteous harvests were signs of divine approval, confirming that the emperor and his government were fulfilling their duties in ways deserving of the Mandate of Heaven. Since the Zhou dynasty, the emperor had sacrificed twice yearly to the *sheji* (社稷 deities of soil and grain/millet). But the gods help those who help themselves: actively performing *quannong* (勸農 promoting agriculture) was part of the Mandate. Imperial officials were expected to study the principles of farming so that they could help the farmers in their charge to improve their methods and raise their output.[17] Peasant households farmed not only to feed themselves but to pay the taxes in millet and cloth that, in the terms of canonical political economy, fed and clothed the state. In return, in addition to provisioning the court, the bureaucracy, and the army, the state used the tax millet accumulated in its granaries to help peasant farmers by regulating prices and by distributing grain in times of shortage. As an example of the scale involved, the Hanjia tax-granary, built in 605 CE in the Sui capital Luoyang, held up to six million bushels, (approximately 600,000 m^3) of millet contained in three hundred huge lime-lined pits.[18]

But by 605, southern rice had already begun to rival northern millet as the source of national wealth. The completion of the Grand Canal in 609 facilitated the transfer of southern tax rice to the northern capital, and by around 800 the southern provinces were overtaking the north in wealth and population. With the fall of the Northern Song in 1127, the millet zone of the old Central States was lost to the empire for over a century. At this point, rice definitively replaced millets as China's iconic crop and its most prestigious food; from now on, the rice regions of the south reigned supreme as the economic and cultural heartlands. Yet symbolically, millet retained an aura of eminence. It remained the grain of imperial sacrifice (see figure 1.2), while renowned agricultural treatises that covered both southern and northern farming still began their sections on cereal crops not with rice but with millet.[19]

Materially, too, millets remained an important—if no longer dominant—resource throughout the north.[20] Farmers grew millets along with wheat and "diverse" or "coarse cereals" (sorghum, buckwheat, soy, and other beans). They sold their wheat to the cities and ate millets themselves—if they could. Sorghum, a large-grained cereal introduced under Mongol rule, was normally used to feed animals or to make liquor; like maize, which arrived in the sixteenth century, sorghum was disdained as a famine food.[21] Northern peasants and the urban poor continued to eat millet when they could, in preference to other "coarse grains."[22] But the preferred staples of better-off households, especially in towns and cities, were local wheat products (*mian* [麵 noodles] or *bing* [餅 flatbreads]) and rice imported from the south. The well-off now consumed millet not as a staple but as a breakfast dish or a nourishing soup for the elderly or for women recovering from childbirth.[23]

Where the late imperial state was concerned, however, millet, with its excellent keeping properties, remained an important grain for provisioning northern populations. Through the 1740s, for instance, the Qing government regularly purchased hundreds of thousands of bushels of millet to stock public granaries, pay salaries, and distribute food relief in Zhili, the capital province. But these supplies were no longer coming from the traditional Chinese millet belt: recently incorporated territories north of the Great Wall—Mongolia and Manchuria—were now the chief suppliers. The old millet zones of the heartlands continued to grow millet but no longer produced sufficient surpluses to supply the national market. Yet it was still to the ancient milletlands of Shandong, Henan, and Shanxi that the court turned for the "tribute millet" used in the all-important twice-yearly ancestral sacrifices made by the emperor at the Temple of the Soil and Grain.[24]

MODERN MILLET: BACK FROM THE BRINK

By the 1920s, when John Lossing Buck and his colleagues at the Nanjing Agricultural Institute surveyed China's farming regions and systems, they found that millet was still widely grown across the hills or mountains of the

FIGURE 1.2
Giuseppe Castiglione, *Gathering of Auspicious Signs*, 1723. The auspicious plants include an unusual double head of foxtail millet (top left). Image courtesy of National Palace Museum, published on Wikimedia Commons: https://commons.wikimedia.org/wiki/File:Gathering_of_Auspicious_Signs.jpg.

north, from Gansu in the west to Shandong in the east, and from Manchuria in the north to the Qinling Mountains in the south.[25] Manchuria exported millet. In the loesslands, the core of the ancient Central States and their millet culture, including the Taihang Range whose villages are exporting their branded millets today, millets clung on tenaciously as subsistence crops, a crucial staple for the poor, as no other cereals would prosper under these harsh conditions. One Hebei villager interviewed by Japanese investigators in 1937 said that his three meals a day consisted of boiled millet and gruels, some vegetables, and bean noodles.[26] Some decades later, Wangjinzhuang households eked out their meager supplies of millet and fuel by adding maize dumplings to the pot; or they prepared *laofan* (捞饭 scooped porridge), where the boiled millet grains were scooped off with a ladle for the family members doing strenuous work on the farm, while those doing lighter work at home drank the liquor from the pot.[27]

Buck's surveys were part of a concerted campaign by the Republican government to modernize agriculture, raise output, and strengthen the nation. Unsurprisingly, official and scientific attention focused on globally important crops. Rice, wheat, and cotton benefited from a national network of breeding programs and experimental stations, and from transnational flows of expertise and material inputs. As crops of merely local importance, millets, in China as in India and Africa, were relegated to the status of what agronomists call "minor" or "orphan crops," neither researched nor developed.[28]

Yet millets were perfect crops for hard times and tough places: up in the arid mountains of Shaanxi, in Yan'an, and the other revolutionary bases, the Red Army lived on millet, along with the local farmers.[29] After Liberation, the upland villages continued to grow millets as they had done for centuries. But from the late 1950s onward, under the pressure of successive campaigns to raise grain output, maize—which has astonishingly high yields compared to all other cereals—steadily nudged millets and other summer crops like sorghum to the margins of local cropping systems and diets.[30]

Scientific breeding programs of the 1960s and 1970s produced new millets with considerably higher yields (albeit requiring chemical inputs). Nevertheless, between 1949 and 2014, the area under setaria shrank by eleven-twelfths, from 9.2 million hectares to 0.7 million hectares; from 1980 to 2010, the amount of millet grain produced fell by two-thirds, from 5.5 to 1.5 million tons. By the 2010s, millets were grown only on hilly and marginal land, in poor and remote areas like the Taihang Range.³¹ Then, suddenly, their yellow millet turned to gold!

GOLDEN PEARLS AND DONKEY PLOWS: COOKING UP A PAST

Today Chinese millets have found burgeoning new markets as healthy heritage foods.³² Drawing on the replenishing qualities attributed to them by imperial medicine, millets are marketed through Taobao and other online platforms and presented in food and health literature as nourishing breakfast dishes as well as postpartum and baby food. As a taste of heritage, they are marketed as basic ingredients for all kinds of traditional cakes and snacks.³³ The labels on the artfully designed packets of millet displayed online declare their contents to be eco-crops, "peasant-grown and donkey-manured," "traditional varieties grown with traditional techniques."³⁴ In other words, they embody China's farming past come to life. But how are those pasts recreated? A comparison of the very different traditions that two of today's leading millet-producing communities, Qinzhou and Wangjinzhuang, have constructed for themselves suggests some of the challenges of successfully incorporating old foods into new foodways.

Qinzhou and Wangjinzhuang are among several marginalized and ostensibly archaic communities in North China experiencing an economic and cultural revival rooted in a renewal of local millet farming. Wangjinzhuang's towering terraces make it a natural tourist destination. Now that the village has good road access, visitors flood in, admiring the stunning vistas, clamoring to have their photos taken with the village's famous donkeys, and devouring bowls of millet porridge in village eateries. Wangjinzhuang farmers market not only their millet but also their maize, Sichuan pepper, chestnuts, and medicinal herbs across the country, relying heavily on the Internet.³⁵ Qinzhou, in contrast, is not particularly picturesque and does not feature as a "must-see" destination on the tourist routes. But it does possess a millet variety so excellent that it was supplied as tribute to the Kangxi emperor three hundred years ago. Leveraging the reputation of this auspicious grain, Qinzhou has vastly expanded its millet acreage and developed a flourishing millet-processing industry.³⁶

In assessing how tradition is mobilized in revitalizing these two millet systems, clearly cultivating the millet crop on the spot is essential in both places. But Qinzhou emphasizes the pedigree of its local millet variety, leaving the local farming and processing system largely invisible. In Wangjinzhuang, it is the visible system of production, the dryland farming techniques and the mixed farming system in which they are embedded, on which the claims to tradition and heritage are founded.

Wangjinzhuang's terraces, located in She County, Hebei, were designated a Nationally Important Agricultural Heritage System (NIAHS) in 2014; in May 2022, the district was granted coveted UNESCO-FAO GIAHS status, defined as follows: "The Globally Important Agricultural Heritage Systems (GIAHS) represent not only stunning natural landscapes but also agricultural practices that create *livelihoods* in rural areas while combining *biodiversity, resilient ecosystems* and *tradition* and *innovation* in a *unique* way."[37] She County's claim to uniqueness is as a "system of dryland terraces," founded on a *wuwei yiti* (五位一体 "quintity") of nested and interdependent resources: stone-donkeys-crops-terraces-villagers.[38]

In her 2019 article "Layer upon Layer," Sigrid Schmalzer analyzes the foundations undergirding Wangjinzhuang's fashioning as a site of agricultural heritage.[39] It seems that the terraces that dominate the landscape today only gradually came to such prominence. One set of terraces in Wangjinzhuang is dated to 1290, and as the county's population grew through late imperial times, through immigration as well as natural increase, here as elsewhere in China farms slowly spread from the valleys up the mountain slopes. But it was during the Maoist era that the landscape was carved into the dramatic form we see today. In the national drive to expand food production, as in many other poor and environmentally fragile regions, the inhabitants of Wangjinzhuang mobilized the only resources they had—their own labor, soil, and stone—to extend the existing terraces into a spectacular landscape of masonry-buttressed strip-fields covering the steep mountainsides from foot to peak. In the 1970s, the village was honored as a model of socialist ingenuity and labor, but it remained economically isolated and poor, dependent on its harvests of millet and maize produced with not only donkey-plows but also motorized hand-tillers.

Even ten years ago Wangjinzhuang's donkeys might well have been on the way out: in most of China, machines have steadily replaced donkeys, mules, and buffalo in the name of efficiency. But the Wangjinzhuang donkeys were saved, along with the wooden plows and harrows they pulled, by the stipulations of eligibility for NIAHS status (see figure 1.3). This system of tillage certainly looked "traditional," and in fact, when it came to one

FIGURE 1.3
A Wangjinzhuang donkey. Photo by Sigrid Schmalzer, reprinted with permission.

of the features of the local farming system deemed unique to the county, namely its capacity for "soil water storage and soil moisture conservation," donkey-drawn plows performed as well (if not better) than mechanical tillers. Furthermore, donkeys conveniently eat the parts of the crops that humans would otherwise discard, converting them into manure for the fields. Donkeys thus substitute for industrial inputs (diesel, chemical fertilizer), promoting an ecologically virtuous feedback system.[40]

Apart from the terraces and the maize, all the principles and practices of Wangjinzhuang's dryland farming system, as well as its crops (millets, beans, chestnuts, Chinese dates, persimmons, medicinal herbs, and the rows of Sichuan pepper trees planted up against the stone walls to catch the heat of the sun) would have been familiar to, and applauded by, the author of the *Qimin yaoshu*. Indeed, many of the farming tools and techniques used in Wangjinzhuang exactly replicate those described in *Qimin yaoshu*, though curiously this ancient and revered precedent is not cited in the abundant popular and scholarly literature on Wangjinzhuang. One *Dagongbao* photo reportage shows a farmer using a traditional plow and harrow that could have stepped right out of any medieval farming manual; another depicts a man harvesting millet in his small terraced field, using the type of sickle that has been in use across East Asia for over two thousand years.[41]

However traditional the sickle he is brandishing may be, the Wangjinzhuang farmer photographed waist-deep among the millet is part and parcel of the creative combination of tradition and innovation intrinsic to the Agricultural Heritage concept: sustainability, resilience, and a viable future, whether environmental or social, depend on adapting to new challenges and creating new opportunities. It is no coincidence that this particular reportage appears in the finance section of *Dagongbao*. The farmer in question, Wang Hulin, is a native of Wangjinzhuang who left to train in microelectronics at the University of Technology in Tianjin. He returned to She County and set up an electronics company that he still directs. But in 2014, after an accident led him to rethink his priorities and dream new dreams, Wang decided to invest his energies and skills in the future of local farming. He began farming himself, in his natal village, and helped set up the She County Terraces Protection and Utilization Association, developing the online marketing that provides local farmers with profitable outlets for their produce. As associate director He Xianlin puts it, the association "unites village cadres, entrepreneurs and organizations of old farmers and craftsmen." In other words, it constitutes the interface between local producers, the state, and "more than 10" *longtou qiye* (龙头企业 dragon-head enterprises) whose chief role in Wangjinzhuang, at least according to He Xianlin, is to develop products and markets for those products.[42]

Dragon-head enterprises are agribusiness companies supported by local government to develop commodity chains, linking farms to markets and often intervening to consolidate and specialize production. Some successfully promote the rural regeneration, poverty reduction, and community building that is their ostensible purpose, while others pursue turnover and profits in exploitative ways evocative of Western agribusiness.[43] With a dozen dragon-heads at work marketing millet, red pepper, and black beans as well as herbs, honey, and nuts, it seems that in Wangjinzhuang individual dragon-heads would have relatively little power to set the terms of local production. Qinzhou, in contrast, appears to be a company town and a one-crop district.

The nineteen townships within Qinzhou do not aspire to NIAHS status and are not categorized as having a specific farming system. Qinzhou's claims to fame and fortune lie not in the uniqueness and traditional nature of its field systems and farming methods, as in Wangjinzhuang, but in the branded identity and quality of the setaria variety it grows. *Qinzhouhuang* (沁州黄 Qinzhou yellow/gold) caught the attention of emperors and was sent to court as a tribute grain. The Kangxi emperor himself (r. 1661–1722) appreciated its delicious flavor, reportedly calling it "king of millets" and "golden pearls." It was renowned for its health-giving properties: sweet in taste, slightly cold in nature, with the effects of clearing heat, diuresis, reducing swelling, and nourishing yin.[44] It is recognized by crop geneticists as one of four traditional setaria varieties famous for their superior flavor. Although breeders have worked on improving and stabilizing the cultivar since the 1980s, *Qinzhouhuang* is one of the few current setaria varieties that has not been subjected to crossing with the "super cultivar" Yugu 1, first released in the 1980s.[45]

The focus in Qinzhou has been on refining this local millet variety as a unique and standardized grain, rather than on refashioning the farming system that produces it. In the early 1980s, local farmers began working with scientists on breeding stable varieties, and in 1989, the Qinzhou Yellow Millet Group (QYMG) "was established as a local flagship company in the processing and sales of [millet] products." Founded by the county, in 2002, QYMG launched as a joint stock company with a staff of around five hundred. Since 2010, county and company have been working together with various leading research institutes to develop "green" but nutritionally enriched varieties for making products like babyfood, currently marketed under the label *Gu zhi ai* (谷之爱 love of millet).[46]

In recent years, millet prices have soared nationally compared to other dryland cereals such as maize. *Qinzhouhuang* is a National Geographic Indication product: it must have been grown within the nineteen townships, using specified varieties, and be certified as having met specific standards. In 2014, about one quarter of the total Group area, the "yellow millet green standardized planting base," was certified as reaching the "organic" standard, at prices three times those for nonorganic millet.[47] By 2016, QMYG's output value was 164 million yuan (USD 25 million), and millet brought in an average of 4,500 yuan annually to farmers. Over 15,000 farmers across Qinzhou had signed contracts with the Group; some switched from maize to millet, others opened up abandoned land high in the hills to grow the valuable crop.[48]

Engaged in every step of the chain, from breeding and nutritional research to processing, product development, marketing, and sales, QYMG offers a classic example of the accelerating trend analyzed by Mindi Schneider and her colleagues whereby a dragon-head enterprise vertically integrates a locality and its product into markets.[49] The "Poverty reduction" webpage on Qinzhou represents the QYMG system as articulating peasant-scale farms with factory-scale processing, depicting a landscape of steep, tiny fields and farmers wielding hoes. The Baidu Baike webpage, however, also features photographs of industrial-scale millet fields.[50] Do the large and small farms coexist peacefully, or is there a trend involved? And how many Qinzhou farmers, if any, chose to grow millet outside the sheltering but controlling umbrella of QYMG? I could find no information about possible renegades.

"The term *agricultural heritage* prompts people to talk about ideas, skills, and objects being handed down through time, but this concept of time tends to be emptied of events and process, what historians and other historically minded scholars privilege."[51] Qinzhou and Wangjinzhuang have each devised their own millet tradition as the basis for a system of rural revitalization that combines the authenticity craved by consumers with the resilience needed by producers. What use does each make of the historical resources potentially at their disposal? It is notable that neither district has mobilized more than a fraction of the long and richly documented history of Chinese millets.

Qinzhou's claim to heritage status is "an object handed down through time," the *Qinzhouhuang* variety with its special qualities. The brand's publicists emphasize the dazzling pedigree of Qinzhou's golden pearls stretching back to Ming tribute lists and on, through Kangxi's lyrical praise, to a list of Gold Medals beginning with the Panama-Pacific Exposition of 1915 in San Francisco.[52] Four centuries of history are antiquity enough for most modern consumers. Even should it occur to the QYMG publicists, it would only distract from *Qinzhouhuang*'s unique connection with place and time to invoke Lord Millet or the millennial history of setaria as national staple. In its standards, furthermore, QYMG does not stipulate any specific techniques or organization of production, "traditional" or otherwise, beyond the use of organic or nonorganic fertilizers—it is place and variety, not method or scale, that assure quality. This bracketing helpfully accommodates the hybridity of Qinzhou's farming system: smallholders and large-scale farms can both sell their millet to the processing plant without undermining the grain's claims to authenticity on national markets. Should large mechanized farms displace small plots worked with hoes, as is becoming increasingly frequent in China's other grain regions, this will not impair *Qinzhouhuang*'s reputation as a quality consumer product.

Wangjinzhuang, by contrast, has no genealogy of glorious awards for its millet varieties. They go unnamed on the packets and are unrecognized beyond the locality, even though the community has now started its own seedbank that includes twenty-two named setaria cultivars along with dozens of vegetables, fruits, and other cereals.[53] Wangjinzhuang's branding depends on its "traditional" farming system, in which millet is represented as just one crop among many rather than the backbone of the system. The local agricultural bureau and the national heritage commission identify the district's cultivation techniques and crop combinations as a unique feature of dryland terrace farming, a system for which they claim seven hundred years of history even though terraces only came to dominate the local landscape in the Maoist era. Understandably, however, tourists and urban consumers are content for history from before the 1950s to fuse into a generic "ancient times" blur, while terraces have been disproportionately prominent in agricultural heritagization since the movement began. In fact, the techniques and crop clusters of Wangjinzhuang are not specific to terraces but belong to a dryland farming system of great antiquity and range, richly documented in *Qimin yaoshu* and organized around millet as the staple cereal. Historically, this system of mixed farming was at its most effective, productive, and resilient when practiced on estates, that is, under conditions of structural inequality and exploitation at odds with the goals of poverty reduction and community revitalization enfolded into Wangjinzhuang's current projects.

Millet farming in Wangjinzhuang has survived centuries of vicissitudes. But how sustainable is it likely to be in the future? Attractive as it may be to tourists and environmental scientists, Wangjinzhuang's quintity of stone-donkeys-crops-terraces-villagers lacks one fundamental dimension of the resilience that is supposedly at the heart of agricultural heritage systems: it cannot survive unless enough villagers play along. Persuading young people to stay in the village, to farm, raise donkeys, and cultivate remote terraces, will not be easy however well the local crops sell online.[54] New-build farmhouses don't accommodate donkeys. Machine tillers save labor but can't access remote fields. Even if tourism and heritage subsidies help maintain a viable population of villagers, the chances of preserving the "traditional" farming landscape and methods of the Wangjinzhuang brand beyond a picturesque but shallow façade seem slender. In any case, as Sichuan pepper is by far the most profitable local crop, the future of millet-growing in Wangjinzhuang seems far less certain than in Qinzhou. If Wangjinzhuang succeeds as a sustainable heritage site, it would not be surprising if its bustling food-stalls turned to purchasing all their millet from other locations, as so many other tourist villages do.

FROM FARMING TO FOODWAYS?

The current millet revival springs from a complex braiding of changes in contemporary Chinese society. Consumer demand for healthy foods, rural tourism, online commerce, the heritage industry, poverty reduction programs, and sustainability research all contribute to growing markets for the grain, and to the regeneration of local millet landscapes supplying those markets. Millet in China today is what Heather Paxson terms an "unfinished commodity"—as food and as crop it encapsulates multiple, heterogeneous forms of value, some in synergy, some in tension—a spectrum of hopes, expectations and affordances, and visions of past and future, that are shaping both production and consumption.[55] So what are millet's likely prospects? How compatible are its culinary, environmental, and social sustainabilities?

Chinese crop-breeders and environmental scientists would like to see millets become a major crop again. They are pressing the government to invest in them as crops of the future, resistant to the drought, salinity, and rising temperatures that increasingly threaten China's food supply. But to work, this strategy would require Chinese consumers to embrace millet not just as a niche product but as a regular staple. Chinese markets for millet as delicacy and health food will likely continue to expand as quality replaces quantity in the food choices of a growing number of consumers. It is even possible that millet could once again become a routine food item—after all, in the last few decades people across China have happily abandoned local cereals in favor of rice, and now are reducing rice consumption to eat more bread. But who would supply the millet, and how would it be grown?

Currently, Chinese millet production is tightly entangled with heritage initiatives. I have suggested that production of millet within the Wangjinzhuang system, for all its environmental virtues, may not be socially or economically sustainable. The Qinzhou system, by contrast, has a demonstrated capacity to expand both output and acreage of millet without damaging its brand reputation or losing its workforce. There are both social and environmental risks associated with the Qinzhou model, but in terms of dietary sustainability, QYMG's expanding range of products turns millet from a mouthful of history into a versatile modern foodstuff, promotes its integration into regular consumer food routines, and thus reduces its dependence on locale and tradition. Much as we all love the Wangjinzhuang donkeys, the Qinzhou industry appears a more solid foundation for millet as a crop, and food, for the future.

NOTES

1. For example, https://pikbest.com/e-commerce/foods-briefing-details-foods-millet-foods-details-taobao-details-simplicity-details-taobao-foods-sha_525166.html (accessed January 25, 2022).

2. Ellen Oxfeld, *Bitter and Sweet: Food, Meaning, and Modernity in Rural China* (Berkeley: University of California Press, 2017), 79; Jing Guan, Jun Gao, and Chaozhi Zhang, "Food Heritagization and Sustainable Rural Tourism Destination: The Case of China's Yuanjia Village," *Sustainability* 11, no. 10 (2019): 2858.

3. The last decade has seen a dramatic surge in millet research, typically coordinated between national, provincial, and local crop science institutes, for example, Xianmin Diao, "Production and Genetic Improvement of Minor Cereals in China," *The Crop Journal, Advances in Crop Science: Innovation and Sustainability* 5, no. 2 (2017): 103–114; Zhenling Cui et al., "Pursuing Sustainable Productivity with Millions of Smallholder Farmers," *Nature* 555, no. 7696 (2018): 363–366. Millet is a fuzzy category, a generic English term for a range of small-grained, hardy cereal species domesticated across Africa and Eurasia, each with their own local names, lores, and histories. Crop scientists investigate millet species separately, but the environmental discourses supported by their research typically advocate for millets in general.

4. Anthony M. Fuller et al., "Globally Important Agricultural Heritage Systems (GIAHS) of China: The Challenge of Complexity in Research," *Ecosystem Health and Sustainability* 1, no. 2 (2015): 1–10.

5. A policy program "intended to bring the forces of production into a pattern of sustainable development [through] a managed transition to a new set of relationships among state, society, economy, and environment"; John Aloysius Zinda and Jun He, "Ecological Civilization in the Mountains: How Walnuts Boomed and Busted in Southwest China," *The Journal of Peasant Studies* 47, no. 5 (2020): 1061.

6. Jakob A. Klein, "Heritagizing Local Cheese in China: Opportunities, Challenges, and Inequalities," *Food and Foodways* 26, no. 1 (2018): 63–83; Zinda and He, "Ecological Civilization."

7. Hou Ji's story is first recounted in the *Shijing* [Classic of poetry] (ca. 1000–600 BCE); Robert F. Campany, "Eating Better than Gods and Ancestors," in *Of Tripod and Palate: Food, Politics, and Religion in Traditional China*, ed. Roel Sterckx (New York: Palgrave Macmillan US, 2005), 98.

8. Alike in their small grains and hardy habit, both setaria and panicum are typical crops of the north, occur in non-sticky and sticky varieties, and were consumed in the same forms. As a staple, steamed as whole grains or boiled as porridge, nonsticky varieties were and are generally preferred, while for cakes and especially for brewing sticky varieties are used. No farmer would confuse setaria with panicum as a crop since the seed heads are unmistakably different, but names were another matter. The *ji* (grain/millet) of Hou Ji denoted setaria in some eras or regions, panicum in others, as did another very common term, *su*. A distinction is always made between the crop-plant and the hulled, edible grain; today *xiaomi* (small grain), as opposed to *dami* (large grain), rice typically refers to setaria and *huang mi* (yellow grain) to panicum.

9. He Hongzhong and Hui Fuping, *Zhongguo gudai suzuoshi* [A history of the cultivation of millet in ancient China] (Beijing: Zhongguo nongye keji chunbanshe, 2015); Hongzhong He et al., "Millet, Wheat, and Society in North China over the Very Long Term," *Environment and History* 27, no. 1 (2021): 127–154.

10. H. T. Huang, *Fermentations and Food Science*, in *Science and Civilisation in China*, ed. Joseph Needham, vol. 6 *Biology and Biological Technology*, pt. 5 (Cambridge: Cambridge University Press, 2000); Constance A. Cook, "Moonshine and Millet: Feasting and Purification Rituals in Ancient China," in *Of Tripod and Palate: Food, Politics, and Religion in Traditional China*, ed. Roel Sterckx (New York: Palgrave Macmillan US, 2005), 9–33.

11. Francesca Bray, "Agriculture," in *Cambridge History of China: Volume Two, The Six Dynasties 220–581*, ed. Albert E. Dien and Keith N. Knapp (Cambridge and New York: Cambridge University Press, 2019), 355–373.

12. Bray, "Agriculture."

13. Jia Sixie, *Qimin yaoshu jiaoshi* [Annotated edition of *Qimin yaoshu*, Essential techniques for the common people], eds. Miao Qiyu and Miao Guilong (Beijing: Agriculture Press, 1982), 5, my translation.

14. See Leung (this volume chapter 2); Françoise Sabban, "'Follow the Seasons of the Heavens': Household Economy and the Management of Time in Sixth-Century China," *Food and Foodways* 6, nos. 3–4 (1996): 329–349; Huang, *Fermentations*.

15. Wang Lihua, *Zhonggu Huabei yinshi wenhua di bianqian* [Changes in the culture of food and drink in medieval North China] (Beijing: Zhongguo shehui kexue chubanshe, 2000), 69–74, 80–83.

16. Jia Sixie, *Qimin yaoshu*, 464.

17. Francesca Bray, "Science, Technique, Technology: Passages between Matter and Knowledge in Imperial Chinese Agriculture," *The British Journal for the History of Science*

41, no. 3 (2008): 319–344; William T. Rowe, *Saving the World: Chen Hongmou and Elite Consciousness in Eighteenth-Century China* (Stanford, CA: Stanford University Press, 2001).

18. He et al., "Millet, Wheat," 143.

19. Notably, Wang Zhen's *Nongshu* [Agricultural treatise] of 1313, and Xu Guangqi's *Nongzheng quanshu* [Complete treatise on agricultural administration] of 1639.

20. He et al., "Millet, Wheat."

21. Pierre-Étienne Will, *Bureaucratie et famine en Chine au XVIIIe siècle*, 1st ed. 1980 (Berlin/Boston: De Gruyter, Inc., 2017), 138 and passim.

22. Lillian M. Li, *Fighting Famine in North China: State, Market, and Environmental Decline, 1690s–1990s* (Stanford, CA: Stanford University Press, 2007), 90–99.

23. Silvano Serventi and Françoise Sabban, *Pasta: The Story of a Universal Food*, trans. Antony Shugaar (New York: Columbia University Press, 2002); Peter Peverelli, "Millet—An Ancient but Rejuvenated Food," *Peverelli on Chinese Food and Culture* (blog), September 16, 2015, https://chinafoodingredients.com/2015/09/16/millet-an-ancient-but-rejuvenated-food/.

24. Will, *Bureaucratie et famine*, 138, 144–145.

25. John Lossing Buck, *Land Utilization in China: A Study of 16,786 Farms in 168 Localities, and 38,256 Farm Families in Twenty-Two Provinces in China, 1929–1933* (Nanking: University of Nanking, 1937).

26. Philip C. C. Huang, *The Peasant Economy and Social Change in North China* (Stanford, CA: Stanford University Press, 1985), 189.

27. Han Zedong, "Danshi piaoyin: Hebei Shexian hanzuo titian xitong de yinshi tixi" ['Poor livelihood': The food system of Shexian County dryland terrace system in Hebei Province], *China Agricultural University Journal of Social Sciences*, no. 6 (2017): 122–123.

28. Diao, "Production and Improvement."

29. In a 1946 interview with the journalist Anna Louise Strong, Mao Zedong declared: "We have only millet plus rifles (*xiaomi jia buqiang* 小米加步枪) to rely on, but history will finally prove that our millet plus rifles is more powerful than Chiang Kaishek's aeroplanes plus tanks"; https://en.wikipedia.org/wiki/Millet_plus_rifles, accessed January 15, 2022.

30. Between 1952 and 1980, the acreage under maize increased by 55.4 percent, and output by 225.2 percent; A. J. Jowett, "China's Foodgrains: Production and Performance, 1949–1981," *Geojournal* 10, no. 4 (1985): 381–383.

31. Diao, "Production and Improvement," 104, 105, 111.

32. Chinese millets are just one example of a wider global trend toward recuperating and commercializing such marginalized traditional staples as Andean quinoa or Indian millets as health foods or superfoods. Comparisons and generalizations are

treacherous, as each revival mobilizes a distinctive, dynamic constellation of resources. Thierry Winkel et al., "Calling for a Reappraisal of the Impact of Quinoa Expansion on Agricultural Sustainability in the Andean Highlands," *Idesia* 32, no. 4 (2014): 95–100; David Meek, "From Marginalized to Miracle: Critical Bioregionalism, Jungle Farming and the Move to Millets in Karnataka, India," *Agriculture and Human Values*, 2022.

33. Peverelli, "Millet." The Baidu page for broomcorn millet has a wonderful set of video clips illustrating the making of millet delicacies; https://baike.baidu.com/item/黄米, accessed March 23, 2024.

34. See note 1 of this chapter.

35. He Xianlin et al., "Shexian hanzuo titian xitong nongye wuzhong ji yichuan duoyang xing baohu yu liyong" [Characteristics and protection of agricultural species diversity and genetic diversity in the dryland terrace system of Shexian County], *Chinese Journal of Eco-Agriculture* 28, no. 9 (June 2020): 1453–1464.

36. "Shanxi Qinzhou huangxiaomi (jituan) youxian gongsi [Shanxi Qinzhouhuang Millet (Group) Co. Ltd.]," October 24, 2019, http://www.xinhuanet.com/energy/2019-10/24/c_1125141637.htm; Shanxi Qinzhouhuang Millet (Group) Co., Ltd., "Shanxi Qinzhouhuang Millet (Group) Co., Ltd., Changzhi, China," eWorldTrade.com, accessed March 11, 2021, https://www.eworldtrade.com/c/qinzhouhuangmillet/.

37. Wenjun Jiao and Qingwen Min, "Reviewing the Progress in the Identification, Conservation and Management of China-Nationally Important Agricultural Heritage Systems (China-NIAHS)," *Sustainability* 9, no. 10 (2017): 1698. https://www.mdpi.com/2071-1050/9/10/1698, emphasis added.

38. The nested system of the quintity is illustrated in He Xianlin, "Characteristics and Conservation Practice of Dryland Terrace System in Shexian County, Hebei Province, China," https://www.giahs-minabetanabe.jp/erahs/assets/pdf/Symposium3-1_HeXianlin.pdf. For further analysis of the system, see, for instance, the six articles by He Xianlin and others in a special issue of *China Agricultural University Journal of Social Sciences* 34, 6 (December 2017).

39. Sigrid Schmalzer, "Layer upon Layer: Mao-Era History and the Construction of China's Agricultural Heritage," *East Asian Science, Technology and Society* 13, no. 3 (September 1, 2019): 413–441.

40. Li Heyao, "Nongshi yu xiangqing: Hebei Shexian hanzuo titian xitong de lü wenhua" [Farming and living: The donkey culture of the dryland terrace agriculture system], *Chinese Agricultural University Journal of Social Sciences* 34, no. 6 (2017): 103–110; Schmalzer, "Layer upon Layer," 219.

41. "Shuishou huixiang suimeng xinnongcun [Landing back home to pursue the dream of a new countryside]," *Dagongbao*, November 5, 2020, http://www.takungpao.com.hk/finance/236133/2020/1125/524712.html.

42. He Xianlin, "Characteristics."

43. Jane Hayward, "Beyond the Ownership Question: Who Will Till the Land? The New Debate on China's Agricultural Production," *Critical Asian Studies* 49, no. 4 (October 2, 2017): 523–545; Qiangqiang Luo, Joel Andreas, and Yao Li, "Grapes of Wrath: Twisting Arms to Get Villagers to Cooperate with Agribusiness in China," *The China Journal* 77 (January 2017): 27–50.

44. For example, "Qinzhouhuang xiaomi," Baidu Baike, accessed March 23, 2024, https://baike.baidu.com/item/沁州黄小米.

45. Xianmin Diao and Guanqing Jia, "Foxtail Millet Breeding in China," in *Genetics and Genomics of Setaria*, ed. Andrew N. Doust and Xianmin Diao (Cham: Springer International Publishing, 2017), 93–113.

46. Peverelli, "Millet"; "Qinzhouhuang xiaomi"; "Soaring Prices of Millet Sees a Shanxi Village Revitalized," China's Poverty Reduction Online, September 26, 2018, http://p.china.org.cn/2018-09/26/content_64087839_4.htm.

47. Shujie Feng, "Geographical Indications: Can China Reconcile the Irreconcilable Intellectual Property Issue between EU and US?," *World Trade Review* 19, no. 3 (July 2020): 424–445; "Qinzhouhuang xiaomi"; "Soaring Prices."

48. "Soaring Prices."

49. Mindi Schneider, "Dragon Head Enterprises and the State of Agribusiness in China," *Journal of Agrarian Change* 17, no. 1 (2017): 3–21.

50. "Soaring Prices"; "Qinzhouhuang xiaomi."

51. Schmalzer, "Layer upon Layer," 430.

52. "Qinzhouhuang xiaomi."

53. Xin Song, Guanqi Long, Ronnie Vernooy and Yiching Song, "Community Seed Banks in China: Achievements, Challenges and Prospects," *Frontiers in Sustainable Food Systems* 5 (April 2021), https://www.frontiersin.org/articles/10.3389/fsufs.2021.630400.

54. Tianyu Guo, María García-Martín, and Tobias Plieninger, "Recognizing Indigenous Farming Practices for Sustainability: A Narrative Analysis of Key Elements and Drivers in a Chinese Dryland Terrace System," *Ecosystems and People* 17, no. 1 (2021): 279–291.

55. Heather Paxson, *The Life of Cheese: Crafting Food and Value in America* (Berkeley: University of California Press, 2012).

2 BECOMING AN EVERYDAY FOOD: SOY SAUCE IN MODERN CHINA (CA. 1800–1930)

ANGELA KI CHE LEUNG

The banality of soy sauce as a condiment in Chinese and East Asian cuisines and its easy accessibility as a global commodity do not naturally inspire the study of its history. The Chinese tend to assume that it is as old as Chinese civilization itself, even after the seminal study on Chinese fermented foods by H. T. Huang in 2000 and its Chinese translation in 2008.[1] When I give talks on the history of soy sauce, describing its first mention in Song texts, audiences in China often react with skepticism at this "late" dating, claiming archaeological findings and older lexes that could suggest earlier existence of soy sauce, or some version of it. On websites of East Asian soy sauce makers, one sometimes finds the claim of soy sauce having some three thousand years of history.[2] Such putative assumption reflects how firmly the idea of soy sauce is ingrained in the imagination of an ancient, uninterrupted Chinese or East Asian tradition. Ishige Naomichi's insightful observation that soy sauce delineates East Asia as a cultural sphere would be strongly backed by this popular perception, while the cultural identifying power of the condiment divests it of its historicity.[3] This chapter pieces together the history of Chinese soy sauce precisely to explain its modern cultural power. It traces the product's trajectory from an obscure traditional condiment for the elite to an everyday food in the late Qing and early Republican period to reveal its fate inextricably linked with that of a Chinese nation striving toward modernity. It was during this relatively recent process of its long history that, I believe, soy sauce acquired the cultural power we are familiar with today.

FROM ELITE FOOD TO POPULAR COMMODITY: A LONG JOURNEY

It is clearly explained in H. T. Huang's thorough study on the history of Chinese fermented foods that the origin of *jiangyou* or *shiyou* (醬油 or 豉油 soy sauce), as we know it today, remains obscure, with the first extant mention of the term *jiangyou* in a thirteenth-century elite recipe text. The earliest extant formula for making soy sauce can be found only in a 1360 dietetic book authored by one of the most influential painters in Chinese history, Ni Zan (1301–1374): "For every official peck of *huangzi* (黃子 fermented yellow bean), have ready ten catties of salt and twenty catties of water. On a *fu* (伏 warm season) day, mix them [in a jar] and incubate."[4] This short formula succinctly mentions the key ingredients in soy sauce: soybeans and the ferment developed on it, salt, and water, though it is unclear whether fermentation was done purely on the beans or on a mixture of beans and grain flour as was practiced in later times. We also know from the description that brewing should begin in the warm season, as indicated by the summer calendrical *fu* day. Despite its succinctness, this fourteenth-century text was a clear indicator of the circulation of soy sauce knowledge and practice among the elite classes, while the term *jiangyou* was increasingly visible in different genres on refined cuisine and living in the Yuan-Ming periods.[5]

However, we have reasons to believe that even with their growing visibility in printed texts, fermented soy foods, including soy sauce, were not an easily accessible commodity for the populace as late as the sixteenth century, even in the most prosperous region of the empire. Tian Yiheng (1524–1591), son of a prominent Zhejiang scholar-official, himself a famous literatus and bon vivant, wrote in the late sixteenth century that various soy foods, including soy sauce "were rare foods for modest peasant families. Those living in isolated hills and valleys might never taste it in their whole life. Those who bought corrupted [products] infested with worms in the market were pitiable." Briefly describing the recipe for making soy sauce that could be manipulated by the well-off to make sauces with fancy flavors, Tian showed that the procedure was familiar to him and his peers.[6]

The condiment's value as a fine food in the pre-Qing period is best illustrated by its descriptions in two related Ming genres—life-nourishing art and *materia medica*, both produced and consumed by the literati class. The condiment was described, for example, in detail in the most important *Bencao gangmu* (本草綱目 *Compendium of Materia Medica*) by Li Shizhen (1518–1593) in long passages in a special chapter on fermented foods, which is also the fourth and last one on *gu* (穀 food crops). The chapter began with *shi* (豉 fermented soybeans) and the sauce that came out of it after brewing with sesame oil, salt, Sichuan pepper, ginger, and scallions, producing a juice *xiangmei juesheng* (香美絕勝 exquisitely aromatic and lovely). Li also described soy sauce that he called *douyou* (豆油 liquid of bean) under the section *jiang* (醬 pastes) made from various legumes in the same chapter.[7] Even though the author discussed the medicinal qualities of some of these fermented foods, the main tone of this section was more about the exquisite taste and aroma of fermented foods made from legumes and grains. As Vivienne Lo and Penelope Barrett have convincingly demonstrated, Li Shizhen was "a gourmet par excellence," and in this impactful and monumental *materia medica*, one can "distinguish broad—if interpermeable—categories of culinary and medical remedies and discourse,"[8] reflecting the literatus's attitude toward food and well-being in this period.

This late Ming literati lifestyle was best illustrated by the *yangsheng* (養生 life-nourishing) genre of the time. In *Zunsheng bajian* (遵生八牋 Eight approaches to nourishing life, 1591) by native of Hangzhou Gao Lian (1537–1642), the most representative of the genre in the Ming, published in the same year as Li Shizhen's *magnum opus*, soy sauce appeared in seven recipes in chapters on food and clothing: to stir fry thin slices of pork, lamb tripe, to marinate crabs and goose meat, to make mustard paste, sesame sauce, and to mix vegetables.[9] One can also add to this list a full section on recipes of making *jiang* and soy sauce in an early sixteenth-century life-nourishing text, and a few more mentions in related genres such as literati jottings, family handbooks, and a mid-seventeenth-century record of Ming court condiments. Based on this obvious but not dramatic development, H. T. Huang estimated that soy sauce started to rival *jiang*.[10] I would add that this dietetic change was largely limited to the well-off and cultured classes.

The dramatic increase of the visibility of soy sauce began only in the early Qing of the late seventeenth century. This development was noted by H. T. Huang in a meticulously compiled table of the number of mentions of soy sauce in extant texts from the early period until the early twentieth century. It shows single-digit mentions in periods prior to the Qing but more than three hundred from the late seventeenth century to the early twentieth century.[11] One comes to a similar conclusion with digital counting, by doing a keyword search with "*jiangyou*" that gives 275 mentions in the *Guji* (古籍

old texts, from early to Republican periods) collection of the Erudition digital database, of which 208 are in the Qing and early Republican periods, with only sixty-seven in pre-Qing texts. Both counts are, naturally, not exhaustive. Major Qing recipe books consulted by Huang not listed in the digital database include *Shixian hongmi* (食憲鴻秘 Guide to the great mysteries of food, 1680) and *Tiao ding ji* (調鼎集 The harmonious cauldron, 1928), which respectively contain forty-six and 585 mentions. Neither count includes rich information in gazetteers and in records on popular life during the Qing, such as the voluminous *Qing bai lei chao* (清稗類鈔 Records of popular customs in the Qing, 1916) that gives 109 mentions of soy sauce, mostly in recipes.[12]

Such hard data substantially strengthens Huang's observation that "By the early years of the Qing, [soy sauce] had attained the status of the most popular condiment in the Chinese food system."[13] This development was highly correlated to the rapid commercialization of soy foods, soy sauce in particular, during the Qing, and more precisely since the so-called high Qing of the eighteenth century.

SOY FOOD COMMERCIALIZATION AND QING POLITICAL POWER

The emergence and subsequent strong presence of sizable urban soy food manufacturers (*jiangyuan* 醬園 or *jiangfang* 醬坊 pickle shops) in the urban landscape was a mid-Qing phenomenon. These shops made all kinds of fermented foods and drinks, but mostly with soy sauce as the branding product. A 1933 gazetteer defined the work of pickle shops as follows: "They can make anything that regulates taste, especially soy sauce."[14] If we do keyword searches using *"jiangyuan"* and *"jiangfang"* in the Erudition gazetteer database, we find 233 mentions, of which twenty-eight appear in gazetteers published between 1796 and 1911, and 205 in those published from 1911 to 1949. Most of the records in post-1911 publications were described as of Qing origin. Moreover, the number of shops is not accurately tallied by the gazetteers, as these publications often register pickle shops only as a business genre, without the statistics.

In official documents and literati writings, one can find sporadic mentions of soy manufacturers in North China, the Jiangnan region, and coastal cities beginning in the Yongzheng period (1723–1735) and increasingly throughout the eighteenth and nineteenth centuries.[15] Pickle shops in major maritime and riverine trade ports grew in significant numbers and began to

FIGURE 2.1
"Soy sauce being drawn by means of a siphon." Image from Elizabeth Groff, "Soy-sauce Manufacturing in Kwangtung, China" (1919). *EAST Collection.* 61. https://digitalrepository.trincoll.edu/eastbooks/61.

establish *gongsuo* (公所 collective guilds) in the 1870s. The earliest, and one of the biggest, was the one established in 1873 in Suzhou by eighty-six preexisting local shops.[16] By the early twentieth century, most counties had more than one shop, and some of the recorded manufacturers claimed their origin in the earlier Qing period or even in the late Ming.[17] The gazetteers show that these manufacturers were mostly located at city fringes, as they needed big spans of land for brewing their products. Sometimes they were grouped together forming business neighborhoods and alleys bearing the name of *jiangyuan*. Pickle shops, now found in many cities, also drew the attention of curious Western visitors in late Qing and early Republican China (see figure 2.1), who took closer looks at the production of soy sauce to find out what the popular liquid really was, as it had once been rumored in Sichuan Province to be boiled-down cockroaches, as reported by Alexander Hosie in the 1890s.[18] By the end of the Qing, pickle shops had become a major component of the Chinese urban landscape, some of which were identified with the cities themselves, such as Liubiju (六必居) for Peking; Yutang (玉堂) for Jining, Shandong; Hu Yumei (胡玉美) for Anqing, Anhui; Dingfeng (鼎豐) for Pinghu, Zhejiang; Feng Wantong (馮萬通) for Shanghai, with emerging brands in southern provinces like Hunan, Fujian, and Guangdong (see figure 2.2).

Republican sources provide more precise information that more accurately depicts the magnitude of the pickle shop phenomenon. A 1926 article on Peking pickle shops gives a count of more than 140 shops of different sizes, and a 1939 article indicates more than forty member shops in the Shanghai pickle shop guild.[19] A 1933 report on the industry in Zhejiang notes 322 pickle shops in the province with larger concentrations of over eighty shops in two counties and anything between a dozen to over fifty shops for the other counties.[20] Many of these shops were established in the Qing. Zhao

Fig. 2. The first drawing of soy.
PLATE III.

FIGURE 2.2
"The first drawing of soy." Image from Elizabeth Groff, "Soy-sauce Manufacturing in Kwangtung, China" (1919). *EAST Collection*. 61. https://digitalrepository.trincoll.edu/eastbooks/61.

Rongguang, a leading food historian in China, undertook the first systematic study of pickle shops in 2005 based on shops registered in Shanghai in the 1930s–1940s. He identifies the names and addresses of 102 shops of the "traditional" artisanal type with prolonged fermentation, and 251 "modern" shops using technoscientific methods including the use of chloric acid to break down the bean protein quickly instead of allowing time for fermentation.[21] This counting, which Zhao still considers incomplete, does show that in Shanghai alone, more than one hundred old traditional pickle shops were doing business in the early twentieth century, many of which were probably established in the Qing. Some of the allegedly "modern" shops might also have been established earlier and changed production methods in the Republican period. The data fully reflects a rapid growth of urban pickle shops from the late eighteenth century and a booming industry at the turn of the twentieth century.

THE GOLDEN AGE OF MANCHURIAN SOYBEAN PRODUCTION AND SOY SAUCE BUSINESSES

The explosion of urban pickle shops after the eighteenth century was the direct consequence of the dramatic increase of soybean supply with the full integration of Manchuria, a major producer of the crop in the Qing empire. Manchuria has the largest span of flat, high-quality, fertile land in the Qing empire, and the climate and the seasonal daylight change of this northern region were most suitable for the growth of some of the best breeds of soybeans.[22] For almost a hundred years after the Manchu conquest of China, there was a ban on mass soybean exportation by maritime trade from Manchuria. Soybeans were levied as banner land tax to be paid in kind as they were the main fodder for horses and camels of the troops. The ban was necessary to preserve the Manchu's military might; it was lifted only when, with soybean production becoming abundant, the Manchu state felt fully confident and secure about its political control of China.[23] In a sense, soybeans facilitated a process of the integration of Manchuria into the Qing Empire.

The process took place over two phases: first, the gradual immigration of Chinese peasants to Manchuria starting in the late seventeenth century that progressively expanded agriculture, especially soybean cultivation, on originally pastoral or forested land,[24] steadily increasing soybean production; and second, the gradual lifting of the bans on soybean exports to China proper via maritime trade starting in the mid-eighteenth century. Both were important political decisions made by Qing emperors during the prime of the dynasty.

The first Qing emperor Shunzhi saw the need to develop agriculture in Manchuria and in a 1653 edict encouraged the recruitment of peasants to cultivate the vast Manchurian land. After successive waves of legal and illegal Chinese immigration under the banner system, by the end of the nineteenth century, Chinese peasants in southern Manchuria had reached three million.[25] Chinese immigration peaked again in the first years of the twentieth century when the Qing state, under great political and financial pressure, officially invited free economic exploitation of the region, opening northern Manchurian pastoral land for agriculture.[26] Sustained increase in soybean production and commodification was the result of the successive waves of Chinese peasant immigration. By the early 1930s, Manchuria produced more than four million tons of beans, or 80 percent of world production.[27] This increase in soybean production was accompanied by the progressive lifting of the ban on Manchurian maritime soybean trade to China proper since the mid-eighteenth century, triggering a rapid influx and circulation of soybeans in China proper, much of it through Shanghai, making it the leading trading hub.[28] The boom further accelerated after 1902 with the end of the "closing off" policy of Mongolian pastures to the north of Manchuria.

Isett provides some figures on the spectacular increase of soybean supply in the nineteenth and early twentieth centuries: annual soybean surplus was under 1.5 million *shi* (around 118,800 tons)[29] in the late seventeenth century and increased to 3 million *shi* in the first half of the eighteenth century, topping four million in 1875 and reaching seven million in 1900. Estimated export increased from 1 to 1.5 million *shi* in the late eighteenth century to three million in 1876, and more than five million *shi* (around 400,000 tons) by 1900.[30] Alexander Hosie, who visited China and Manchuria in the last decade of the Qing regime, estimated in 1904 that over 600,000 tons of soy were exported annually from Manchuria to China's southern provinces and Japan.[31] Another estimate of Manchurian bean export in 1907 put the amount to China to be around 279,000 tons, and 476,150 tons to Japan.[32] By the 1920s, Manchuria was producing 60–70 percent of China's soybean export and 80 percent of the world's soy output.[33]

The sustained growth of the maritime soybean trade between Manchuria and China since the mid-eighteenth century was a key factor in the explosion of soy food commodification in all parts of China, and, with it, new urban dietary traditions based on the everyday use of soy-based condiments, in particular soy sauce. Part of that story was the spectacular rise of Shanghai, ultimately overshadowing Suzhou as the leading city in Jiangnan. This change was largely a result of Shanghai's central position in the eastern maritime trade in the nineteenth century, a process well studied by Fan Jinmin.[34] The Shanghai *douhang* (豆行 soy trade guild) established in 1765 was a landmark, and by the mid-nineteenth century, the guild at Cuixiu Tang (萃秀堂 Cuixiu Hall) in the City God Temple Garden became the biggest and most powerful of the twenty-one merchant guilds in the city by 1843, and a key stakeholder of the city's governance.[35] Its site is still a tourist attraction today. Manchurian soybean traders along China's eastern coast formed a tight network and subsequently acquired a dominant position in municipal governance in the most prosperous part of China in the late imperial period.[36]

The importance of Manchurian beans for the soy food business in the Jiangnan region can be illustrated by the situation in Jiangsu in the 1930s. According to a 1936 provincial report, Jiangsu Province produced 562 million kg of soybeans, while the demand was 629 million kg, resulting in a shortage of 66.8 million kg.[37] A significant portion of the shortage could come from the need for soy sauce. A survey in 1938 on urban soy sauce consumption in the province tells us that the average annual per capita consumption was 8.85 kg.[38] As soy sauce contains about 40 percent soybeans by content,[39] the annual per capita consumption of soy sauce alone would be 3.54 kg. With an urban population in Jiangsu Province of around 5.38 million at the time,[40] the total amount of soybeans needed to satisfy the urban soy sauce market would be around 19 million kg. If we also include the demand for soy sauce from Jiangsu's non-urban population (approximately twenty-three million), assuming arbitrarily that they needed 50 percent less soy sauce than their urban counterpart, then the total non-urban demand would be 50.3 million kg, totaling 69.3 million kg for the whole province. This figure exceeds the soybean shortage (66.8 million kg) for the whole province in 1936 and represents about 10 percent of the 629 million kg needed by the province. This rough estimate, though inexact, does indicate the importance of Manchurian soybeans in sustaining the growth of the soy food business in Jiangsu Province alone, which had a national reputation for the high quality of its soy sauce.

Manchurian bean exports also spurred the development of soy sauce businesses in southern China where local soybean production was negligible in the late nineteenth century. Guangdong merchants managed to obtain three-fifths of the soybeans exported through Niuzhuang to Hong Kong, most of which were consumed in the province. By the 1920s, 50–60 percent of exported Manchurian beans went to central China through Shanghai, and 30–50 percent went to the southern provinces of Fujian and Guangdong. An anthropological report on a Guangdong soy sauce manufacturer southwest of Canton in 1919 confirms that the soybeans used in the production were the best yellow beans from Niuzhuang.[41] Manchurian beans reaching all parts of China proper since the late eighteenth century greatly increased and stabilized the soybean supply, thereby enabling the soy food businesses to flourish. As the soy sauce production involved not only soybeans but also a considerable amount of taxable salt, soy sauce as a commodity became an important source of revenue for the provinces and the state, and it would play an important role in modern China's state building.

CULTURAL POWER OF SOY SAUCE AS EVERYDAY FOOD

Like sugar, brilliantly studied by Sidney Mintz, soy sauce steadily infiltrated into multiple aspects of people's everyday life in late imperial and modern China, and it became "symbolically powerful, for its use could be endowed with many subsidiary meanings."[42] This is revealed in the frequent mentions of soy sauce in recipes and other genres, such as account books, tax registers, popular novels, gazetteers, magazines and newspapers, tax records, personal diaries, letters, and ritual texts. It also became an identity food for the region, the nation, the native place, and the family. Soy sauce became a food that created connections and made distinctions.

The rapid and nationwide commercialization of soy foods paradoxically accentuated the value of traditional homemade sauces as a gift of intimacy. Late Qing diaries and family instructions by scholar-officials often include mention of soy sauce being given as a gift by junior family members to their seniors. The family instructions by Zeng Guofan (1811–1872), the powerful Hunan Confucian scholar-official, are particularly revealing. In an entry on the third day of the eighth month of 1866, he wrote, "Women [in the household] should refine their skills in making *xiaocai* (小菜 accompanying foods) like fermented bean curds, soy sauce, pickled vegetables, fine vinegar, and bamboo shoots, etc. They should make these often and send some to me. The domestic principles of serving one's parents, uncles, and aunts are emphatic on such tasks. But no need to send foods purchased in shops."[43] Similarly, a 1903 entry in the diary of the influential scholar-official Weng Tonghe (1830–1904) described how he received one pot of soy paste and one pot of soy sauce from his nephew, delivered by the latter's cook. Li Ciming (1830–1895), a mid-ranked scholar-official, recorded receiving jars of soy sauce sent to him by his niece in the capital where he worked, and by his younger sister in the third month and again in the third leap month of 1886.[44] Clearly homemade soy sauce symbolized for Zeng, Weng, and their peers the everyday chores and pleasure that sustained the well-being of a respectable Confucian family. It nourished intergenerational and gender relations within the household. Their words also reveal an important aspect of soy sauce making of the time: that commercialization of soy foods did not replace or preclude traditional homemade ones, prepared mainly by women (in contrast to commercial products produced exclusively by male workers to avoid female pollution during fermentation). The wide circulation of recipes of homemade sauce throughout the period until the latter half of the twentieth century shows that, for a long time, homemade traditions went hand in hand with the pickle shop phenomenon.[45] However, only homemade sauce had the power of fostering intimate family relations; the gifts received by Weng and Li mentioned previously were also likely to be homemade by younger women in the family.

Commercial soy sauce, on the other hand, had become in the nineteenth century a budgeted item for public lineage expenses. In the 1868 account books of the Jue lineage in Shicang, Zhejiang, several expense sheets list purchases of soy sauce by a lineage member who traveled to the nearest urban center to attend the imperial examination.[46] These show that, first, rural and township households were also consumers of commercial soy foods sold mainly in urban centers; and second, such purchases could be for public events of the lineage, including important rituals. In the 1911 account sheets of a funeral ritual in Shexian in Anhui, soy sauce was listed twice.[47] Late Qing family instructions on ancestor worship also systematically listed soy sauce as one of the four basic food items put in four little dishes offered to ancestors on the altar, together with tea, liquor, and rice in northern traditions, or with vinegar, oil, and pickled foods in southern ones. The coexistence of homemade and commercial soy sauces in the nineteenth and early twentieth centuries seemed to have broadened and deepened the cultural power base of the condiment.

Soy sauce by this period also seemed to have acquired a defining power through its smell and taste. The mathematician and scientist Zheng Fuguang (1780–1853) philosophized in 1842 about foods' good and bad smells and tastes. Comparing soy sauce with shrimp sauce, both tasty and made by *oufu* (漚腐 decomposition), Zheng wrote, "Soy sauce is rid of all the bad smell when done, which is why everyone loves it, while the bad smell of shrimp sauce remains, so that some hate it and others love it [for the *xian* (鮮 fishy) taste]."[48] Soy sauce indeed seemed to have reached the status of the most popular condiment in China.

However, there was actually no "standard" Chinese soy sauce but a proliferation of a wide range of regional soy sauces. The complexities of the production methods dependent on the ingredients, environment, climate, ferments, incubation time, and so on had produced regional soy sauces with different tastes and qualities. By the early nineteenth century, even the average consumer had developed a sensitivity to the differences in taste of regional sauces, which defined the distinct tastes of regional cuisines. The diary of Bi Huai, a scholar from Tongxiang, Zhejiang recorded in 1836 a particularly memorable dinner he had with a close friend while on his way to the capital for the imperial exam. The two shared a pork stew eaten with the soy sauce of his native place. Being away from home, he wrote, they were both fully satiated with *guxiang fengwei* (故鄉風味 the taste of home).[49] The previously mentioned scholar-official Li Ciming, working in the capital at the time, also noted in his diary the pleasure he had in receiving in the third month of 1887 a jar of sauce from his native place, Shaoxing, sent by a friend.[50] The commercialization of soy sauce by this time revealed, and perhaps reaffirmed, the different regional taste preferences that became an important issue for food experts in the twentieth century,[51] further amplifying the condiment's power as a regional identity food.

Soy sauce also became a major food item listed in Qing diplomatic conventions. Systematically listed as *jiangyou* or *qingjiang* (清醬 light soy sauce), it was a "must" in the standard daily provisions for foreign envoys, ambassadors, and translators visiting China, from Korea to the northeast, to the Indian subcontinent to the southwest of China. The condiment was typically listed with different quantities of meat, fish, or poultry; vegetables; and so on to showcase Chinese hospitality.[52]

Soy sauce in the early twentieth century also embodied China's desire for a scientific modernity. With the growing economic and cultural importance of the condiment and its overwhelming popularity, it was under great pressure to modernize and industrialize toward the turn of the twentieth century. The *Jingshi gaodeng shiye xuetang* (京師高等實業學堂 Peking Higher College on Industrial Study), modeled on Meiji industrial colleges, was established in Peking in 1904 and supported research on traditional foods, including soy sauce and fermentation as an aspect of applied chemistry, one of the four disciplines taught at the college by Japanese- or Western-trained teachers.[53] One of the missions of the college was indeed to transform China's *gongyi or shougongyi* (工藝 or 手工藝 traditional handicrafts) with new science and technology to spearhead China's industrialization.[54] Early foundational research included investigating and comparing various regional sauces.[55]

For Chen Taosheng (1899–1992), one of the first and most impactful Chinese food scientists, who graduated from the Peking Higher College on Industrial Study in 1922 with a degree in chemical engineering, industrialization of the condiment implied reducing the risks in fermentation and shortening fermentation time with new biochemical techniques, mostly developed by the Japanese.[56] Chen and his peers assiduously experimented with these techniques in state and university laboratories throughout the 1920s and 1930s until the outbreak of the war with Japan in 1937.[57] Chen combined his laboratory work with site visits of regional sauce manufacturers and found the award-winning (1934) Guantou (琯頭) sauce of his hometown of Fuzhou, Fujian Province, the most promising model for industrialization, as the fermentation of only soybeans (without wheat flour) took less incubation time yet produced a product of superior quality.[58] To counter Japanese competition, Chinese researchers focused on the study and application of premade starters, based on the mold *Aspergillus oryzae* developed by the Japanese to trigger a safe, controlled fermentation process. The work of Chen and his peers, however, was largely disrupted by the war with Japan and the civil war that followed.[59]

CONCLUSION

Chinese soy sauce as an everyday food came into being only after the eighteenth century during the height of Qing power. In this respect, the story is reminiscent of the history of sugar, with its entanglement of empire and nation formation as a new food enters the lives of the working classes.[60] Soy sauce, of course, was not *new* to the Chinese in the eighteenth century as was sugar to the Europeans in Mintz's story. But it was transformed from a traditionally elite condiment into a commodity for everyday consumption, without significant technological change until the 1920s when new biochemical techniques were introduced in food making.[61] Moreover, soy sauce's transformation of the late-Qing politico-economic landscape owes much to the unprecedented supply of Manchurian soybeans made possible by the massive Chinese migration in Manchuria as well as astute maritime traders.

The process of intensive commercialization since then not only broadened soy sauce's economic impact at the national level but also created or even expanded its cultural power as an identity food, highlighting different tastes and smells made with specific local ingredients and technologies, and as an enhancer of intimate social relations and official diplomatic contacts. As an everyday food, it created identity by making distinctions and strengthened family structure and continuity by being shared as a gift to respected elders and ancestors. Soon, soy sauce making became a test case of China's ability to retain national dignity and to modernize when challenged by imperial powers.

But the story of soy sauce is an open-ended one. Today we are witnessing the incredible post-industrial venture of soy sauce making in every part of East Asia where its values as an identity food, a health food, a heritage food produced by an ethical crop system, and an authentic food are being reimagined and recreated more enthusiastically than ever before. Its magic in molding our post-industrial everyday life is as intriguing as its fermentation process.

NOTES

The research of this paper is supported by two RGC funded projects: 1. GRF (HKU 17612218: The Birth of East Asian Modern Entrepreneurship: The Case of Soy Sauce Making, 1880–1960" 2018–2021; 2. CRF C 7011-16G "Making Modernity in East Asia: Technologies of Everyday Life, 19th–21st Centuries," 2017–2021.

1. H. T. Huang, *Fermentations and Food Science*, in *Science and Civilisation in China*, ed. Joseph Needham, vol. 6: Biology and Biological Technology, pt. 5 (Cambridge: Cambridge University Press, 2000), 358–374. See also Hong Guangzhu, "Doujiang he dou jiangyou qiyuan chutan" [A preliminary study on the origin of soy paste and soy sauce], in *Zhongguo shipin kejishi gao* [Preliminary studies on Chinese food history] (Beijing: Zhongguo shangye chubanshe 1985), 90–111. Huang basically agrees with Hong's views.

2. See the two biggest Chinese soy sauce makers: the Hunan soy sauce maker Jiajia (https://www.163.com/dy/article/FQVS8A520522CR6J.html) and Haitian in Guangdong (https://m.sohu.com/n/488147146/).

3. Ishige Naomichi, *Sekai no tabemono: Shoku no bunka chiri* [The world's foods: the cultural geography of diets] (Tokyo: Asahi Shimbunsha, 1995), 13–14. Ishige argues that soy sauce defines east Asia as a region culturally distinct from Southeast Asia where fish sauce is the main condiment.

4. Huang, *Fermentations and Food Science*, 362, quoting Ni Zan's *Yunlin Tang yinshi zhidu ji* [Dietary system of the Cloud Forest Studio].

5. One can add to this list a recipe recorded by Han Yi of the late fourteenth century in his dietetic book *Yi Ya yiyi* [legacy of Yi Ya]; see also Huang, *Fermentations and Food Science*, 372.

6. Tian Yiheng, *Liuqing ri zha* [Daily jottings] *juan* 26:204 (*Guji* collection of the Erudition digital database; henceforth "*Guji* collection"). Tian's description mentioned the use of wheat flour with beans to make sauces that could have wolfberry or rose flavor.

7. Li Shizhen, *Bencao gangmu*, vol. 2 (1885; Taibei: Wenguang Tushu, 1982), *juan* 25:876–890.

8. Vivienne Lo and Penelope Barrett, "Cooking up Fine Remedies: On the Culinary Aesthetic in a Sixteenth-Century Chinese *Materia Medica*," *Medical History* 49 (2005): 395–422, esp. 420 and Lo, "Pleasure, Prohibition, and Pain: Food and Medicine in Traditional China," in *Of Tripod and Palate: Food, Politics and Religion in Traditional China*, ed. Roel Sterckx (New York: Palgrave Macmillan, 2005), 163–185.

9. Gao Lian, *Zunsheng ba jian* [Eight sections on life-nourishing] (Beijing: Renmin weisheng chubanshe 1994), 421, 425, 427–428, 436–437.

10. Huang, *Fermentations and Food Science*, 373.

11. Huang, *Fermentations and Food Science*, 373.

12. Of the three hundred mentions in the Gazetteer collection of the Erudition digital database, only two are pre-Qing, in the last years of the Ming; *Qing bai leichao* in the Database on Chinese Historical Sources of the Institute of History and Philology, Academia Sinica.

13. Huang, *Fermentations and Food Science*, 374.

14. *Guan xianzhi* [Gazetteer of Guan County in Sichuan Province] (1933).

15. *Shanzhi ketan* [Conversations in the Mountain Studio, early eighteenth century], *juan* 5:28 (*Guji* collection) on a Hangzhou pickle shop in the Kangxi period; two pickle shops in Peking were said to be established during the Kangxi period, "Beijing zhi jiangyuan ye" [Pickle shops in Peking], *Xing Hua* 23/35 (1926): 23.

16. "Suzhou fu wei jiangfang ye chuangjian gong suo" [Establishment of the pickle shop guild by the Suzhou prefecture], *Mingqing Suzhou gongshang ye beike ji* (Suzhou: Jiangsu renmin chubanshe, 1981), 260–261.

17. A few manufacturers in big urban centers like Beijing, Tianjin, and Shanghai claimed their origins in the early Qing, and even the Ming, which cannot be verified. Most typical was the case in Beijing, see "Beijing zhi jiangyuan ye" 1926 op. cit.

18. Alexander Hosie, *Three Years in Western China: A Narrative of Three Journeys in Ssu-ch'uan, Kuei-Chow, and Yun-nan* (London: George Philip & Son, 1897), 164.

19. "Beijing zhi jiangyuan ye"; "*Jiangyuan ye*" [Pickle shops], *Shangye yuebao* (Shanghai) 19, no. 5 (1939): 9.

20. Fang Peng, "Pinghu niangzao chang bainian licheng" [The hundred-year journey of the pickle shop in Pinghu] in *Wushi cunqiu hui shi pian—Pinghu wenshi ziliao* (Pinghu town: Zhongguo renmin zhengzhi xieshang huiyi Zhejiang Sheng Pinghu Shi weiyuanhui wenshi ziliao weiyuanhui, 1999), 93–99; *Zhejiang Sheng shiye zhi* [Record of Industries in Zhejiang Province] (1933), quoted in Gui Qiang and Zi Xiawei, *Pinghu Lao-Dingfeng jiangyuan jingyuan yanjiu* [A study of the management of the Old Dingfeng pickle shop in Pinghu] (Shanghai: Shanghai cishu, 2017), 211–212.

21. Zhao Rongguang, "Zhongguo lishi shang de jiangyuan yu jiangyuan wenhua shulun—yi Shanghai shi de chuantong yu xinshi jiangyuan ziliao duibi wei zhu" [A discussion on pickle shops and the culture of pickle shops in history mainly based on the traditional and modern pickle shops in Shanghai], *Yinshi wenhua yanjiu*, 4, no. 16 (2005): 9–19.

22. Lin Hanming, *Yidou yi shijie* [One bean, one world] (Hong Kong: Joint Publishing Co. 2023), 195–207.

23. Katō Shigeshi, "Dongbei dadou doubing shengchan de youlai" [Origins of soybean and soy meals produced in Manchuria], trans. Wu Jie, in *Zhongguo jingji shi kaozheng* [Studies in Chinese economic history] vol. 2 (Beijing: Zhonghua shuju, 2012), 985.

24. The irreversible environmental impact of this history is given in David Bello, *Across Forest, Steppe, and Mountain. Environment, Identity, and Empire in Qing China's Borderlands* (Cambridge: Cambridge University Press, 2016), esp. chapter 3.

25. Ding Yizhuang et al., *Liaodong yimin zhong de qiren shehui* [Immigration and Eight Banner Society in Liaodong] (Shanghai: Shanghai kexue yuan chubanshe, 2004), 8, 214.

26. Diao Shuren and Yi Xingguo, *Jin sanbai nian Dongbei tudi kaifa shi* [Development of Manchuria in the recent 300 years] (Changchun: Jilin wenshi chubanshe, 1994), 103, 110; Liu Xiaomeng, "Qingdai Dongbei liumin yu Man-Han guanxi" [Migrants in Qing Manchuria and Manchu-Han relations], *Qingshi yanjiu* no. 4 (2015): 1–22.

27. James Kung and Nan Li, "Commercialization as Exogenous Shocks: The Effect of the Soybean Trade and Migration in Manchurian Villages, 1895–1934," *Explorations in*

Economic History 48 (2011): 580; Sakura Christmas, "Japanese Imperialism and Environmental Disease on a Soy Frontier, 1890-1940," *Journal of Asian Studies* 78 (2019): 822.

28. Fan Jinmin, "Qingdai zhongqi Shanghai chengwei hangyun ye zhongxin zhi yuanyin tantao" [Study on Shanghai becoming a maritime trade center in the mid-Qing], *Anhui shixue* 1 (2013): 29-38.

29. One *shi* is equivalent to 79.2 kilograms.

30. Christopher Isett, *State, Peasant, and Merchant in Qing Manchuria, 1644-1862* (Stanford, CA: Stanford University Press, 2007), 226-228, 231.

31. Alexander Hosie, *Manchuria. Its People, Resources, and Recent History* (London: Methuen & Co., 1904), 244.

32. Lei Hui'er, *Dongbei di douhuo maoyi* (1907-1931) [Manchurian soybean trade 1907-1931) (Taipei: Guoli Taiwan shifan daxue lishi yanjiusuo zhuankan, 1981), 30.

33. Kung and Li, "Commercialization as Exogenous Shocks," 569; *New York Times*, August 19, 1928, Section N, 48.

34. Fan Jinmin, "Qingdai zhongqi Shanghai."

35. Fan Jinmin, *Ming Qing Jiangnan shangye di fazhan* [Commercial development in Jiangnan in the Ming-Qing periods] (Nanjing: Nanjing Daxue Chubanshe, 1998), 153-154; *Shanghai xian xuzhi* [Sequel to Shanghai Gazetteer] (1918), *juan* 3.

36. Xu Tan, "Qianlong-Daoguang nianjian de bei yang maoyi yu Shanghai de jueqi" [North Sea trade in the Qianlong-Daoguang period and the rise of Shanghai], *Xueshu yuekan* 43, no. 11 (2011); Fan Jinmin, "Qingdai qianqi Shanghai de hangye chuanshang" [Maritime trade and traders in early Qing Shanghai], *Anhui shixue* 2 (2011); Zhang Bo, "Zhidu tiaozheng yu Qingdai Dongbei douhuo maoyi geju de bianqian" [Institutional adjustments and the evolution in Manchurian soybean trade], *Tianjin Daxue Xuebao* 11 (2008).

37. "Jiangsu sheng ge xian er-shi-wu nian xiaji zhuyao zuowu zhongzhi mianji ji chanliang guji biao - dadou" [Cultivation area and production of major summer crops in Jiangsu counties in 1936, the case of soybean], *Jiangsu sheng jianshe yuekan* 4, no. 3 (1937): 9-12.

38. Ma Junya, *Hunhe yu fazhan: Jiangnan diqu chuantong shehui jingji de xiandai yanbian, 1900-1950* [Fusion and development: modern changes in Jiangnan traditional social economy, 1900-1950], (Beijing: Shehui kexue yanjiu suo chubanshe, 2003), 327-328.

39. Alexander Hosie, *Report by Consul-General on the Province of Ssuch'uan* (London: Harrison and Sons, 1904), 19-20. Hosie reported that the Sichuan maker took twenty-eight catties of soybeans to make seventy catties of soy sauce.

40. Zhu Kezhen, "Lun Jiang-Zhe liang sheng renko zhi midu" [On population density of the Zhejiang and Jiangsu provinces], *Dongfang zazhi* 23, no. 1 (1926): 91-112.

41. Lei Hui'er, *Dongbei de douhuo maoyi*, 5; Elizabeth Groff, "Soy-Sauce Manufacturing in Kwangtung, China," *Philippines Journal of Science* 3, no. 15 (1919): 310.

42. Sidney Mintz, *Sweetness and Power: The Place of Sugar in Modern History* (New York: Viking 1985), 186.

43. Zeng Guofan, *Zeng Wenzheng gong jiaxun* [Family instructions by Zeng Guofan] (1879), *juan* 2:49 (*Guji* collection).

44. Weng Tonghe, *Weng Wengong gong riji* [Diary of Weng Tonghe], manuscript, p. 8744; *Xun xue zhai riji* [Dairy of Xunxue studio, 1889–1894], chapter *xin* 2:2945 (*Pudie* [Diary and genealogy] collection of the Erudition digital database).

45. This tradition continued in Taiwan and China in the 1970s.

46. *Shicang qiyue* [Contracts and local documents in Shicang township], series 5, volume 4 (Hangzhou: Zhejiang daxue chubanshe 2010), 174, 192, 198.

47. The cost of soy sauce in this case was equivalent to more than that of two eggs. http://dfwx.datahistory.cn/ (database of Shanghai Jiaotong University).

48. Zheng Fuguang, *Feiyin yuzhi lu* [A record to clarify broad and obscure knowledge], 1842:53 (*Guji* collection).

49. Bi Huai, *Gongche riji* [Diary of my days in officialdom], 1911 edition: 35 (*Pudie* collection).

50. Li Ciming, *Xunxue zhai riji* [Diary of Xunxue studio], chapter Xin bis: 3085 (*Pudie* collection).

51. According to Hong Guangzhu, Chinese condiments are geographically distinctive on account of different natural environments and material resources: sweet in the south, salty in the north, sweet and salty in the east, acidic and spicy in the west and southwest, and a complex mixture of different tastes in the center. The differences, for Hong, are due to different natural environments and material resources. See "Zhongguo yinshi wenhua de dili he lishi Beijing [Geographic and historical background of Chinese food culture], *Zhongguo yinshi wenhua*, ed. Nakayama Tokiko (Beijing: Zhongguo shehui kexue chubanshe, 1992), 222–227. Qing consumers often claimed specific tastes within these macro regions.

52. *Qinding Libu celi* [Regulations of the Imperial Ministry of Rites), Daoguang edition (ca. 1844); *Chouban yiwu shimo puyi* [Annex to history of management of barbarian affairs (1821–1874)], Republican edition.

53. See the memorial by Censor Zhao Qilin (1859–1935), "Qing ban Jingshi gaodeng shiye xuetang zhe" [Memorial for the establishment of the Industrial College in the Capital, 1904], in *Dacheng hui conglu* 45 (1934): 25a–27b.

54. James Reardon-Anderson, *The Study of Change: Chemistry in China 1840–1949* (Cambridge: Cambridge University Press, 1991), 159. Handicrafts represented 67.8 percent of the Chinese industrial sector in as late as 1933.

55. Wu Chengluo (1892–1955), a leading chemist and teacher of some leading food experts on fermentation, did a study on soy sauce in 1924 comparing regional products. See "Zhongguo jiangyou zhizaofa" [Methods of making Chinese soy sauce], *Gongda zhoukan* 6 (1924): 2–4.

56. For the history of fermentation science in modern Japan, see Victoria Lee, *The Arts of the Microbial World: Fermentation Science in Twentieth-Century Japan* (Chicago: University of Chicago Press, 2021).

57. Ma Chunhuan, "Zhongguo jindai weishengwu gongye di kaituo zhe he dianji ren Chen Taosheng [The pioneer and founder of modern Chinese microbial industry, Chen Taosheng], *Zhongguo keji shilao* 4 (1983): 1–42.

58. Chen Taosheng, *Gaodeng niangzao xue* [Advanced fermentation science] (Shanghai: Shangwu yinshuguang 1953), 348–352.

59. See my chapter "Soy Sauce in Crisis: China's First Engagement with Technoscience (1900–1950)," in *Crafting Everyday Food: Technology, Tradition and Transformation in Modern East Asia*, eds. A. Leung and H. Stevens (Honolulu: University of Hawai'i Press, 2025).

60. Mintz, *Sweetness and Power*, chapter 2.

61. Premade ferment starter was scientifically developed by the Japanese in the late nineteenth century. See Victoria Lee, "Mold Cultures: Traditional Industry and Microbial Studies in Early 20th-Century Japan," in *New Perspectives on the History of Life Sciences and Agriculture*, ed. Denise Phillips and Sharon Kingsland (New York: Springer, 2015), 231–252; Chinese scientists studied the technique for industrial use only in the 1920s and 1930s.

3 BEEF IN CHINA: A HISTORY IN EIGHT DISHES

THOMAS DUBOIS

How should we write the history of beef in China? To start, we'll need to peer into the pasture, the slaughterhouse, and the retail network to see how beef in China was actually made and sold at different points in China's long history. We'll also need to understand what people thought about beef—whether they considered it nutritious, prestigious, immoral, or uniquely modern. Finally, we will need to see how all these factors have changed over time.

Most ways of understanding food history rely on these two faces of commodity and culture. Some works have focused exclusively on one aspect or the other, tracing the forces behind the global circulation of cod, salt, coffee, tea, or beer,[1] or seeking to understand the ideas of health or status that drove people to accept, reject, or fetishize foods like milk or tofu.[2] Margaret Visser's *Much Depends on Dinner* combines the two, exploring both the origins and meanings of seven items that make up an ordinary meal.[3] In this volume, Francesca Bray's chapter on millet (chapter 1) and Angela Leung's on soy sauce (chapter 2) do something similar, tracing how each of these two foods were made and how each one came to prominence in Chinese diets and imaginations.

This chapter adds a third facet: cuisine. By this, I mean the actual cooking, the steps that turn a commodity such as millet, soy sauce, or beef into food. This shift in perspective moves the consumer from the end of the story to the middle, extending the process of transformation into the kitchen. Doing so highlights both social considerations, like tradition and hygiene, and practical ones about cost and difficulty. Giving full attention to cooking reveals the overlap of industrial and domestic spaces, and the influence of branding and advertising on culinary consciousness.

To illustrate the complex interaction of culture, commodity, and cuisine, this chapter presents the history of beef as a series of eight iconic dishes, each embodying a distinct facet in the ways that beef was made, sold, imagined, and consumed. Moving from the fourteenth century to the present day, we will see beef dishes that reflect the techniques and limits of premodern production and preservation, the cultural forces of dietetic modernity and Cold War symbolism, and the commercial realities of the new market economy, including the introduction of new business models and new approaches to production and branding.

PREINDUSTRIAL CATTLE CHAINS: FAZHI STEWED BEEF

Through most of the fourteenth-century bandit novel *Heroes of the Water Margin*, the book's heroes are either fighting or feasting. The tavern meal eaten by legendary warrior Wu Song captures the spirit of the latter:

> Sitting down and resting his spear, Wu Song called out, "Bring me a drink!" The innkeeper placed three bowls, a pair of chopsticks, and a plate of cooked food before Wu Song and filled his bowl to the brim with wine. Wu Song grabbed the bowl and drank it off, saying, "That wine has some kick! How about something to fill my stomach?" The innkeeper replied, "We only have cooked beef." Wu Song replied, "All right, cut me off two or three *jin*." The innkeeper went to the back and cut off two *jin* of beef, which he brought out on a large plate and placed in front of Wu Song before refilling the bowl of wine.[4]

As the evening drags on, Wu continues to consume inhuman amounts of liquor and meat, growing belligerent as the innkeeper tries to cut him off. When he finally staggers out of the inn, Wu encounters a tiger, which he kills with his bare hands. Even now, ordering "two *jin* of beef" (斤 one kg in contemporary exchange) is a way of announcing that one's evening of drinking is going to get wild.

For many modern readers, this scene will seem incredible, less the drunken tiger abuse that ended the evening than the two *jin* of beef that started it. Draft animals like cattle were vital to Chinese agriculture, and the slaughter of healthy animals was at times banned. Even without legal enforcement, many were simply revolted at the thought of eating an animal that lived alongside and loyally helped the farmer. But these restrictions were not absolute. They excepted animals that were lame, old, or otherwise unfit for work. Even during times of agrarian distress, when the ban on slaughter was fully enforced, cattle were used for temple sacrifices.

Indeed, many considered beef to be wholesome and nourishing, a point made by the novel itself. When, in another scene, Wu Song correctly suspects foul play lurking in his plate of steamed dumplings, the proprietress laughingly changes the subject, saying, "In this day and age, do you really think that there could be human or dog meat in your dumplings? For generations, my family has served nothing but pure beef!"[5]

Literary snippets like these show only episodic evidence of what people thought about beef. To see what they actually ate, we need to look to the early twentieth century—a time when social surveys confirm that the city of Beijing alone was consuming the equivalent of fifteen thousand head of cattle per year.[6]

Where did all those animals come from? For Beijing, the answer was a dedicated livestock trade that each year drove thousands of animals from Mongolia to the Great Wall crossing at Zhangjiakou, and thence along the "golden road" into the city. This profitable trade was dominated by established trade firms like Dashengkui (大盛魁) that had the resources to mount massive caravans to the grasslands, drive the animals back to the interior, and hold them at grazing grounds to await a favorable price.[7] The daily slaughter and butchering at Beijing's Cattle Street attracted dozens of specialized tradespersons to "collect the blood and skin, sell the meat, peel off the membranes, gather up the fur, [and] others who pick over the small pieces, cut through the chest and the stomach, and extract oil from the bones."[8]

In other places, beef came from old farm animals. Cattle were raised across China on marginal land like the dry hills of central Shandong or the reedy Yellow River floodplains in Henan. Animals once valued for their labor eventually became food. Despite the moral compunction some may have felt, surveys of cattle dealers confirm that animals who had reached the end of their working lives were overwhelmingly sold for slaughter rather than being released into the wild.[9]

These cattle were tough, and so was their meat. For this reason, recipes of the time generally favor wet cooking methods such as stewing. We see this preference in the eighteenth-century cookbook *Tiaoding ji* (調鼎記 Flavoring the pot), which, out of hundreds of recipes, includes a small section on beef, including an account of how such cuts might have been prepared:

Fazhi niurou (法製牛肉 Proper preparation for beef)

Take four *jin* of good beef, cut into sixteen pieces, wash, and pat dry. Gently rub in a half *jin* of good bean paste and one to two *liang* of fine salt, and simmer over a low fire. Add spices [to the cooked meat], to improve the color and taste.[10]

This barebones presentation is more template than recipe. It is left up to the cook to decide what cut to choose and how much water and which spices to add. What the recipe does specify is technique: large chunks of beef are slathered with bean paste and boiled until tender, cooled and sliced thin, and served alongside a dish of spices, a forgiving technique that makes good work out of even very tough meat. This was likely the preparation of the two *jin* of beef in the *Heroes of the Water Margin*, or indeed anywhere else where work cattle were consumed.

PREINDUSTRIAL PRESERVATION: PINGYAO PICKLED BEEF

If one solution to distance in the preindustrial beef chain was to walk live cattle to market, the other was to create a product that could be stored and transported without refrigeration. There were many ways to preserve meat, mostly variations of salting and pickling, either dry or wet. Salting is ideal for fatty pork or duck, which dries without desiccating. But beef could also be air dried in thin slices or cured whole as in Yunnan's *ganba* (幹巴 dried beef), each method preserving and enhancing the taste of the meat. Beef was also commonly wet-brined, a technique used to preserve all sorts of animal protein. The Yuan-era recipe for fermenting a whole gutted fish in a wet solution of salt, spices, and a starting agent of red rice was fairly typical of this method.[11] Like dry curing, wet methods produce a distinct taste and texture, transforming raw meat into a prized specialty product.

Pingyao in central Shanxi is traditionally known for two things: commerce and cattle. During the Qing, Pingyao was a node in the Shanxi merchant network and part of a cattle-producing region that extended into northern Henan and western Shandong. Pingyao had deep traditions of eating beef, butchering their own cattle, and sharing the meat with neighbors. Arriving to his new post, late-Qing magistrate Wang Peiyu noted with obvious dismay that "It is the custom in Pingyao to slaughter cattle to sell meat. This is an old and pervasive practice."[12] Wang went on to ban the sale of beef, likely to little effect.

Pingyao's specialty product was pickled beef, a delicacy that was said to have been invented when fifth-century invaders punished the region by slaughtering the local work cattle. According to the story, the commander ordered the confiscated beef stored in troop provisions of salted vegetables, later discovering that the process had greatly improved the taste. By the Ming, pickled Pingyao beef was widely traded across north China, propelled by demand from the diaspora of Shanxi merchants and the emergence of specialized trading companies such as Xingshenglei (興盛雷), Zilicheng (自立成), and Longshengwang (隆盛旺). By the mid-Qing, there were six hundred dealers of pickled beef in Pingyao alone.

As a prized local product, Pingyao beef was praised for its color, texture, and taste, but what really makes Pingyao beef stand out is its cultural cachet. Given that pickling meat is one of China's oldest culinary arts, the account of its accidental discovery is very unlikely to be true. But that does not lessen its meaning. Stories like the tale of Empress Dowager Cixi stopping to praise the local delicacy while on the run from the Boxer suppression propelled the reputation of Pingyao beef, which even featured in a folk song about local products during the 1950s.[13] In the same way that nearby farmers have reinvented the fame of millet to appeal to ecologically minded consumers (see Bray, chapter 1), makers of Pingyao beef would revive these stories in the 1980s as the product sought new status and new markets.[14]

INDUSTRIAL EXPORT CHAINS: GYŪNABE

The early twentieth century initiated a revolution in China's beef chains, creating high-value, high-efficiency industrial channels that combined rail transport, modern slaughter, and dedicated markets. To see the result of these changes, we can look to Japan and specifically to the dish of *gyūnabe* (牛鍋 beef hotpot).

During the early twentieth century, many Chinese cities invested in improving meat production. Aiming to clean up the streets, ensure the collection of the slaughter tax, and promote a healthy meat supply, late Qing reforms mandated the creation of municipal slaughterhouses in Shenyang, Tianjin, and Beijing. In Shanghai, where Western observers had once expressed horror at the unsanitary state of slaughtering and butchery, reforms enacted by the Shanghai Municipal Council cut down on fraud and malpractice in the meat trade and funded the 1935 construction of a massive three-story slaughtering facility.[15]

Colonial railways were among the first foreign investors to develop China's beef industry for overseas markets. The Russian-owned China Eastern Railway connected cattle producers in Hulunbuir to a modern slaughtering industry in Harbin and built ice houses and specialized transport that allowed grassland beef to reach markets in Russian Amur and Vladivostok. German railways connected cattle-rich western Shandong to state-of-the-art slaughter and processing facilities in the colony (1898–1914) of Qingdao.[16] German planners also crossbred Shandong cattle with imported Simmenthal, creating a new breed that was literally made for beef. Instead of ropey work animals, the new market for high-quality beef demanded tender cattle, fattened on green waste.

After Japan took possession of Qingdao in 1914, the Shandong industry switched to serving the Japanese market. Japan itself had already embraced beef as a symbol of dietary modernity, emblematic of the country leaving behind its farinaceous Asian past and emulating the protein-rich diets of the West. On New Year's Day of 1872, the young Meiji emperor had kicked off the new era by publicly eating meat. That act, along with the enthusiastic promotion by reformers such as Fukuzawa Yukichi, gave dishes like *gyūnabe*, beef sliced thin and briefly stewed in miso broth, an aura of progressive destiny and national salvation. To supplement Japan's own relatively small beef production capacity, merchants began importing beef from Russia and Korea. However, Japanese diners preferred the beef from the Qingdao slaughterhouses and paid a premium over other imports, creating a new export industry for specially bred and raised beef cattle. Although these were by no means draft animals, the symbolic significance of Japan growing in strength by devouring the continent still evoked the ire of Chinese nationalists who blamed China's agrarian backwardness on the slaughter of domestic work cattle for foreign bellies.[17]

THE SOCIALIST GOOD LIFE: BRAISED POTATOES WITH BEEF

On a visit to Hungary in 1964, Soviet leader Nikita Khrushchev famously told a group of journalists that "socialism is a dish of beef goulash and potatoes." Khrushchev wasn't there to hand out recipes, but rather was expressing support for Premier János Kádár's brand of non-doctrinaire "goulash socialism" that aimed, above all, to increase material living standards. The metaphor simultaneously represented the blending of ideas and the goal of every household having a steaming pot of stew. A clear rejection of both Stalinism and the idea of endless revolution, Khrushchev's statement was roundly criticized by the Chinese press.[18] This story (or at least the quote) remains well known to most Chinese born before 1970 and shows the powerful imagery of food as China lurched toward a new political direction. At the symbolic center lay Khrushchev's goulash, which was translated for Chinese readers simply as *"tudou shao niurou"* (土豆 燒牛肉 braised potatoes with beef).

During the early 1950s, China adopted a love of all things Russian; from women's fashion to customer service, the Soviet Union became synonymous with the socialist good life.[19] Food represented a powerful focus of aspiration. As Soviet Ukraine rebuilt its dairy industry, Chinese newspapers gushed about the rivers of milk, even though dairy consumption was hardly common in most of China. Even as other private restaurants were being closed or folded into "joint management," Beijing's Moscow Restaurant remained one of the capital's iconic dining locations. When Khrushchev's Soviet Union was denounced as revisionist, the reputation of iconic foods suffered accordingly. Satirizing what he saw as the cowardice of Soviet rapprochement with the United States, Mao Zedong closed a 1965 poem with the unmistakable imagery of Khrushchev's well-known culinary metaphor:

> A tripartite pact was signed, under the bright autumn moon two years ago. There'll be plenty to eat.
> The potatoes are braised, now add the beef.
> Stop talking nonsense.
> Look, the world is being turned upside down.[20]

Yet despite the widespread notoriety of *"tudou shao niurou,"* the dish itself was not common. While 1950s cookbooks did contain recipes for soy sauce braised beef with potatoes, beef itself became increasingly scarce over the decade, as domestic production was redirected to foreign markets and the early pressures of collectivization pushed down the size of the national cattle herd. Yet scarcity only enhanced the dish's symbolic value. Years later, braised beef with potatoes had not lost its allure. First posted as a journalist to the Chinese embassy in West Germany, Wang Shu (who would later serve as deputy foreign minister) recalled the thrill of traveling through Eastern Europe and trying actual goulash in homes, restaurants, and at roadside rest stops, where it was served without ceremony in an unassuming cup. Wang praised the dish as "cheap and tasty, served fast so we could be on our way."

But even as new memories accumulated, Wang could not help recalling anew the story about Khrushchev's goulash socialism each time he ate the dish.[21]

REFORM-ERA ENTREPRENEURS: TINNED BRAISED BEEF

For all its problems, the planned economy was, in many ways, a boon to the long-term development of China's cattle production, prompting infrastructural investments in pastoral irrigation, veterinary care, feed processing, and breed improvement.[22] Following a crash after the Great Leap Forward, China's herd of work cattle gradually recovered, thanks in part to a strict ban on slaughter. For the planned economy, cattle remained a strategic productive resource; their labor belonged to the collective, and ultimately to the state. In most of the country, slaughtering an animal required special permission from the collective, which retained possession of the meat, bones, and hide.

The 1980s brought three major changes. Deng Xiaoping's economic reforms gradually decollectivized ownership of productive assets, including animals. With cattle now the property of households, herd numbers immediately rose. The second was a race for development, in which local governments were tasked to work with state-owned enterprises and new township-village enterprises to develop industries that could earn income either abroad or in China's growing domestic consumer market. The third was the mechanization of agriculture. As small tractors replaced livestock as a source of labor, county governments faced the decision of what to do with the thousands of draft animals still being kept in strategic reserve. Some counties returned these animals to farmers, while others sold them to neighboring regions. A few worked with local entrepreneurs to develop beef industries.

The result was in many ways the opposite of the industrial streamlining of the 1930s, as thousands of entrepreneurs, often with local government backing, scrambled to get their beef to market. Established cattle-producing regions like western Shandong enjoyed high-level support. Shandong beef was promoted at trade fairs, and by the National Export Agency. By the late 1980s, frozen Shandong beef was again being shipped to Japan. But most producers faced significant technical barriers to entry. With almost no capacity for cold shipping or storage, the first challenge was preservation.

Canning was among the easiest and least expensive options. In 1987, the county-level Kalaqin Zuoyi (喀喇沁左翼 Horchin Left Wing) in Inner Mongolia spent 150,000 yuan on a small-scale canning line to process its large cattle surplus for export to Iran.[23] Initially, almost all canned beef was destined for export; according to a 1982 report, the product was basically unheard of in China.[24] By the 1990s, dozens of factories from Harbin to Kunming were producing small runs of canned beef, sold in tins that were beautifully decorated but otherwise essentially indistinguishable from each other. Only a few producers had the quality and reach to build a brand, especially for coveted overseas markets. The most famous of these was state-owned Meilin (梅林 Maling), which exported canned products to overseas customers in Southeast Asia, the Middle East, and the Soviet Union.[25]

As China's domestic market grew, the era's indefatigable local entrepreneurs expanded into new lines of beef products. Inner Mongolia and Sichuan developed new snack markets for dried beef. In 1988, the Pingyao County Food Factory made a 200,000-yuan investment in a canning line and briefly produced tinned luncheon meat. Before long, they had moved to shelf-stable vacuum-packed foil pouches, which preserved the texture of the product, and were sold alongside such similarly packed delicacies as Dezhou chicken and Yangzhou goose.[26]

Consumers prized these new products for their novelty and portability. Both canned and vacuum-packed beef could be heated right in the package, using the hot water that was always in supply from tea boilers. Like powdered milk or instant noodles, these portable products were an ever-present feature of long train trips and crowded dormitories. Over time, hundreds of small producers of tinned and vacuum-packed beef were winnowed down to a few well-known brands, such as the Guanyun brand from the Pingyao Beef Company. Driven in part by the revelation of the widespread substitution of horse or donkey meat at the unregulated lower end of the market, consumers increasingly sought out trusted names.[27] By 2007, the Guanyun brand had emerged as one of China's most valuable.[28]

UPSCALING A BRAND: LANZHOU BEEF NOODLES

During the early 1980s, Chinese officials encouraged a dietary shift from pork to beef and sheep, since grazing animals did not consume grain.[29] Yet, while the consumption of beef steadily grew, fresh beef remained a niche purchase, in part because of the time and effort it required to cook at home. Instead of raw beef, people would instead buy a finished product, such as beef that had been cooked in an industrial kitchen, ready to be sliced and served cold with garlic and vinegar.[30] Or else they would eat beef in restaurants, which had the time and equipment to prepare long-stewed beef, the original Chinese version of beef and potatoes, or even a Hong Kong-style beef curry. A more daily-use meal might include noodles topped with stewed or sliced beef. Of these dishes, the most famous was *Lanzhou niurou lamian* (蘭州牛肉拉麵 Lanzhou beef noodles).

Lanzhou beef noodles, as the name suggests, originated in Lanzhou, supposedly invented in 1915 by a shop owner named Ma Baozi.[31] By the 1990s, Lanzhou beef noodles were everywhere, made fresh in market stalls by noodle masters who pulled one piece of stretchy dough into two, then four, then eight, and then dozens of strands that were quickly boiled and served with broth, onions, radish, coriander, and a few thin slices of cooked beef.

Besides the taste, one of the attractions of Lanzhou beef noodles was the price. In early 1990s Jinan, a bowl of Lanzhou beef noodles was exactly one yuan, an impossibly low price that reflected the minimal cost of operations. In bus stations and roadside markets nationwide, Lanzhou noodle stands had a similar setup: a wide pot of water heated over a lump of pressed coal dust, a table for working the noodles, and a few basic materials for dressing the finished dish. Like the streetside sellers of beef or mutton kebabs, the purveyors of these stalls were often Muslims from western China. (A 2011 survey found that 70 percent of the sellers in Shanghai came from Qinghai.[32]) The visible display of minority ethnicity in Lanzhou beef noodle stalls was more than just a show for the customers. Workers and owners were often recruited from the same region or even the same village, committed to bringing money back home. The model was wildly successful, allowing stands to survive by keeping costs extremely low and ensuring a steady supply of resources from home, often including the beef itself.

Over time, the dish inevitably faced pressure to move upmarket. In 2010, Lanzhou beef noodles received official recognition of geographic indication (GI), meaning that, like Champagne wines, Parma ham, or Pingyao beef, its place of origin was now a legally protected brand.[33] The larger phenomenon of Lanzhou beef noodles seeking a brand entailed two additional processes. The first was official promotion of food tourism, a tactic that had already been used with great success in places like Sichuan. Government promotion of Lanzhou beef noodles as an iconic item of cuisine included recognition by the usual awards, as well as such touristic efforts as the building of a Lanzhou

Beef Noodle Museum that paired the experience of the dish with the exotic image of China's far West.[34] The second was a series of measures aimed at upscaling the business model. These included state-sponsored training for cooks and the promotion of chains to displace the large number of individual small-scale proprietors (estimated in 2015 at fifty thousand shops nationwide).[35] In emulation of the "dragon-head" model promoted throughout other industries, state planners increasingly sought an iconic brand to serve as the cultural standard bearer for taking Lanzhou beef noodles global.[36] Yet although a few major chains (e.g., Jinding 金鼎, Mayoubu 馬有佈, and Sadamu 薩達姆) of Lanzhou beef noodles have indeed emerged, none have been able to displace the small-scale family proprietor.[37]

FRANCHISE FEVER: BIG MAC

In 2019, three professors from a small technical university in Hebei published an article decrying Western fast food as unhealthy, exploitative, and addictive, nothing less than a "new opium."[38] Such criticism was not new, nor was the call for China to create its own fast food.[39] Decades earlier, when KFC, Pizza Hut, and McDonald's were just entering China, Chinese competitors immediately followed. Among the earliest of these was Mr. Lee California Beef Noodle King, which opened its first store in Beijing in 1986.[40] Distinct from a bowl of open-air market noodles, California Beef Noodle King replicated much of the appeal of the then-scarce Western franchises, offering a quick and easy meal in brightly lit, colorful surroundings. Despite the restaurant's supposed California roots, the main attraction was a fairly standard dish of *hongshao* (紅燒 braised) beef noodles—a taste not so different from beef stewed with potatoes or the soy sauce-cooked beef in tins. Rather, what was new was the operational model. Like its fast-food competitors, California Beef Noodle King precooked much of its food in a central kitchen, to be reheated and served in franchised outlets.

The efficiency of this model has proven hard to resist. Since the 1990s, a potent combination of forces—urban planning that turned street markets into indoor malls, ever more sophisticated cold chains that allowed for cheaper and more reliable centralized food production, and a proportional increase in dining out (a broad category often abbreviated as FAFH, "food away from home") as part of the daily food intake—have all allowed replicable food franchises to overwhelm individually operated restaurants at all price points and across food sectors, from bubble tea to roast duck.[41] But the most visible and the most transformative front in China's franchise frenzy has been in fast food, in which no name is more iconic than McDonald's.

The advance of the Golden Arches into China represents the country's modern food revolution in microcosm. When China's first McDonald's opened in 1990 in Shenzhen, it was mobbed with customers attracted by the sheer novelty of the place. As the restaurant spread to other cities and became a more familiar fixture of daily life, people made it their own, just like they would any other public space.[42] This normalization of the experience was followed by a more fundamental change—the price fell; that is, prices rose only slightly at a time when the average national income increased ninefold. McDonald's rode the wave of FAFH, growing by 2020 to 3,700 outlets nationwide. Nevertheless, it is something of a stretch to compare the success of the restaurant to opium addiction. If the fifty million Big Macs sold in 2015 were shared equally among China's urban population alone, each person would barely get a bite.[43]

The transformation of a McDonald's meal from a special treat to an unremarkable part of daily life is the exact opposite of the upmarket aspirations of foods like Lanzhou beef noodles. Along with being quick and convenient, the Big Mac is cheap, reflecting the cost-saving innovation of mass-scale, centralized food production. This dependable, infinitely replicable efficiency is the irresistible force of "McDonaldization," and as China's current franchise frenzy will attest, its impact has been felt well outside of hamburgers.[44]

BRANDING THE LOCAL: CHAOSHAN BEEF HOTPOT

China's 2001 entry into the World Trade Organization introduced a wave of new food imports that benefited consumers but presented a severe challenge to domestic producers. Imported foods were often both cheaper and higher quality than local ones, a problem that grew more visible as a series of safety scandals drove Chinese consumers to seek out foreign products.

China's beef producers felt the full impact of this new competition. Over the 1990s, Chinese beef production had grown steadily in quantity but not in quality, producing a situation where the domestic supply of low-quality beef actually outstripped demand. After 2001, imports outcompeted Chinese-made beef in both price at the low end and in quality at the high end, and the high end was what was growing. With supermarkets rapidly replacing fresh open-air markets and new restaurants increasingly aiming at upscale diners, the field came to favor imported grain-fed beef that used its foreign origin as a brand: precut steaks from Australia, stew meat from Ireland, and *wagyu* from Korea.[45] Survey after survey confirmed that consumers wanted higher-quality produce and were, moreover, willing to pay for it.[46]

Chinese producers eventually began making their way up the value ladder. Especially after the food safety scandals of 2008, consumers enthusiastically sought out new quality-driven sectors like organic produce. Some of China's meat producers embraced the opportunity by trying to create their own distinct brand image around place. Unlike the cultural heritage of Pingyao beef or Lanzhou beef noodles, this evocation of locality places the value on the unspoiled quality of the fresh meat, including the supposed medicinal properties of meat from animals grazed on the Hulunbuir grasslands or Guizhou's Guanling Mountains,[47] often enhanced by an exoticized image of minority husbandry practices.[48] Place-as-brand reflects not only on intense promotion by local governments, but also new initiatives like the 2018–2022 *xiangcun zhenxing zhanlüe guihua* (鄉村振興戰略規劃 Rural Rejuvenation Strategic Plan) that encourages rural brands to add value beyond simple price competitiveness.[49] Yet despite all these efforts, few of the nation's beef producers have had particular success distinguishing themselves on reputation alone.

Rather, place-based branding that has largely remained the domain of finished dishes can lay claim to both a culinary heritage and a productive one. The restaurant setting is particularly important for beef, which is especially suited to FAFH. More than grilled meat, beef in the restaurant sector means *huoguo* (火锅 *hotpot*), a *gyūnabe*-style dish of thinly sliced meat, cooked at the table in a bubbling pot of broth. Estimated in 2022 at 612,000 stores, hotpot is a large and diverse sector.[50] In terms of quality and cost of ingredients, hotpot can be very cheap or very expensive. Chain restaurants like Haidilao and Xiabu Xiabu distinguish themselves primarily on price, while more expensive restaurants highlight exotic Japanese or Korean *wagyu* beef.

Within this very large and competitive sector, Chaoshan-style beef hotpot has used locality to carve out a niche. More than the flavor of the broth or the quality of the animals, the uniqueness of Chaoshan hotpot is how the beef is prepared. Rather than the usual method of shaving curls off of a frozen block of meat, the beef in Chaoshan hotpot is prepared fresh, ideally within three hours of slaughter. Because the meat is never refrigerated, different cuts retain their individual taste and texture. Local diners understand and appreciate these subtleties. But even diners who are not familiar with what makes Chaoshan hotpot culinarily unique will still recognize the *name* of Chaoshan hotpot, giving vendors an important edge in facing mass-market chain rivals.

CONCLUSION

This chapter has sprinted through a complex commodity history, stopping briefly to introduce eight distinct moments of commercial, social, political, and technological significance. By building the story around finished dishes, it has made the case that production does not stop at the point of retail, but rather continues into the home or industrial kitchen, and eventually to the dinner table. No mere afterthought to production, cooking is the difference between an undifferentiated commodity and anything we would recognize as food. The decision of what and how to cook reflects and culminates all of the forces and processes that came before.

Although each of these eight dishes emerged at a specific moment in time, the history of cuisine itself is hardly linear. Separated by a hundred years, the pickled beef of the late Qing and the tinned beef of the 1980s were both responses to the same problem—the need to make good use of a perishable product that is in temporary oversupply. Pingyao beef, Lanzhou beef noodles, and Chaoshan hotpot each show efforts to bind food to locality, using vehicles ranging from folktales to government-sponsored heritage programs to lay claim to culinary technique while creating a cultural cachet that enhances the market value of local specialties. Finally, Japanese *gyūnabe*, Khrushchev's beef and potatoes, and the iconic Big Mac all show the cultural value of food taking on a life of its own, the reputation sometimes arriving well in advance of the food itself.

Finally, the culinarily curious will be glad to know that all of our eight dishes are available today, each one having outlived and evolved beyond the particular circumstances that fostered its creation. Such is the plasticity of culinary meaning. Processing techniques like pickling, drying, or canning still add value that is distinct from their original function of preservation. The social meaning of Japanese *gyūnabe* or a fast-food hamburger remains significant, although very different from what it initially had been. The story of beef shows how cuisine adds new layers of cultural value to itself over time, as custom is relabeled as heritage and the pull of novelty gives way to that of nostalgia.

NOTES

1. Mark Kurlansky, *Cod: A Biography of the Fish that Changed the World* (New York: Penguin Books, 1997).

2. Jia-Chen Fu, *The Other Milk: Reinventing Soy in Republican China* (Seattle: University of Washington Press, 2018); Veronica S. W. Mak, *Milk Craze: Body, Science, and Hope in China* (Honolulu: University of Hawai'i Press, 2021).

3. Margaret Visser, *Much Depends on Dinner: The Extraordinary History and Mythology, Allure and Obsessions, Perils and Taboos of an Ordinary Meal* (New York: Grove Press, 1986).

4. Shi Nai'an, *Shuihu zhuan* [Heroes of the Water Margin] (Taibei: Sanmin shuju, 2016), *juan* 23.

5. Shi Nai'an, *Shuihu zhuan, juan* 27.

6. Tao Menghe, *Beiping shenghuofei zhi fenxi* [Analysis of living costs in Beiping] (Beijing: Shangwu yinshuguan, 2011), 54–55.

7. "You Mengshang Dashengkui" [Traveling Mongol trade house Dashengkui], *Neimenggu wenshi ziliao* 12 (1984): 125–131.

8. *Beijing Niujie zhi shu "Gang zhi"* [Gazetteer of the Beijing Cow Street "Gang zhi"] (Beijing: Xinhua, 1991), 34.

9. On moral and legal restrictions, see Vincent Goossaert, *L'interdit du boeuf en Chine: agriculture, éthique et sacrifice* [The ban on beef in China: Agriculture, ethics and sacrifice] (Paris: Collège de France, Institut des Hautes Etudes Chinoises, 2005); E Liu, "Qingdai 'zaisha maniu' lü yanjiu" [Research on the Qing-era "cattle and horse slaughter" law], *Lishi dang'an* 3 (2015), 67–75. Surveys appear in Liu Qinli and Du Bosi, "Minguo shiqi niufanzi de shangye fengsu yu maoyi wangluo: Liyong waibao fangtan de fangshi shiyan" [Commercial customs and trade networks of Republican-era cattle brokers: An experiment in outsourcing ethnographic research], *Minsu yanjiu* 2 (2022): 23–34.

10. For a brief explanation of *Tiaoding ji* and culinary sources, see Thomas David DuBois. "Ten Centuries of Chinese Food Writing: What Do We Do with All These Recipes?," *Food, Culture & Society* (2023), https://doi.org/10.1080/15528014.2023.2235164.

11. *Pujiang wushi zhongkuilu: gufa zhicai, yincang de chuniang shidan* [Instructions for the housewife: Ancient foods, the menu of a hidden-away woman cook] (Shanghai: Shanghai wenyi chubanshe, 2021), 44.

12. 1882 Pingyao County gazetteer quoted in Lei Bingyi, *Pingyao niurou* [Pingyao beef] (Taiyuan: Shanxi chubanshe, 2011), 9.

13. The song "In praise of local products" (*kua tuchan*) begins with the line "Beef from Pingyao, Taigu cakes."

14. Lei Bingyi, *Pingyao niurou*, 5–9; Wang Zhen, "Shanxi Zhonghua laozihao xilie zhi wu—Pingyao niurou" [Chinese heritage brands in Shanxi—Pingyao beef], *Pinpai* 6 (2008).

15. W. J. Blackwood, "Meat and Milk Inspection in Shanghai," in US Department of Agriculture, ed., *Fifteenth Annual Report of the Bureau of Animal Industry for the Year 1898* (Washington, DC: Government Printing Office, 1899), 205–212.

16. Thomas David DuBois, "Many Roads from Pasture to Plate: A Commodity Chain Approach to China's Beef Trade, 1732–1931," *Journal of Global History* 14, no. 1 (2019): 22–43.

17. Liu Xingji, *Zhongguo gengniu wenti* [China's draft cattle problem] (Jiangsu: Shiyebu Zhongyang zhongchuchang, 1935).

18. *On Khrushchov's Phoney Communism and Its Historical Lessons for the World* (Beijing: Foreign Languages Press, 1964).

19. Karl Gerth, *Unending Capitalism: How Consumerism Negated China's Communist Revolution* (Cambridge: Cambridge University Press, 2020).

20. Mao Zedong, "Nian Nujiao: niao er wenda" [Two birds: a Dialogue] (1965), *Shikan* [Poetry journal] 1 (1976).

21. Shu Wang, "'Tudou shao niurou' gongchan zhuyi de youlai" ['Beef and potatoes' the root of communism], *Dangshi zongheng* 4 (2006): 29.

22. Xie Chengxia, *Zhongguo yang niuyang shi (fu yanglu jianshi)* [History of cattle and sheep husbandry in China (appended short history on deer raising)] (Beijing: Nongye chubanshe, 1985), 125–135.

23. Sun Wenguo, "Niu xiang jianqi guantouchang" [A cow county founds a cannery], *Xin nongye* 15 (1987): 27.

24. Su Huishan, "Qiantan Qinchuan niu de rouyong jiazhi jiqi rouyong tujing" [Research on establishment and promotion of the brand of Guizhou Guanlian beef], *Zhongguo huangniu* 2 (1988), 51–54.

25. Wang Baodong, "Chukou niurou guantou ying zhuyi de jige wenti" [Some problems to be aware of regarding canned beef for export], *Zhongguo shangjian* 12 (1995), 17.

26. Cun Wai, "Guanyun niurou de pinpai fazhan de lu" [Road to developing the Guanyun beef brand], *Nong chanpin jiagong* 4 (2007): 71–73.

27. Lao Dong, "Zhenjia zhi you—Pingyao niurou shijian ji sikao" [Worries over real and fake—thoughts on the Pingyao beef incident], *Xianfengdui* 5 (2004): 30–31.

28. Zhao Yiliang, Mu Zhemin, "'Guanyun' pinpai shengzhi 4000 wan" [Guanyun brand appreciates 40 million], *Shanxi ribao*, July 26, 2007.

29. Jacob A. Hoefer and Patricia Jones Tsuchitani, *Animal Agriculture in China: A Report of the Visit of the CSCPRC Animal Sciences Delegation* (Washington, DC: National Academy Press, 1980), 69.

30. Brooke Edwards, et al., *The Sino-Australian Cattle and Beef Relationships: Assessment and Prospects* (Ultimo NSW: The Australia-China Relations Institute (ACRI) 2016), 32.

31. Chen Xiaoli, "Lanzhou niurou mian pinpai de wenhua tazhan wenti shenxi" [Deep analysis of the problems with lanzhou beef noodle brand's cultural expansion], *Sheke zongheng* 27, no. 7 (2012): 125–127.

32. Chen Xiaoli, "Lanzhou niurou"; Liu Jia, "Lanzhou niurou mian chanyehua jingying diaocha fenxi" [Investigation and analysis of the industrial management of Lanzhou beef noodles], *Hezuo jingji yu keji* 16 (2011): 84–86.

33. Chen Xiaoli, "Lanzhou niurou."

34. Li Ruixue, "Lanzhou lamian pinpai zhanlüe sikao" [Strategic thinking on the branding of Lanzhou pulled noodles], *Gansu keji zongheng* 38, no. 6 (2009): 124–126.

35. Wang Guoqun, "Canyin qiye jingying zhong de wenti yu sikao—yi Lanzhou niurou mian liansuo wei li" [Thoughts and problems on the management of restaurant management—the case of Lanzhou beef noodle chains], *Dongfang qiye wenhua* 11 (2015): 12–13; Liu Jia, "Lanzhou niurou mian."

36. Yan Xinyan, "Lun Lanzhou qingzhen yinshi wenhua tese ji qi minzu pinpai suzao celüe—yi Lanzhou niurou mian wei li" [On the ethnic characteristics and brand creation strategies of Lanzhou halal foods], *Shipin anquan daokan* 27 (2019): 52–53.

37. Liu Jia, "Lanzhou niurou mian"; Ma Qiming, "Lanzhou niurou mian wenhua de chuancheng yu fazhan" [Transmission and development of Lanzhou beef noodle culture], *Fazhan* 12 (2017): 84–86; Qiu Zhaolei, "Qiantan Lanzhou lamian de fazhan xianzhuang wenti ji shenglüe cong chenggong qiye Jinding Lanzhou diyimian tanqi" [Discussion of the development of Lanzhou noodles from the perspective of the most successful enterprise Jinding Lanzhou noodles], *Beifang jingmao* 11 (2015): 55–57.

38. Wang Chenxi, Liu Zijian, and Wang Wen, "Yang kuaican zhengzai qinshi woguo de laodongli" [Western fast food is eroding China's labor strength], *Shipin anquan daokan* 23 (2019): 26.

39. Zhu Jianping, "Jiasu fazhan juyou Zhongguo tese de Zhong shi kuaican shizai bixing" [Imperatives for accelerating the development of Chinese style fast foods], *Shipin keji* 12 (2003); Guonei maoyibu, *Zhongguo kuaicanye fazhan gangyao* [Overview of China's fast-food industry], Document 96 (1997), *Zhongguo falü shujuku*, http://fgcx.bjcourt.gov.cn.

40. Zhang Guangliang, "Meiguo jiazhou niurou dawang—yinqi bu zhengdang jingzheng an" [The unfair competition case of California beef noodles], *Renmin sifa* 8 (1995): 47–49; On China's struggle to build a fast-food industry, see Hattie Liu, "Can China Save Its Struggling Fast-Food Industry?" *The World of Chinese*, July 21, 2018. https://www.theworldofchinese.com/2018/07/quick-fix/.

41. Junfei Bai, et al., "Food away from Home in Beijing: Effects of Wealth, Time and 'Free' Meals," *China Economic Review* 21, no. 3 (2010): 432–441.

42. Yunxiang Yan, "McDonald's in Beijing: The Localization of Americana," in *Golden Arches East: McDonald's in East Asia*, ed. James L. Watson (Stanford, CA: Stanford University Press, 1997), 39–76.

43. Yu Cheng, "Big Mac Is Favorite Burger at McDonald's China," *China Daily*, August 4, 2018.

44. George Ritzer, *The McDonaldization of Society* (Thousand Oaks, CA: Pine Forge Press, 1993).

45. Shuwen Zhou, "Formalisation of Fresh Food Markets in China. The Story of Hangzhou," in *Integrating Food into Urban Planning*, eds. Yves Cabannes and Cecilia Marocchino (London: UCL Press, 2018), 247–263.

46. Fred Gale and Kuo Huang, *Demand for Food Quantity and Quality in China*, U.S. Department of Agriculture Economic Research Report No. (ERR-32) (2007).

47. Su Shaohui, "Guizhou Guanlian niurou pinpai jianshe yu tuiguang celüe shenjiu" [Research on establishment and promotion of the brand of Guizhou Guanlian beef], *Nanfang nongye* 13 (2019), 79–81; Li Wei, Wang Xinyuan, "Hulunbeier niu yang rou chanye pinpai jianshe tanxi" [Analysis of branding in the Hulunbuir beef and mutton industry], *Hulunbeier xueyuan xuebao* 29, no. 03 (2021): 75–78.

48. Megan Tracy, "Pasteurizing China's Grasslands and Sealing in *Terroir*," *American Anthropologist* 115, no. 3 (2013), 437–451; Yuan Demin, et al., "Pinpai niurou shengchan de chaju yu duice" [Gaps and corrective measures in the production of branded beef], *Yangzhi jishu guwen* 1 (2008): 99.

49. Su Shaohui, "Guizhou Guanlian niurou"; *Xiangzhen zhenxing zhanlüe guihua 2018–2022* [Rural revitalization strategic plan, 2018–2022], http://www.moa.gov.cn/ztzl/xczx/xczxzlgh/.

50. Meituan/Dazhong dianping yanjiuyuan, 2022 *Zhongguo huoguochanye fazhan dashuju baogao* [China hotpot industry development big data report] (n.p.: Meituan/Dazhong dianping yanjiuyuan, 2022).

II SYSTEMATIZING EXPERTISE: FOOD, SCIENCE, AND TECHNOLOGY

4 ABSORBING VITAMINS: HOW A NUTRITIONAL PARADIGM WAS REINVENTED IN REPUBLICAN CHINA

HILARY A. SMITH

In 1919, a peculiar item appeared in the pages of the *Zhonghua yixue zazhi* (中華醫學雜誌 *China Medical Journal*), the periodical of a professional association founded just a few years earlier to promote Western-style medical knowledge. The article's title proclaimed that *"weitaming"* (維他命 "vitamins")—written in quotation marks, indicating that the term was novel—"are an essential factor in foods." Its author began by identifying rich sources of vitamins, including "barley, unpolished rice, fermented things, and heart muscle" as well as liver. And then things got strange. The author offered a version of an English-language joke that he had, with charming sincerity, turned into a meditation on the power of the vitamin. The joke was that a doctor had recommended cod liver oil, a highly regarded supplement, to a sickly farmer who'd come to him with tuberculosis. Three years later, the two met again by chance and the farmer praised the doctor's prescription, saying, "I followed your advice. I wanted to eat dog-liver oil but I couldn't get any. So I hunted down a dog, cut out his liver and ate it." As a result, he claimed, "my disease has been successfully, suddenly cured." The doctor was astonished. He said, "I didn't direct you to eat *dog*-liver oil. What I wanted you to drink was *cod*-liver oil." The author then observed, "When I first heard this I thought it was a joke, but looking at it from today, maybe dog liver"—like cod liver—"has lots of 'vitamin' efficacy in it that would make his tuberculosis seem to disappear."[1]

This article, one of the earliest recorded mentions of the word "vitamin" in Chinese, conveys the aura of power and mystery that surrounded this foundational concept of modern nutrition science when it first entered China. Presented as an abstract, health-promoting substance, it combined the mystique of animal vitality with the promise of a panacea, something that might cure not only the few vitamin-deficiency disorders that had been discovered but even tuberculosis, then one of the most widespread and deadliest diseases around.

Outside of China, the vitamin was making a similar splash. The English word had been coined seven years earlier, and the concept of an essential element in foods that was not protein, carbohydrate, or fat but that seemed equally necessary to keep eaters alive had compelled nutrition scientists to revise their paradigms of what healthy eating required. Without vitamins, they determined, human bodies could not thrive. The discovery fueled a paradigm shift in nutrition science, which up to that point had focused on establishing the *minimum* provision of food necessary to keep the poor alive or to keep the working class productive. The "new nutrition," by contrast, focused on *optimum* diets that would encourage flourishing and development; it often framed diet in national terms, implying that faulty eating had contributed to the struggles of weaker, poorer nations.[2] This new paradigm made its way to China in the 1920s and 1930s, mostly through the activity of Chinese nutrition scientists trained abroad. In China, the new nutrition, though favored by a Western-oriented, science-centered Guomindang government, became but one element of a complex landscape of ideas about healthy eating. The vitamin took on new meanings, sometimes assimilated to ideas about deficit and surfeit originating in classical Chinese medicine or in folk knowledge.

This chapter examines the rise of the vitamin and its integration into the vibrant, varied intellectual life of Republican China (1912–1949). It shows how preexisting notions about dietary supplementation both accommodated and resisted this new nutrition-science concept. The story of the vitamin in modern China contributes to a growing body of literature that is reconfiguring the place of Western scientific knowledge there, showing that it constituted only one among many forms of knowing—one that was, itself, changing shape in response to its environment.[3] This was especially true of nutrition science, which was still in its infancy worldwide.

The science and technology of *food* in modern China have centered on persistent questions about scarcity and sufficiency, as Sigrid Schmalzer's contribution to this volume (chapter 5) highlights for the Mao era. In that later period as in the one I discuss here, many saw science as a way to overcome the dangers of food inadequacy in a large population. But as Schmalzer's, Fan Yang's, and my essay all demonstrate, in China, foodways-related science and technology have sometimes diverted attention from systemic problems responsible for a great deal of human suffering. Below, we will examine the emergence of the vitamin in the context of early nutrition science internationally and in China, and then explore the ways in which nutrition-science concepts interacted with other approaches to deficiency and supplementation prevalent in the Republican period.

THE DISCOVERY AND INVENTION OF VITAMINS, 1912–1950

The vitamin was to nutrition science what the antibiotic was to medicine. Just as the development of antibiotics in the 1940s made it seem that modern medicine might eliminate infectious disease altogether, the development of vitamin supplements starting in the 1930s gave hope that humankind was growing ever closer to conquering malnutrition. With the proper inputs of vitamins A, B, C, and D, the thinking went, serious cases of deficiency disease could be prevented, as could the suboptimal health caused by less-serious deficits.

It was the Polish chemist Casimir Funk who first coined what would become the word "vitamin"—*vitamine*—in 1912, with *vita* (from "vital") indicating the thing's importance in maintaining life, and *amine* reflecting Funk's conviction that these things all shared a common molecular structure. They were all, he supposed, organic bases or *amines*. When he coined the term, however, the molecular structure of the substances he referred to as *vitamines* was unknown. The definition of a vitamin was therefore imprecise. All scientists knew was that these substances could prevent or cure the symptoms of a new class of diseases that Funk called "deficiency diseases," including beriberi, scurvy, and pellagra. Thus, the original vitamin, Vitamine B, was not a particular chemical compound; instead, it was *the substance in rice polishings that prevents beriberi from developing*. "B" originally stood for beriberi.[4]

Soon other conditions joined the list of deficiency diseases, including night blindness, rickets, and some forms of sterility. As scientists worked out the molecular structure of the substances that protected against and cured these conditions, they realized that not all *vitamines* were, in fact, amines. In 1920, Jack Cecil Drummond, a physiologist at University College, London, therefore suggested dropping the final "-e" so that the word ended in "-in" instead, since chemists often used the ending "-in" to name "a neutral substance of undefined composition"—a nebulous description indeed. To keep track of these discoveries, Drummond suggested lettering the substances alphabetically according to the order of their discovery, with the idea that the letters would serve as placeholders "until such time as the factors are isolated, and their true nature identified."[5] Vitamin B's "B" for beriberi was grandfathered in, and the alphabetical system began from the next vitamin to be discovered, the anti-night blindness vitamin A. The anti-scurvy, anti-rickets, and anti-sterility factors then became vitamins C, D, and E, respectively.

Complications began to crop up, however. By 1925, vitamin B had begun to unravel; the anti-beriberi substance originally discovered seemed to contain at least two different chemical compounds, one that prevented beriberi and another that worked against pellagra. Over the course of the 1930s, the substance was broken down further and other active components emerged. Vitamin F, so named in 1923, was decommissioned by 1930, losing its vitamin status and becoming a group of "fatty acids" instead. By 1935, when vitamin K appeared, its discoverers discarded Drummond's alphabetic placeholder system and returned to something like the original rationale. "K" was not the next letter in the sequence, but referred instead to what the substance did: it was an anti-hemorrhagic factor that helped blood coagulate, and K stood for Koagulering in the key researcher's native Danish.[6] After vitamin K, no further vitamins gained a foothold in the nutrition science literature. The flamboyant biochemist Albert Szent-Györgyi announced a "permeability controlling" vitamin P in 1936, but by 1950, vitamin P had been rejected. The Federation of American Societies of Experimental Biology recommended that the term be dropped altogether since scientists had been unable to show that the vitamin really existed.[7]

Vitamins F, G, H, and a handful of the numbered Bs followed vitamin P into oblivion as scientists determined either that they were not necessary for *human* health (sometimes a missing dietary factor caused symptoms in a laboratory guinea pig but not in a human being) or that they were identical to a previously named vitamin. Thus, by 1950, the peculiar gap-ridden set of letters and numbers that we still have today had more or less fossilized.

INTRODUCING VITAMINS TO CHINA: THE ROLE OF THE NUTRITION SCIENTISTS

The age of vitamin discoveries corresponds neatly to China's Republican period. The *vitamine* and the Republic came into being in the same year, 1912, and in 1950 when scientists were drumming the last major vitamin discovery, vitamin P, out of the books, the Chinese Communists were likewise drumming the last vestiges of the Republic off the mainland. In China, news of the vitamin arrived not long after the formulation of the concept in Europe and America, as the 1919 article quoted in this chapter's introduction attests. At that time, however, nutrition science had not yet taken shape as a discipline and a profession in China; discussions of the vitamin were therefore undertaken by enthusiastic amateurs with eclectic educations, rather than by those who had had extended training in science. It is not surprising, then, that the vitamin described in that 1919 article was a kind of undifferentiated, potent force. The author did not distinguish, as scientists were by then doing, water-soluble from fat-soluble vitamins, or anti-beriberi from anti-night blindness from anti-scurvy vitamins. As noted previously, he did not limit the efficacy of vitamins to curing vitamin deficiencies but suggested that they had the power to check tuberculosis.

We should not dismiss such prescientific formulations as irrelevant to the story of nutrition science in China. As Eugenia Lean has argued, the "heterogeneous engagement with manufacturing, science, and commerce" that characterized China in this period was not "merely a 'transitional' stage in which China was simply a newcomer on an inevitable and desirable course toward modernity."[8] When Chinese scholars who had attained doctoral-level nutrition-science training abroad returned to China beginning in the mid-1920s, they did not enter an environment lacking well-developed ideas about food and health. Nor was the vitamin completely unknown there. The concepts of the vitamin that had already taken shape in magazines and advertising informed a popular understanding that the work of nutrition scientists did not erase or supplant.

One piece of evidence that demonstrates this dynamic is the proliferation, and persistence, of multiple Chinese terms for the English word "vitamin." While English speakers settled quickly on *vitamines* and then vitamins—terms invented and propagated by scientists—in Chinese a variety of names emerged. The earliest ones, judging from searches in databases of Republican-era journals and magazines, were transliterations that started to appear at the very end of the 1910s. Most common was *weitaming* (維他命), but *weitaiming* (維太命) and *weitamin* (維塔民) also appeared. These felicitous renderings not only approximated the sound of the English word "vitamin" but also conveyed, at least in a broad sense, the meaning of the concept. *Weitaming* means "safeguarding his life (or lifespan)," while *weitaiming* means "safeguarding the greatest lifespan." *Weitamin* implies something like "preserving the pillar of the people."

More scientific terms came later, starting in the mid-1920s. These were translations rather than transliterations, pairing a modifier that represented life or vitality with the character *su* (素). *Su* was the same character used in *yuansu* (元素), the term that denoted chemical elements, reflecting nutrition science's roots in chemistry laboratories. These names, too, were multiple. A number of different modifiers were chosen to precede *su*, including *shenghuo* (生活), *huoli* (活力), *weisheng* (衛生), *shengming* (生命), and *shengji* (生機), representing vigor, vibrancy, longevity, strength, and more. Clearly, the scientists crafting a translation to represent "vital-element" had a rich vocabulary of vitality to choose from. *Weitaming* and *weitaiming* seem to have become the preference for commercial use; there are many advertisements for things so identified in newspapers such as *Shen bao* (sometimes written *Shun Pao*), but fewer for *weishengsu* (維生素). *Weishengsu* became the term preferred by the scientists. Both *weitaming* and *weishengsu* continue to be used today in Chinese, with something like those same associations.

It is no coincidence that the scientific term, *weishengsu*, began to appear in print more frequently in the mid-1920s, because it was at that time that nutrition science was beginning to emerge as a profession in China. In 1912, there had been no field of nutrition science in China; nutrition science had had no departments, no professors, no research institutes, and no scholarly journals. Even the institutions that would later host these departments and scholars did not exist yet: places like Peking Union Medical College, Yenching fluorescence University, Central Medical College in Nanjing, and the Henry Lester Institute in Shanghai. Most of these institutions were founded later in the 1910s (apart from Henry Lester, founded in the 1930s). They started offering biochemistry and physiology courses in the 1920s, and it wasn't until the beginning of the 1930s that formal nutrition-science programs began to grow out of biochemistry and physiology departments. It took the displacement and disruption of the anti-Japanese war in the early

1940s to bring into being the first research institutes specifically devoted to nutrition science. The nation's first national nutrition conference was held in 1941 in Chongqing; the first Chinese scholarly association dedicated to nutrition science, the Zhongguo Yingyang Xuehui (中國營養學會 Chinese Nutrition Society), was established in 1945; the first Chinese journal of nutrition science appeared in 1947.[9]

Almost all of the leading figures in early Chinese nutrition science spent significant time in North America. They earned graduate degrees from such institutions as Ohio State University and Indiana University (Zheng Ji, also known as Libin T. Cheng, 1900–2010), MIT and Harvard (Wu Hsien, 1893–1959), and McGill and the University of Pennsylvania (Hou Hsiang-ch'uan, 1899–1982). Even those who did not earn degrees in North America nevertheless spent time there undertaking further study and research.

Considering how connected Chinese nutrition scientists were to laboratories and people in North America, it is unsurprising that they shared the same interests and concerns.[10] In particular, Chinese scientists took up vitamins and avitaminoses as one of their major programs of research, publishing their work in the *Chinese Journal of Physiology* and the *Chinese Journal of Nutrition*. The purpose of much of this research was to figure out how vitamin-sufficient Chinese diets were. The journals are replete with articles assessing the vitamin content of particular Chinese herbs or foods and articles reporting on the vitamin content of the livers of deceased Chinese subjects. Scientists also worked hard to disseminate knowledge about vitamins to the public, presenting it as the most important bit of knowledge that nutrition science had to offer. For example, in 1929, the organic chemist Zheng Zhenwen (1891–1969) published a book called *Yingyang huaxue* (營養化學 *Nutritional chemistry*), one of the earliest books published on nutrition in China. In the preface, Zheng celebrated the discovery of vitamins (here, *huolisu* 活力素), declaring, "Those who aren't experts in this specialty should research it, and ordinary people should pay attention to it in order to deliberately and practically advance the average person's nutrition."[11] It quickly became clear to the reader that vitamins were, for Zheng, the central concept in nutrition science, as six out of his eight chapters focused on vitamins. Another early Chinese book on nutrition was Wu Hsien's *Yingyang gailun* (營養概論 *Introduction to nutrition*), published in the same year, 1929. In it there is a chapter on *yingyang bu liang zhi zhuangtai* (營養不良之狀態 conditions of malnutrition). What sorts of conditions did Wu connect with poor nutrition? Almost entirely vitamin deficiencies. He wrote at the outset that consuming excessive nutrients *could* cause malnourishment, but he averred that that was unusual, explaining why he discussed only deficits in the chapter. Descriptions of conditions caused by severe deficits of vitamins A, B1, B2, C, D, and E filled almost the entire chapter. It included lurid photographs of a pigeon with a bent-back neck said to be suffering from a version of beriberi after being fed white rice exclusively, a woman and children with rickets caused by severe vitamin D deficiency, and a sluggish, dull-eyed dog showing the effects of a vitamin A deficit.[12]

Was this assaying of Chinese plants and livers, and spreading the gospel of the vitamin to the public, actually improving the overall well-being of Chinese people? Did Chinese people actually have a vitamin deficiency problem? The evidence does not support a resounding yes. For one thing, popular articles discussing vitamins often gave the English names for avitaminoses and sometimes explained their symptoms, suggesting that they didn't expect most readers to have encountered these conditions. In an article published in 1929, Wu Hsien mentioned *jiaoqi* (腳氣 beriberi) and *goulou* (佝僂 rickets) but provided the English names for these diseases in parentheses, highlighting their foreignness.[13] Many of the published studies concluded that the Chinese foods and diets investigated were not, in fact, nutrient deficient. And those that concluded that they were based these conclusions on Western standards of nutritional sufficiency. They were not basing their conclusions on observing whether or not Chinese people were actually showing symptoms of vitamin deficiencies.

In 1937, the nutrition scientist Lan-chen Kung published a meta-analysis of the studies that had been done on Chinese diets up to that point. When it came to vitamins, she noted, "An evaluation of the adequacy of vitamins in the Chinese diet is rather difficult." She went on to suggest that the consensus of nutrition scientists was that Chinese diets were probably vitamin deficient, though her language suggests that this was questionable. She noted that their conclusions were based on "'Paper analyses' and animal experimentation," not on "quantitative experiments on human subjects."[14]

If Wu's articles for middle-class urban readers and Kung's studies, which often focused on college women, give us an impression of nutrition in a relatively privileged stratum of Republican society, the 1937 report that a team of Chinese nutritionists submitted to the League of Nations suggests that vitamin deficiencies were not rampant among the rural poor, either. This report was prepared by a team headed by Hou Hsiang-ch'uan, then a researcher with the Henry Lester Institute in Shanghai, for the League's Mixed Committee for the Study of Problems of Nutrition. It reads, "beri-beri [vitamin B1 deficiency] has never been widespread among rural populations . . . According to the data we have collected, pellagra [vitamin B3 deficiency] does not play an important part in any of the Asiatic countries . . . For the sunny countries of Asia, rickets [vitamin D deficiency] does not constitute a problem of any importance."[15]

That is not to say, however, that vitamin deficiency disorders were not a major problem for *some people* in China. Scattered reports appear in the scientific literature of vitamin-deficiency disorders, particularly beriberi, among factory workers in Shanghai, students at boarding schools, hospital patients, prison inmates, the army, and refugees. What did all of these people have in common? The Chinese team's report to the League of Nations identified it: beriberi in China, they said, was associated with "towns, armies, shipping, prisons, schools and large undertakings—in short, places where catering is collective."[16] In other words, most vitamin-deficiency sufferers were being fed by some kind of centralized authority as cheaply as possible. Vitamin knowledge might have improved the health of institutionalized populations by enabling authorities to continue to feed people cheaply without inducing clinical nutritional deficiency disorders. Ironically, however, vitamin discoveries may also have facilitated many people's malnutrition in a broader sense, by focusing narrowly on keeping people alive and free of obvious symptoms of disease rather than on enabling them to eat in balanced, moderate, and fulfilling ways.

For the majority of Chinese people in the 1920s and 1930s, who did not live in an institutional setting such as the army or a refugee camp, vitamin-deficiency disease seems to have been unusual. Nevertheless, nutrition scientists insisted that the country as a whole was suffering from suboptimal health induced by nutrient-poor diets. In their published work, they had to explain away the apparent strength and stamina of Chinese laborers. Such people might appear normal—or even remarkably strong—but their appearance was misleading.[17] Inadequate diet, according to the scientist Wu Hsien, was a national or racial problem specific to the Chinese.[18]

This shared perception was no doubt reinforced by the international image of China as the "Sick Man of Asia" and as a "Land of Famine."[19] China had indeed experienced a series of devastating famines in living memory, including two notoriously severe famines in north and northwestern China in the years 1920–1921 and 1928–1929, exactly when the first generation of Chinese nutrition-science researchers was beginning to publish. The Chinese press reported on these famines extensively, so the scientists could hardly have been unaware of them.[20] As Chinese nutrition science became institutionally established in the late 1930s and 1940s, continuous warfare ensured that the nutritional challenges of refugees and soldiers would attract attention in the discipline's formative years, further accentuating nutrition scientists' early emphasis on scarcity, deficit, and the need for growth. Vitamins were not a solution to recurrent famines or to war-induced dietary disruption, of course, but they were the tool readiest to hand among nutrition scientists, whose desire to revive and strengthen China enhanced their zeal for dietary reform.

INTRODUCING VITAMINS TO CHINA: THE ROLE OF DRUG COMPANIES

Meanwhile, a parallel understanding of vitamins was developing in the newspapers and magazines of Republican China, where drug companies advertised their wares. Advertisements for "vitamins" (typically *weitaming* 維他命) frequently appeared in newspapers throughout the period. It is unclear what these branded medicinals contained or how they were produced, but their advertising copy drew on both the new charisma of science and established ideas from classical medicine. Because the vitamin promised something like universal illness prevention, and could switch easily among foreign and Chinese, scientific and folk registers, it fit well into the burgeoning commercial drug market that Sherman Cochran has described for this period. As Cochran notes, although the entrepreneurs who ran the most successful Chinese pharmaceutical companies sometimes "had no Western partner or Western financial backing, had never been in the West, and had never studied Western pharmacology or received any Western education," they nevertheless often presented their companies and products as Western in origin.[21] Their most popular products, such as Ailuo Brain Tonic, Humane Elixir, and Man-Made Blood, tended to be broad in application and vague in composition, much like vitamins.

A typical example of register-melding in vitamin advertising is the "Three Virtues Vitamins" ad that appeared in the newspaper *Shen bao* in 1929 (figure 4.1). A banner across the top trumpets that the product was "Invented by the German Robert Koch"—a clear falsehood, considering that the great bacteriologist died in 1910 before vitamins were discovered, but one that reflects Koch's global preeminence as a man of science. The text along the right side of the advertisement weaves together both scientific and classical concepts. On the one hand, it claims that the vitamins improve *"xinchen daixie"* (新陳代謝 metabolism) and increase *"dikang yuanli"* (抵抗原力 resistance and innate strength), using technical neologisms describing physiology and immunity. But at the same time, it indicates that the product can *"yingyang jingshen qi xue"* (營養精神氣血 nourish nerves, *qi*, and blood). Nerves were a fixation of early twentieth-century Western medicine in China as elsewhere, but *qi* and blood had a long history as vital substances of concern in classical Chinese medicine.[22] Then, too, the name of the drug company, San De (三德 Three Virtues), resonates with Chinese culture and religion; although the precise referents of the term "three virtues" differed from one context to another, it featured in both Buddhist and Daoist texts and had been mentioned in the ancient *Book of Documents* and *Rites of Zhou* as well. Tellingly, the banner at the bottom of the advertisement proclaims the product "twenty-five times more effective than fish liver oil."

Vitamins inherited the mystique of cod liver oil, a substance that, like vitamins, had entered Chinese discourse through contact with Western countries, but had done so decades earlier. Health-seekers in Western countries had apparently been dosing themselves with cod liver oil for more than a century.[23] But though it was a popular and widespread remedy, cod liver oil did not win Western scientists' wholehearted endorsement before the advent of vitamins. It was not that scientists thought it did not do any good, but they thought it did not do any *specific* good; if it had any effect at all, they thought, it was just because it was a source of fat, and any other kind of oil or fat would have had the same positive effect on health. After vitamins were discovered, however, researchers learned that fish liver oil was rich in vitamin A, and thus the popular enthusiasm for the stuff seemed to be more precisely, scientifically vindicated.

This was the pattern in China as well. A 1919 advertisement by the English pharmaceutical company Burroughs Wellcome for its Kepler brand cod liver oil with malt extract (figure 4.2) shows a box of the oil haloed with characters touting its panacean functions. The product is a *"da bupin* (大補品 great tonic), famous around the world, that lengthens life, benefits longevity, relieves weakness, and shores up what is collapsing." The term *bu* (補 to supplement or tonify) had long been used in classical Chinese medicine for drugs thought to invigorate patients who, due to illness, senescence, or other

FIGURE 4.1
An advertisement for "Three Virtues Vitamins" (San De weitaming) in *Shen bao* (November 20, 1929): 12. Image courtesy of Chinese Modern Newspapers database (www.dhcdb.com.tw).

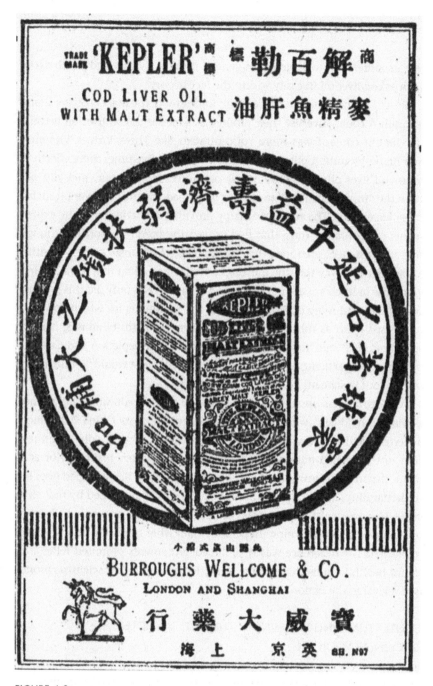

FIGURE 4.2

An advertisement for "Kepler" Cod Liver Oil with Malt Extract ("Jiebaile" maijing yuganyou) in *Shen bao* (November 8, 1919): 16. Image courtesy of Chinese Modern Newspapers Database (www.dhcdb.com.tw).

factors, suffered from depletion. A general-purpose remedy and prophylactic such as cod liver oil fit easily within the *bu* category.

By 1933, when vitamins were better understood, the makers of Star Brand Vitamin A could advertise their product as superior to cod liver oil because its vitamin content was more concentrated; like Three Virtues Vitamins, Star Brand Vitamin A pills were touted as "twenty-five times more effective" than cod liver oil.[24] Moreover, their promoters pointed out, they did not have the "fishy stench and greasiness" that made taking cod liver oil such a bitter experience. The advertising copy provided a description of the experiments on laboratory mice that had proven the product's worth. Although this advertisement used more precise language, and foregrounded scientific knowledge in a way that the Burroughs Wellcome ad had not, it was no less grandiose in its claims about the product's efficacy. Not only did this product work faster and more effectively for "all the symptoms for which cod liver oil is used, such as tuberculosis, rickets, adenosis, night-blindness, corneal softening, etc.," but "for those suffering from bodily depletion and damage, or weakness after being ill," taking Star Brand Vitamin A would "immediately bring about [a state of] health."[25]

As Cochran has observed, it was not only "highly educated and cosmopolitan individuals" who introduced Western ideas to China.[26] Products advertised and sold as vitamins, which in most cases had no direct connection to the foreign-trained nutrition scientists, research institutes, or academic departments introduced in the previous section, likely shaped popular understandings as much as or more than the articles published by that elite cadre. Nor, we might observe, was the faith in vitamins' power that advertising copy expressed any more hyperbolic than what nutrition scientists had published. The difference was that the advertisements promised relief and robust health to individuals, while Wu Hsien and his fellow scientists promised revival to the nation.

CONTEXTUALIZING VITAMINS IN THE DIET: THE ROLE OF CHINESE MEDICINE

Another important influence shaping Republican perceptions of vitamins and supplementation were practitioners of classical Chinese medicine. As Bridie Andrews and Sean Hsiang-Lin Lei have shown, in the 1920s and 1930s, there was no comprehensive, coherent philosophy or set of practices that one could point to as representing the way of Chinese medicine.²⁷ That was equally true of dietary ideas and practices, for which there was no unitary Chinese medical approach. In general, however, the classical canon emphasized balance and moderation and focused on the context of meals more than their content. Texts such as the *Yellow Emperor's Inner Canon*, compiled in the first century BCE, organized foods into a five-part matrix that encompassed the rest of the natural world as well: flavors, seasons, weather, planets, numbers, musical notes, colors, organs of the body, orifices, and emotions could also be assigned one of five categories, called wood, fire, earth, metal, and water. The five-phase matrix informed considerations of dietary balance, and encouraged eating attuned to multiple aspects of a person's environment and constitution. Maintaining such a balance required moderation, and classical medical writers prescribed restraint and regularity in eating as part of a larger pattern of self-discipline meant to foster health. Early Chinese doctors' idea of a diet-related disorder was, accordingly, one whose symptoms stemmed from eating irregularly, and particularly from eating excessively.²⁸

This perspective differed from the nutrition-science notion that the optimal diet involved maximizing one's intake of good foods. Any food could be good in one context and bad in another in the view of Chinese medicine. For a physically weak person convalescing from a wasting illness, cow's milk could be excellent nutritionally, but for a robust person with a rich diet, it could be detrimental. There was no absolute scale of good and poor according to which foods could be ranked in this understanding. In contrast, articles in nutrition-science journals often included tables that ranked foods according to their vitamin content, awarding pluses, double-pluses, and triple-pluses for higher concentrations.

Despite their differences with nutrition scientists, however, Republican-era Chinese-medicine practitioners also contributed to shaping the vitamin. After all, their emphasis on balancing sour, bitter, pungent, salty, and sweet made scientists' concern with balancing protein, carbohydrates, fats, minerals, and vitamins seem natural, even if there was no one-to-one correspondence between the two sets of categories. Many of them also proved quite willing to adopt and adapt elements of new scientific knowledge into their own thinking about food and digestion. The terms for vitamins, and even the modern term *yingyang* (營養 nutrition), almost never appeared in Chinese-medicine journals of the 1920s and 1930s, but the authors of articles in these journals did pay attention to how Western-style physiologists described the process of digestion and attempted to reconcile it with what classical texts said about how food gets transformed inside the body.

Consider, for example, the special 1935 issue of *Yijie chunqiu* (醫界春秋 *Annals of the world of medicine*), a journal of Chinese medicine, devoted to diseases of the stomach. Contributors described in detail the mechanics of digestion that scientists had elucidated: the anatomy of the esophagus, stomach, and intestines; the way the stomach moves; and the chemical composition of gastric juices. Their main concern, however, when it came to disease-causing eating, was as much *how* a person ate as *what* they ate. Shi Yiren (1896–1966), a well-known Chinese-medicine doctor from Wuxi, wrote, "When your spirit is extremely happy your gastric juices flow abundantly, so you can eat more. When you are worried, your gastric-juice secretion is weak, so you eat less. (Ancient medicine transmitted [this wisdom as] wood can conquer earth)." He asserted that while parasites or bacteria caused some types of stomach disease, other types emerged "because food and drink are not eaten in a regular manner, and fruit and oily things are randomly tossed in."[29] In his contribution, Chen Botao similarly counseled attention to the circumstances of eating. He wrote, "Chew completely to make digestion more convenient for your stomach. Eating an appropriate amount will benefit the tension and relaxation of your stomach. Do not tightly bind your belly. Do not force yourself to eat when you are angry or annoyed. Do not snack between meals." Practice this, he declared, and "your stomach *qi* will naturally be robust."[30]

Contributors to this issue were concerned about depletion and deficiency, but it was not vitamin depletion they worried about, and the way to correct depletion was not to consume more of any particular food. Instead, they emphasized the danger of depleted stomach and spleen *qi* leading to poor digestion and consequently to disease. What depleted spleens and stomachs? The same sort of thing that depletes overall energy: overwork. Eating a lot, or eating quickly, or eating irregularly asked too much of the digestive organs. It wore them out, and over time, this caused symptoms to appear.

Although their proposed etiology of deficiency differed from that of the nutrition scientists, and they focused on individuals rather than the nation or the race, Chinese-medicine practitioners, too, perceived depletion as a common problem in the life of modern Chinese. They believed that the changes of modern life had rendered citizens enervated and impotent.[31] These factors helped lay the groundwork for Chinese eaters to accept the gospel of the vitamin even more readily. There is no reason why a person suffering the effects of depletion, even if she presumed it had been caused by exhausting the spleen and stomach, would not turn to Three Virtues Vitamins to give her nerves, *qi*, and blood a boost.

CONCLUSION

The vitamin was the centerpiece of a body of knowledge, in some ways quite foreign, that nutrition scientists brought to China in the 1920s through 1940s. Chinese nutrition scientists spoke a different language, and not just the Western languages in which they had earned their degrees at universities half a world away. Even in their native tongue, they described bodies and foods in terms of chemical constituents, trained their gaze on the population rather than the individual, and spoke statistically of the aggregate. They sought to learn about the diets of humans by strictly controlling those of mice. They had little to say about the context of eating, the social and psychological and environmental factors that had been central to classical Chinese dietetics.[32]

In other ways, however, the vitamin that so concerned Chinese nutrition scientists also resonated broadly among their compatriots. In fact, the eclectic commercial and intellectual atmosphere of Republican China shaped the Chinese vitamin just as much as the foreign universities had done. China's burgeoning pharmaceutical marketplace turned the vitamin into a potent panacea, suited to ailments both modern and of long standing, both narrow (such as rickets) and general (such as weakness or fatigue). Meanwhile, the emphasis on depletion in classical medicine primed Chinese consumers to understand and accept the concept of a vitamin deficiency, and readily believe, as nutrition scientists insisted, that such deficits were an ever-present danger. The vitamin in China, therefore, did not belong exclusively to Western nutrition science but informed and was informed by other forms of knowledge about eating and health.

Even today, some versions of Chinese dietary advice reflect that mutual shaping. Like nutrition scientists who focus on chemical constituents and active ingredients, some Chinese-medicine practitioners today promote the vitamin-like "superfood" qualities of certain herbs and plants featured in classical dietetics, such as gouji berries or ginseng.[33] The recent work of one of the founders of Chinese nutrition science, Zheng Ji, demonstrates this intertwining even more dramatically. Zheng lived to the remarkable age of 110 and spent much of his later life writing about *yangsheng* (養生 self-cultivation) and longevity. In 2008, when he was 108 years old, he published a book titled *Bulao de jishu—bai sui jiaoshou yangsheng jing* (不老的技術—百歲教授養生經 *The technique of anti-aging: a hundred-year-old professor's classic of self-care*). In some respects, his book reads like a page torn from a 1930s nutrition-science textbook: it espouses a diet that is balanced in the way a budget is balanced, with intake equaling output. It is a diet in which 70 percent of the calories come from carbohydrates, 20 percent from fats, and 10 percent from protein. It consists of mostly vegetable protein but also some animal protein. It provides adequate amounts of vitamins A, B, C, D, E, and K, as well as the chemical elements that a body needs in small amounts, such as sodium and phosphorus.

On the other hand, the healthy diet that Zheng espouses is also bland and eaten at regular times in regular amounts; does not lean too much toward spicy, salty, sweet, or any other strong flavor; and is chewed thoroughly, swallowed slowly, and eaten silently in a cheerful mood. It is moderate, as the eater always stops before they are full. It includes appropriate amounts of warm boiled water. In the winter, it includes medicinal foods to *buzhong yiqi* (補中益氣 supplement the center and boost *qi*).[34] Zheng combines recommendations that could have come from nutrition science anywhere on the planet with *yangsheng* ideas that are distinctively Chinese. In that regard, his recent work reflects, at a distance of many decades, the pluralistic environment in which his discipline first developed.

NOTES

1. Tian Niao, "'Weitaming' wei shipin zhong zhi yi yaosuo'" [Vitamins' are an essential factor in foods]," *Zhonghua yixue zazhi* 5, no. 2 (1919): 101.

2. One of the most frequently cited expressions of the "new nutrition" is Elmer Verner McCollum's *The Newer Knowledge of Nutrition: The Use of Food for the Preservation of Vitality and Health* (New York: MacMillan, 1919).

3. Some recent works reconfiguring the place of Western science and medicine in early twentieth-century China are: Bridie J. Andrews, *The Making of Modern Chinese Medicine, 1850–1960* (Honolulu: University of Hawaii Press, 2015); Sean Hsiang-lin Lei, *Neither Donkey Nor Horse: Medicine in the Struggle over China's Modernity* (Chicago: University of Chicago Press, 2014); David Luesink, William H. Schneider, and Daqing Zhang, eds. *China and the Globalization of Biomedicine* (Rochester, NY: University of Rochester Press, 2019).

4. Casimir Funk, "The Etiology of the Deficiency Diseases," *Journal of State Medicine* 20, no. 6 (1912): 341–368; Kenneth J. Carpenter, "A Short History of Nutritional Science Part III (1912–1944)," *Journal of Nutrition* 133, no. 10 (2003): 3023–3032.

5. Jack Cecil Drummond, "The Nomenclature of the So-Called Accessory Food Factors," *Biochemical Journal* 14, no. 5 (1920): 660.

6. Henrik Dam and Johannes Glavind, "Vitamin K in Human Pathology," *The Lancet* 231, no. 5978 (March 26, 1938): 720–721.

7. A. Szent-Györgyi, "From Vitamin C to Vitamin P," *Current Science* 5, no. 6 (December 1936): 285–286; "Clouds over Rutin and Vitamin P," *The Lancet* 256, no. 6640 (December 2, 1950): 690–691.

8. Eugenia Lean, *Vernacular Industrialism in China: Local Innovation and Translated Technologies in the Making of a Cosmetics Empire, 1900–1940* (New York: Columbia University Press, 2020), 8 and 18.

9. Zheng Ji, *Zhongguo zaoqi shengwu huaxue fazhan shi, 1917–1949* [History of the early stage of development of biochemistry in China, 1917–1949] (Nanjing: Nanjing University Press, 1989).

10. Zhang Li, "Peter P.T. Sah and the Synthesis of Vitamin C in China and Europe," *East Asian Science, Technology, and Medicine* 20 (2003): 92–98.

11. Zheng Zhenwen, *Yingyang huaxue* [Nutritional chemistry] (Shanghai: Commercial Press, 1929).

12. Wu Xian, *Yingyang gailun* [Introduction to nutrition] (Shanghai: Commercial Press, 1928).

13. Wu Xian, "Yingyang yu jiankang" [Nutrition and health], *Yixue zhoukan ji* 2 (1929): 28–31.

14. Lan-chen Kung, "The Chinese Diet," *Nutrition Notes* 8 (December 1937): 7.

15. Mixed Committee on the Problem of Nutrition, "Final Report of the Mixed Committee of the League of Nations on the Relation of Nutrition to Health, Agriculture and Economic Policy" (Geneva, 1937), 317–320.

16. Mixed Committee on the Problem of Nutrition, "Final Report," 317–320.

17. See Jia-Chen Fu, *The Other Milk: Reinventing Soy in Republican China* (Seattle: University of Washington Press, 2018), 83–87.

18. Hsien Wu, "The Chinese Diet in the Light of Modern Knowledge of Nutrition," *Chinese Social and Political Science Review* 11, no. 1 (January 1927): 56–81.

19. Ari Larissa Heinrich, *The Afterlife of Images: Translating the Pathological Body between China and the West* (Durham, NC: Duke University Press, 2008); Walter Mallory, *China: Land of Famine* (New York: American Geographical Society, 1927).

20. Kathryn Jean Edgerton-Tarpley, "From 'Nourish the People' to 'Sacrifice the Nation': Changing Responses to Disaster in Late Imperial and Modern China," *The Journal of Asian Studies* 73, no. 2 (2014): 447–469.

21. Sherman Cochran, *Chinese Medicine Men: Consumer Culture in China and Southeast Asia* (Cambridge, MA: Harvard University Press, 2006), chapters 3 and 4.

22. On nerves and neurasthenia in Republican China, see Emily Baum, *The Invention of Madness: State, Society, and the Insane in Modern China* (Chicago: University of Chicago Press, 2018), chapter 4. On blood in modern Chinese medicine, see Bridie Andrews, "Blood in the History of Modern Chinese Medicine," in *Historical Epistemology and the Making of Modern Chinese Medicine*, ed. Howard Chiang (Manchester: Manchester University Press, 2015), 113–136.

23. Ruth Guy, "The History of Cod Liver Oil as a Remedy," *Western Medical Observer* 84, no. 4 (1924): 185.

24. "Jiazhong weitamin" [Vitamin-A Star Brand], *Yiyao xue* 10, no. 9 (1933): 30.

25. "Jiazhong weitamin."

26. Sherman Cochran, "Marketing Medicine and Advertising Dreams in China, 1900–1950," in *Becoming Chinese: Passages to Modernity and Beyond*, ed. Wen-hsin Yeh (Berkeley: University of California Press, 2000), 90.

27. Andrews, *The Making of Modern Chinese Medicine*; Lei, *Neither Donkey Nor Horse*.

28. Hilary A. Smith, "Beyond Indulgence: Diet-Induced Illnesses in Chinese Medicine," *Historia Scientiarum* 27, no. 2 (2018): 233–253.

29. Shi Yiren, "Wei zhi yanjiu" [Stomach research], *Yijie chunqiu* no. 100 (9 no. 4) (April 15, 1935): 1–3.

30. Chen Botao, "Weibing yangsheng lun" [A discussion of cultivating life for stomach disease], *Yijie chunqiu* no. 100 (vol. 9, no. 4) (April 15, 1935): 35–36.

31. Baum, *The Invention of Madness*; Hugh Shapiro, "The Puzzle of Spermatorrhea in Republican China," *positions* 6, no. 3 (1998): 551–596.

32. See the critiques of Gyorgy Scrinis in *Nutritionism: The Science and Politics of Dietary Advice* (New York: Columbia University Press, 2013).

33. Volker Scheid, "From Civilizing Foods for Nourishing Life to A Global Traditional Chinese Medicine Dietetics: Changing Perceptions of Foods in Chinese Medicine," in *Moral Foods: The Construction of Nutrition and Health in Modern Asia*, eds. Angela Ki Che Leung and Melissa L. Caldwell (Honolulu: University of Hawaii Press, 2019), 241–261.

34. Zheng Ji, *Bulao de jishu: Bai sui jiaoshou yangsheng jing* [The art of anti-aging: A hundred-year-old professor's classic of self-care] (Taipei: Gao bao, 2008), chapter 2.

5 TASTE 100 HERBS: MATERIAL SCARCITY AND LOCAL PLANT KNOWLEDGE IN THE MAO-ERA CAMPAIGN FOR NATIVE PESTICIDES

SIGRID SCHMALZER

In August 1960, in the midst of the worst famine in world history, the Kaifeng Municipal Science and Technology Committee published a reference volume for local production of fertilizers and pesticides (figure 5.1).[1] The book was handwritten and mimeographed on low-grade recycled paper that was soft, gray, fibrous, and speckled with bits of darker pulp or sometimes even pieces of bark or twig. The pages—both the words themselves and the material on which they were inked—spoke to scarcity and resourcefulness. The text served the urgent priority of food production: wild plants, together with every other available resource, would be combined according to knowledge distilled from traditional Chinese doctors and modern chemists, and then applied to crops to boost growth and forestall disease and insect pests. The book's paper, too, followed the principle of making do with what was available: as land once used to grow bamboo for papermaking was converted to food production,[2] mills turned to inferior sources for their raw materials. In places with severe shortages (including the region around Kaifeng), rural people ate bark, twigs, and some of those same wild plants described in such handbooks, sometimes poisoning themselves in the process. During those years, the lines between food, medicine, poison, and even paper became more blurred than usual.

FIGURE 5.1
The cover of a handbook on native fertilizers and pesticides, published in Kaifeng, 1960. Photo by Jessica Johnson; book in author's collection.

This chapter explores the *da gao tunongyao yundong* (大搞土农药运动 mass campaign to manufacture native pesticides), first launched in 1958 to support food production during the Great Leap Forward. The campaign was part of the PRC leadership's effort to overcome material scarcity and transform agriculture along both modern and revolutionary lines. It relied on and celebrated local knowledge of the myriad uses of wild and cultivated plants, systematizing this knowledge for the first time in history to help local governments manufacture botanical, "native" pesticides. Of special importance was local people's knowledge of the medicinal properties of plants, including their traditional classification in terms of *xingwei* (性味 character of taste/smell): bitter, spicy, numbing, acrid, stifling, and sour.

Inserting this topic into the larger history of foodways in modern China is an unusual move, but it is an important one if we are to grapple conscientiously with the complexities of the Mao era and its relationship to the rest of modern Chinese history. The question of food in Mao-era China usually begins and ends with the question of food availability—or, more to the point, unavailability. Indeed, failing to center the reality of hunger and famine would constitute a grave injustice against the people who suffered from them. However, the overwhelming narrative of deprivation, together with the broader repudiation of radical politics, has prevented recognition of the relevance of the Mao era—and perhaps especially the Great Leap Forward—to the questions animating conversations about food and agriculture in China today. In his study of the culinary traditions of Guangzhou, for example, Jakob Klein has demonstrated that despite the very real food shortages experienced during the Mao era, "attempts to create a socialist food culture not only relied upon but even came to celebrate and solidify preexisting culinary practices and institutions," the legacies of which can be traced in post-Mao Guangzhou cuisine.[3] In this chapter, I similarly argue that the Mao era represents important continuities in a history that spans the pre-1949 and post-1978 periods, but I shift the inquiry from urban to rural society and select a case that highlights other knowledge forms—including agricultural and medicinal—central to the production of food.

The Kaifeng handbook, along with many other materials from the native pesticides campaign, provides an illuminating window onto modern Chinese foodways precisely because it brings into focus the complex relationships among scarcity and abundance; modernization and tradition; technoscience and local knowledge; state and rural society; plants and people; nourishment and poison; and food, agriculture, and medicine. These themes resonate strongly with the concerns of alternative food and agriculture movements today, and they demonstrate compelling continuities in the history of Chinese foodways throughout the modern period, as evidenced by many of the other chapters of this book. The dominant narrative jumps over the discredited Mao era and appeals directly to imperial-era "tradition." In contrast, this chapter demonstrates the pivotal role played by the Great Leap Forward within the larger history of Chinese foodways, in which state, scientific, and rural actors have sought to address the dilemmas of industrialized agriculture by systematizing local knowledge. In the process, it underscores the need to respect the enormity of what Mao-era historical actors were attempting in the transformation of food production, to recognize the consequences for rural people struggling to satisfy their most fundamental food needs, and to do justice to the positive role played by diverse actors in mobilizing local plant knowledge.

SCARCITY

The native pesticides campaign was one small part of Chairman Mao's effort to create full-fledged communism within the span of just a few years. Marxists had typically foreseen a long process of development under state socialism before society would achieve the ideal, "from each according to their ability, to each according to their needs"; in 1958, with the launch of the Great Leap Forward, Mao proposed that the revolutionary spirit of the masses could propel China to that stage far more quickly. But spirit and organization alone would not be enough: the Leap required the transformation of material production. This meant the rapid development of industry, which relied on a massive increase in food (and especially grain) production—enough not only to feed China's laborers but also to send to the Soviet Union in exchange for machinery, oil, and other industrial inputs.

Increasing food production required the rapid modernization of many aspects of agriculture. Mao specifically itemized eight necessary areas for improvement, of which crop protection (controlling plant diseases and insect pests) constituted one. But the same scarcity that China's leaders sought to overcome with the Great Leap Forward posed a limit on crop protection as on so much else: China lacked the quantities of chemical pesticides that developed countries were increasingly employing to protect their harvests, and it could not rely on foreign countries to supply them.

The pressure to accelerate agricultural production led to badly conceived technological changes, outrageous harvest projections, downright lies about surpluses, and far greater extraction than many rural communities could bear. At the same time, the Leap's utopian vision led to the establishment of canteens where commune members could eat to excess without payment—a claim to abundance that belied continued struggles with scarcity and would appear tragically short-sighted when famine set in.[4] These are the most well-known aspects of the Great Leap Forward. But there was far more to it than that. To achieve targets and demonstrate revolutionary zeal, people around the country also tapped existing knowledge and developed innovative strategies to maximize production with limited resources.

Local knowledge of wild plants—seemingly a limitless resource requiring no inputs—presented a particularly fruitful area to explore. Depending on the species, wild plants supplied fiber (for clothing and paper), oil (including for food, soap, candles, and paint), starch (for food), tannin (for processing leather), rubber, resin, pesticides, and of course medicine. A 1958 article in *Bulletin of Biology* detailed all of the above uses along with the plants that could supply them and their availability; it covered pesticides in the same section along with medicinal herbs. Indeed, looked at from some angles, pesticides were a kind of medicine: the Chinese term for pesticides, *"nongyao"* (农药 literally "agricultural medicine"), includes treatments for plant diseases and sometimes also fertilizers. With respect to using wild starchy plants for food, the article noted, "If we can make comprehensive use [*chongfen liyong* 充分利用] of wild starches, not only can we increase peasant income, but we can also save the country much grain." Wild plants could supply substitutes for maltose, tofu, bean thread noodles, lotus root paste, processed cakes, and other foodstuffs. For example, according to the article, China had just that year substituted 100,000 tons of acorns for 70,000 tons of food crops, in the process raising peasant income by 12 million yuan.[5]

In May 1958, an article in *Insect Knowledge* noted that the national supply of chemical pesticide for that year was 480,000 tons, and they had 3.1 million pesticide sprayers; given the "high tide" of the Great Leap Forward these were not enough to satisfy demand. Hence the call for every area to adopt the policy of "simultaneously pursuing native [*tu* 土] and foreign [*yang* 洋] methods to go big in native insecticides."[6] Over the course of 1958, the participating localities reported the establishment of more than 2.4 million native pesticide factories, manufacturing more than five hundred types of pesticides mixed into more than three hundred compound sprays, and applying 16 million tons of spray to the fields.[7] The scarcity of chemical pesticides was the incessant drumbeat driving the native pesticide campaign, and this scarcity was portrayed as an opening to develop native self-reliance.

In truth, the supply of wild plants was by no means unlimited. Rather, the call to harvest wild plants for use as native pesticides quickly ran up against other priorities. As one article from June 1958 noted, some of the plants were needed more urgently for human medicine; some were needed for oil, fiber, or starch; and others were valuable as timber. Arguments had already begun to erupt as to how the different needs could be balanced. In fact, figuring out how such resources could be allocated rationally was itself a question that required scientific research.[8] A publication from Da County, Sichuan—a frontrunner in the campaign—similarly highlighted the supply problem and noted that the county had begun cultivating plants useful for native pesticides in the "ten margins": that is, the edges of fields, mountains, ditches, rivers, homes, graves, and other places where more plantings could be squeezed in.[9] And so, what had been trumpeted as a limitless bounty pointed instead, again, to scarcity.

Moreover, and as we will see below, the food substitution project that seemed so positive in 1958 would in a few years take on a much less pleasant flavor. Still, we would be wrong to dismiss these ideas out of hand: exploring new ways of using available resources was a fundamentally rational response to the crucial problem of scarcity, and it remains a touchstone for resource conservation and management in China to this day. Accounting for such efforts and recognizing the magnitude and persistence of the development challenge China has faced helps us move away from a smug dismissal of this painful history, and toward an empathetic recognition of complexity, along with an acknowledgment of the many shared priorities of the Great Leap era and today.

KNOWLEDGE

The scarcity of chemical pesticides presented an immensely practical reason to ramp up a policy orientation that, ever since the early 1940s in the revolutionary base area of Yan'an, had already enjoyed great ideological significance for the CCP: the development of *tu* (土 local or native) science to serve as a complement to *yang* (洋 foreign) science, as in the slogan *"tuyang jiehe"* (土洋结合 combine *tu* and *yang*). Thus, in the campaign for native pesticides, *tu* referred not only to the pesticides themselves but also to the type of science being practiced. In this respect, it is important to note the multivalent associations of *tu* and *yang*, beyond those implied by the similar pair *zhong* (中 China) and *xi* (西 the West). While both binaries point to the differences between "Chinese" and "Western," *tu* (literally "soil") further implies rural and mass-based, in contrast with *yang* (literally "ocean"), which implies academic and elite.[10] To those familiar with contemporary Chinese food culture, zhong may be more familiar in discourse on cuisine, while *tu* appears in more "earthy" contexts of locally raised heritage breeds—as in *tu zhu* (土猪 native pigs). *Tu* science was thus not just Chinese but also local, earthy, frugal, and peasant-based science.

Tu science celebrated making do with locally available materials, and this, of course, was what the native pesticide campaign was all about. But for the frontrunners of the campaign in particular, another aspect of *tu* science was also essential: local knowledge. The national Ministry of Agriculture did not release its encyclopedia of native pesticides until September 1958—months into the campaign. Until then, cadres tasked with developing native pesticides had to consult more accessible authorities, and they quite logically turned first to traditionally trained doctors, pharmacists, and veterinarians. The treatment of crops that fed people had historically been linked to the treatment of people themselves, and wild plants were the foundation of that knowledge. For example, rice blast disease in Chinese is *dao wenyi* (稻瘟病 rice febrile epidemic) or *dao rebing* (稻热病 rice hot disease), suggesting a kinship with the human ailments known in traditional Chinese medicine as *wenbing* (温病 warm-factor disorders).[11] As we will see below, herbs useful in treating warm-factor disorders in people are also used in treating rice blast disease. Some of the traditional medical experts consulted by the cadres reported having experience treating crops—for example, using *tonggen* (桐根 Paulownia roots) to prevent damage from cabbage moths, the legume *Millettia pachycarpa* to combat cutworms, and *Stemona* vines to control corn borers. They also knew the proper preparation methods and the timing of applications.[12] One article published in *Insect Knowledge* pointed out that even "old monks" had knowledge of wild plants that should be tapped for the campaign.[13]

Materials on native pesticides often cited the long history of their use in China. As one article in *Insect Knowledge* declared, "Several thousand years ago, in the process of their productive labor, our ancestors unearthed and used abundant native pesticides. They accumulated a great deal of experience, which in the current plan of combining *tu* and *yang* constitutes an extremely valuable scientific heritage."[14] One of the most powerful botanical insecticides promoted during the campaign was rotenone, which was found in chicken-blood vine among other legumes of southern China, and which one 1958 article noted with pride was first discovered by "Chinese laboring people" and first recorded in 1557 in the "great Chinese scientist Li Shizhen's *Materia Medica*." The plants themselves were originally used as a piscicide (they easily kill fish that can then be consumed by humans without apparent ill effects); the chemical rotenone had subsequently been isolated and employed worldwide as an insecticide.[15]

In fact, however, there was no focused, systematic attention to botanical pesticides as a field of knowledge, at least in the written record, during the imperial era. The famous and comprehensive series *Science and Civilisation in China* devotes forty-eight pages to the subject but acknowledges that the evidence is "scattered." The majority of examples it considers from imperial-era texts do not actually refer to agricultural uses of the herbs at all, but rather to killing lice, fleas, intestinal worms, and other human afflictions. Indeed, one of the sources it relies on most heavily is a volume published in 1959 by the Joint Office for Scientific Research on Native Pesticides, discussed in detail in the next section.[16]

Nonetheless, the connection to agricultural heritage created something of a mystique around the use of wild plants to produce native pesticides, and sometimes authors drew on classical language to convey this sentiment. Evoking the Song dynasty poet Su Shi's Daoist-inflected *Red Cliff Rhapsody*, one article opined, "China's native pesticide resources are extremely abundant: we can take them without end, use them without exhausting them (*qu zhi bu jin, yong zhi bu jie* 取之不尽，用之不竭)." The author further called on the legendary ruler Shennong, whose most celebrated contribution to Chinese civilization was the identification of medicinal plants: "So the masses say: Everyone acts as Shennong, going everywhere and tasting one hundred herbs, changing the useless into the useful, changing poisonous plants into agricultural medicine [i.e., pesticides]."[17] Such references are familiar fare in agricultural heritage discourse today, as found in Bray's chapter in this volume (chapter 1): their significance in Great Leap-era discourse bears remembering.

Traditional medicinal knowledge was not limited to an inventory of useful herbs: it also provided principles to help determine which to combine and which to separate. According to some sources, it was essential to identify the classification of the herb in terms of the *xingwei* (性味 character of taste/smell) of traditional medicine, specifically whether a given plant was bitter, spicy, numbing, astringent, stifling, or sour. On the basis of these classifications, herbs would be separated for storage. While it was fine to store bitter herbs together with spicy herbs, numbing herbs should be kept separate from spicy ones.[18]

People also recommended employing the principle of *shiba fan yao* (十八反药 eighteen conflicting medicines) to identify toxic combinations—for example, onions and honey. While this principle had originally been developed to help people avoid ingesting *fan* (反 conflicting) foods and medicines whose combination would produce toxins in the body, the knowledge could also be turned around and used the other way. The afterword to a native pesticide handbook published in Da County detailed which combinations increased toxicity and thus effectiveness (e.g., numbing plus bitter, or spicy plus bitter), which decreased it (e.g., numbing plus spicy—the famous *mala* combination characteristic of Sichuan cuisine), and which had no effect (e.g., bitter plus stifling).[19]

An article by the Da County Department of Commerce explained that to increase moistness or suspension in a compound, they added material based on the *xingwei* of each herb. "For example, add 2–3 percent honey locust to spicy herbs, 4–5 percent chinaberry to bitter herbs, etc." (The "etc." here implied that readers should know how to apply this principle to herbs with other characteristics, suggesting an assumption that such medicinal knowledge was indeed widespread among rural people.) The authors further noted the importance of an herb's *xingwei* in the manufacturing process: some herbs (especially those with numbing *xingwei*) were sensitive to heat. So, for example, when preparing garlic for use against rice blast disease, it was important to use cool water.[20]

In the spirit of *tu* science, some texts suggested that ordinary peasants, not just doctors or pharmacists, possessed valuable knowledge about wild plants. One article in *Insect Knowledge* explained, "Our ancestors began using native pesticides in ancient times and accumulated thousands of years of rich experience that they have passed down to the people. All peasants have experience identifying, collecting, and using poisonous plants. They don't need equipment; all they need is to distinguish the *xingwei* (sour, spicy, bitter, astringent, numbing, stifling) to determine toxicity." The author went on to cite the example of an "old peasant" in Anhui Province named Feng Ruizhai who used an herb called *mao'er yan* (猫儿眼 or 甘遂 "kitten eyes" or *gansui*) to control the disease known as wheat rust, with an efficacy of 80–100 percent. "These kinds of people are everywhere," the article emphasized.[21]

Another article highlighted an example from the Dong nationality region of Baise in Guangxi, where commune members created a compound made of *chafu* (茶麸 tea bran), a plant known in the Dong language as *guoxin ye* (果心叶 fruit-heart leaves), ashes, and an herb known as "chicken blood" (known elsewhere as "chicken-blood vine" and by numerous other names); it was found to be 95 percent effective against rice borers. The article underscored, "There are similar examples all over the country."[22]

The mobilization of rural people and the application of traditional medical knowledge (framed as the accumulated wisdom of laboring Chinese people over the centuries) in local pesticide factories: this was quintessential *tu* science. But the native pesticide campaign was expected to create a "material basis for the uniting of *tu* and *yang*" and so "open a new road for technological revolution."[23] Where *yang* science came in was the systematic organization and dissemination of local experiences, the chemical analysis of herbs, and the testing of the pesticides by professional researchers.

The Joint Office for Scientific Research on Native Pesticides, founded in 1958, conducted research directly on the pesticides and their targets. So as not to lose time over the winter and spring, they bred insects in the lab and tested the pesticides there—recognizing that the conditions were different from those in the field but maintaining nonetheless the usefulness of the information produced. They tested 404 types of native pesticides on mosquito larvae. In each case they tested a tincture made with water and another made with alcohol, and they also compared the same species collected from different locations, though the reasons for the sometimes dramatic differences in the samples' efficacy were difficult to ascertain.[24]

One of the chief problems was the difficulty of systematizing knowledge gathered from diverse localities. As the Joint Office emphasized, local factories had developed different manufacturing methods. In one factory, a plant might be pressed, while in another it might be soaked, boiled, or ground. This could have a significant impact on potency. They further noted that most of the materials were wild plants, and many of those were unfamiliar; even if the plants themselves were known to the scientists, the local names would likely be obscure.[25] This is a fundamental problem in translating between local and global knowledge systems; the rich profusion of local dialects in China made this a very familiar dilemma for scientists in many fields.

Academic work on native pesticides manifested most visibly in the encyclopedic handbooks known as *tunongyao zhi* (土农药志 records of native pesticides). These books typically adopted a strikingly practical tone: unlike many other kinds of books, they presented very little in the way of explicit ideology, confining the political discourse to a preface or two. However, the books nonetheless reflected Maoist ideology in their very makeup. Most significantly, the volumes drew on information gathered in different localities from around China and often specified in each case where information on the different herbs had come from; this exemplified the relationship between *yang* and *tu* and resonated more broadly with the Maoist philosophy of the mass line, in which the wisdom of the masses would be systematized to create more generalizable and extendable knowledge.[26]

And yet, there were limits on just how earthy, how grassroots, how *tu*, published materials on native insecticides got. These limits are particularly noticeable in the faintness of the traces left by traditional medical knowledge—particularly the pattern of references to *xingwei*, to the eighteen conflicting medicines, and to the consultation of practitioners of traditional medicine. The prominent examples appear almost exclusively in materials relating to the campaign as it unfolded in Sichuan.[27] While it is possible that this reflects a regional strength in traditional Chinese medicine, I think it is more likely a result of how early the campaign work began in Sichuan, before much guidance had emerged from central authorities: it was thus more of a local effort, and officials were truly making do with local resources.

The afterword of a handbook published by Da County (Sichuan) in May 1959 strongly emphasized the importance of knowing the *xingwei* of the plants so that they could be combined effectively, but the descriptions of the plants in the main body of the text did not include this information. Similarly, the book published in Kaifeng in 1960 included information on *xingwei* in one table charting the properties of various plants; but a second table contained no information on this point—in that one, the information was copied from an authoritative guide produced by the Agricultural Ministry in September 1958, which contained no mention of *xingwei*, Chinese doctors, or anything else relevant to traditional Chinese medicine.[28] Tellingly, the most *tu* form that research on native pesticides took—drawing on knowledge of local plants passed down through the generations by medical practitioners and farmers—did not become inscribed in the most influential sources produced by central authorities.

Scholars have often seen the processes of systematization and rationalization as risking the erasure of local knowledge and flattening of traditional epistemologies. In contrast, Lili Lai and Judith Farquhar have offered a compellingly optimistic analysis of the state-sponsored systematizing of national minority medicines in contemporary China. In their view, the "rationalizing process" that "has emerged as a modernizing imperative" has simultaneously "generated new charismatic forms of healing effectiveness and authority."[29] As we will see in more detail in the following section, the systematizing of *tu* plant knowledge in the native pesticides campaign could be read as generating "new charismatic forms"; still, the erasure of local actors bears watching.

GARLIC

A concrete example will afford a more tangible sense of how the handbooks worked in practice. *Dasuan* (大蒜 garlic *Allium sativum*) has occupied an important place in Chinese food culture since it was imported from western Asia during the Han dynasty.[30] Its medicinal properties were also recognized then, and by the time Li Shizhen compiled his *Bencao gangmu* (本草纲目 *Compendium of Materia Medica*, 1578), garlic had acquired numerous applications, from snake bites to warm-factor disorders. Consistent with this foundation, in handbooks on native pesticides, garlic was commonly recommended to treat rice blast (or "rice heat disease").

The Ministry of Agriculture's September 1958 book had this to say in its entry on garlic:

> A member of the lily family. Its bulb and leaves have a very strong odor. Can be used as medicine.
>
> In Guangxi they take 3 *jin* of peeled garlic bulbs, mash it into a paste, and put it in 100 *jin* of clear water to soak for half an hour to control *dao rebing* (稻热病 rice blast). After spraying 1–2 times, rice blast will decrease by 30–50 percent. It is more effective than Bordeaux [a common fungicide mixture made from copper sulfate, lime, and water] and is also effective against brown spot disease. In Anhui, they mash one *jin* of garlic and soak it in 10 *jin* of water to control stem rust in rice.
>
> In Taizhou, Zhejiang, they have experimented with taking 5 *jin* of mashed garlic bulb and soaking it in 100 *jin* of water to control rice blast, with very good results. In Cixi, Zhejiang, Comrade Li Minghua of the Bureau of Agriculture has conducted preliminary experiments taking 100 *jin* of mashed garlic and adding 20 *jin* of water to control angular leaf spot in cotton, with results approximating those for chemical sprays.
>
> In Changshu, Jiangsu, they squeeze half a *jin* of liquid from 1 *jin* of garlic, then add 3 *liang* (150 g.) of camphor for every 1 *jin* of garlic juice. After mixing thoroughly, they soak it in 4 *jin* of water. In controlling cotton aphids it is 82.2 percent effective.

In contrast, a provincial-level volume published by the Fujian Agricultural Tool Reform Office in November 1958 included a longer description of the plant itself and an illustration, but no discussion of practices from other places.

Some handbooks, including the handwritten one produced in Kaifeng mentioned in the introduction, copied the Agricultural Ministry's entries word for word; the meticulous hand-copying of the Kaifeng volume suggests that the Ministry's handbook was not available in sufficient numbers, such that redistribution even in this laborious way using scarce fiber resources was considered valuable.

When the Joint Office for Scientific Research on Native Pesticides published its eagerly awaited volume in May 1959, it offered somewhat longer entries for each plant, including different sections on morphological characteristics, distribution, growing environment, composition (complete with chemical formulas for the effective agents), uses, and preparation methods and targets of control. In the last section of the garlic entry, it listed eight methods and targets, most of which were attributed to specific provinces or counties. This format was chosen for imitation by some other handbooks, for example the one published in Da County, Sichuan.

Both the similarities and the differences among these materials deserve notice. Most obviously, the books shared the common purpose of providing accessible information necessary for the production of native pesticides. Some presented detailed descriptions of the plants' morphology, a characteristic of modern field guides not shared with traditional medicinal encyclopedias like Li Shizhen's Ming-Dynasty *Bencao gangmu*. In other ways, some of the handbooks are quite reminiscent of those older texts—for example, in the organization of each entry into categories of information and in the listing of documented applications, in each case attributed to a specific source. Though many borrowed liberally from one another, these were by no means simple reproductions. Each individual text involved efforts to tailor available information to the perceived needs of local users, and each also involved decisions made by the individual authors and editors assembling the content regarding the most user-friendly format and most appropriate language.

In 1960, the journal *Chemistry World* reported on experiments isolating the chemical components found in garlic and demonstrating their antimicrobial effects in treating crop diseases. According to the authors, by moving from the plant itself to the chemicals it contains, the research fulfilled a key goal of the era: "*tuzhong chuyang*" (土中出洋 produce the *yang* from out of the *tu*). That is, with simple and locally available ingredients, create substitutes for expensive materials that China would otherwise be dependent on foreign sources to supply.[31] While the more common slogan remained "combine *tu* and *yang*," the sense that *tu* was a steppingstone (needs must, faute de mieux) to *yang* is palpable in many of the published materials. This helps explain the evidence the handbooks supply as to the integration of *tu* and *yang*: while we find conscious efforts to retain traces of the origins of local knowledge, there is at least as much deliberate incorporation of both the trappings and substance of professional science. And we see the ways in which certain aspects of *tu* science—in particular, the contributions of traditional doctors, monks, and other suspect actors—were erased as the knowledge became systematized and generalized in these texts.

POISON

Like so much else about the Great Leap Forward, the campaign for native pesticides began with optimistic fanfare in 1958 but became ever more complex and dissonant over the subsequent years. Nowhere is this more apparent than in the many accounts of poisonings, when desperate people turned to eating noxious wild plants, including some used as botanical pesticides.

The Great Leap–era sources exhibit a tension between the ostensibly natural and wholesome character of native pesticides and their powerful toxicity. Sources frequently pointed to the safety of botanical pesticides and their beneficial effects on soil health and fertility. As one article explained, the safety of botanicals was especially significant when treating fruits and vegetables that are eaten as is—with foreign pesticides, harvest would have to wait one or two weeks before harvesting and eating, while crops treated with native pesticides could be harvested and consumed immediately.[32] And yet, Great Leap–era sources often highlighted the risks involved in manufacturing native pesticides. While some articles made sanguine references to local people smelling and tasting the herbs to determine their *xingwei*, others specifically warned against haphazardly smelling or tasting and encouraged people to wear a mask and goggles when working with the plants—and even to place masks on livestock when processing the pesticides. They also sometimes offered first-aid recommendations in case of accidental poisoning: for example, if someone were poisoned with a plant classified as numbing, they should use a plant classified as spicy to rinse out their mouths and follow by ingesting rice or mung beans.[33]

Poisonings sometimes occurred for other, more tragic reasons. One was the ingestion of toxic plants in an effort to consume what the state referred to as *dai shipin* (代食品 food substitutes). According to PRC historian Gao Hua, this term was first used in 1955 when peasants in Guangxi Province collected wild fruit and tree bark to eat, in place of the grain that had been taken by the state under the Unified Purchase and Sale Policy. In 1960, in the face of increasingly impossible-to-ignore reports of food shortages, Mao called for more food to be grown and more so-called food substitutes to be found. Commonly promoted food substitutes included the algae chlorella; artificial starch made by grinding stalks, roots, and other plant parts into a powder to be added to dough for steamed buns; and artificial meat produced with bacteria cultures. Perhaps ironically, the Insect Research Institute of the Chinese Academy of Sciences, which was more often concerned with the protection of crops from insect pests, recommended collecting and eating insects to bolster nutrition.[34]

The increasingly desperate search for food substitutes also led to alarming rises in cases of accidental poisoning. Sometimes the plants in question were in fact food crops but required special processing to remove toxins. Such was the case with cassava, which is the source of tapioca starch but contains high quantities of cyanide in its raw form: a telephone report from Fujian Province to the central authorities in December 1960 blamed cassava for poisoning 2,071 people with 294 fatalities: presumably, desperate hunger led people to consume the plant raw.[35]

While cassava was not to my knowledge used as a pesticide, other plants blamed for poisonings were. Three such plants were the *huaishu* (槐树 scholar tree *Sophora japonica*), *chunshu* or *chouchun* (椿树 or 臭椿 tree of heaven *Ailanthus altissima*), and *cang'er* (苍耳 Siberian cocklebur *Xanthium strumarium*).[36] A report on the famine from Shandong Province in January 1959 documented that after consumption of scholar tree and tree of heaven, "the entire body would become swollen." In the worst instances, "the body could no longer retain excess fluid, which then erupted through the skin, oozing out with a yellowish color."[37] In April 1960, following reports of numerous poisonings from cocklebur seed, the Ministry of Health attempted to ban its promotion—the promotion, and the poisonings, continued nonetheless.[38]

The horror of famine also led people to suicide, and intentionally ingesting pesticide was one frequently cited method.[39] It is unclear whether any of these cases involved native pesticides; the greater potency of chemical pesticides would make them far more likely resources for this ghastly endeavor of last resort. But the same principle of scarcity that propelled the native pesticides campaign in the first place may have led some to try botanical alternatives, especially if they were readily available in a local factory.

These are all horrible fates to contemplate, and nothing lightens the tragedy. At the same time, it is reasonable to expect that the number of accidental poisonings would have been considerably greater if not for the knowledge about wild plants that was widely held in rural communities—as documented especially in the early stages of the native pesticide campaign, before *yang* knowledge had caught up with *tu* knowledge. Where people resorted to eating poisonous plants, it was probably after the edible plants had already been consumed. Where the shortages did not quite reach that stage, many people no doubt survived because they knew which parts of which plants were poisonous and which were safe to eat.

LEGACY

Recent years have seen a surge of interest in what is now being called *Zhongyi nongye* (中医农业 Chinese medicine agriculture).[40] In 2016, I met with a proponent, Zhu Weiping, and visited two of the experimental sites he established with the organic agriculture company that he helped run. He had not heard of the Great Leap–era native pesticide campaign, nor did he seem inclined to explore the connection. This did not surprise me. The tragedy of the Great Leap famine and the more general reputation of the Great Leap as a colossal failure make it an unlikely source of inspiration for hopeful scientists and entrepreneurs in China today. And yet, there is no question that the Chinese medicine agriculture of the present belongs to the same historical narrative as the campaign we have examined in these pages. Indeed, it has far more in common with the Great Leap campaign than with the imperial-era history it is more inclined to evoke. The richest sources produced on "traditional" knowledge of botanical insecticides were produced in the Great Leap Forward, while only scattered records exist for the imperial era.

A key difference between today's Chinese medicine agriculture and the native pesticides campaign of the Great Leap Forward lies in the relationship of each episode to industrialized agriculture. In 1958, the PRC sought to transform Chinese agriculture as part of the larger goal of creating a modern, industrial nation. The lack of chemical pesticides presented one of many obstacles related to the overarching conundrum of scarcity; here, as in so many other areas, Chinese farmers, workers, and officials were urged to embrace "native" materials and knowledge—at times characterized as a great "treasure-house" and at other times (perhaps more honestly, given overwhelming evidence of the nation-state's modernizing priorities) as merely a stop-gap measure. And, of course, during the famine the scarcity of chemicals paled in comparison with the scarcity of food itself: so scarce that algae and acorns were promoted as substitutes, and people resorted to tree bark and wild plants when nothing else was available.

Today, industrialized agriculture dominates the Chinese landscape, and chemical pesticides have, since the 1970s, ceased being a scarce resource and have instead increasingly become a happy abundance or a scourge, depending on one's point of view. The environmental and health consequences of this transformation have inspired an impressive array of movements for more "natural" food production—fueled in large part by the increasingly abundant disposable income of the urban middle and upper classes. That these movements include not only "organic" and "ecological" principles, but also the "traditional" principles of Chinese medicine, speaks to the commercial value of heritage in contemporary China—reflected also in the marketing of "heritage" breeds of meat animals and "heritage" varieties of grains and vegetables.

The native pesticides campaign of the Great Leap Forward and present-day alternative food movements thus occupy different positions within the larger historical arc of agricultural modernization in China. Throughout this history, we see state agents, scientists, and rural people actively mobilizing diverse forms of knowledge in their efforts to find nourishment and healing for people and their crops. The stark differences between the eras should not prevent us from recognizing the complexity of the challenges that Mao-era historical actors faced, the sophistication of their responses, or the familiarity of many of their fundamental concerns.

NOTES

1. Kaifengshi kexue jishu weiyuanhui and Kaifengshi kexuejishu xiehui, eds., *Tunongfei tunongyao ziliao huibian* [Compilation of resources on native fertilizers and native pesticides] (no publishing info, 1960).

2. Jacob Eyferth, *Eating Rice from Bamboo Roots: The Social History of a Community of Handicraft Papermakers in Rural Sichuan, 1920–2000* (Cambridge, MA: Harvard University Asia Center, 2009).

3. Jakob A. Klein, "'For Eating, It's Guangzhou': Regional Culinary Traditions and Chinese Socialism," in *Enduring Socialism: Explorations of Revolution and Transformation, Restoration and Continuation*, ed. Harry G. West and Parvathi Raman (New York: Berghahn Books, 2009), 46.

4. Hanchao Lu, "The Tastes of Chairman Mao: The Quotidian as Statecraft in the Great Leap Forward and Its Aftermath," *Modern China* 41, no. 5 (September 2015): 539–572, 543.

5. Qiao Gongjian and Yin Zutang, "Shinian lai yesheng zhiwu liyong fangmian de chengjiu" [Achievements in the last ten years related to the use of wild plants], *Shengwuxue tongbao* 1959, no. 10: 443; Hebeisheng weishengting liangshi tinghe, ed., *Yecai he daishipin* [Wild plants and food substitutes] (no publishing info, preface dated 1960).

6. Benkan bianweihui, "Kunchongxue zai nongye fengchan zhong de juda chengjiu" [Great achievements of entomology in producing bumper harvests], *Kunchong zhishi* 1958, no. 5: 199–200.

7. Lü Youlan, "Zongjie tuiguang tunongyao de shida haochu" [Ten benefits of extending native pesticides], *Kunchong zhishi* 1959, no. 1: 5–6.

8. Tunongyao kexue yanjiu lianhe bangongshi, "Dagao tunongyao de chengjiu ji jinhou fazhan" [Achievements and future development of the going big in native pesticides [campaign]], *Kunchong zhishi* 1958, no. 6: 251–252.

9. Sichuan Daxian zhuanqu kexue yanjiusuo, ed., *Tunongyao zhi* [Records of native pesticides] (Sichuansheng difang guoying Daxian yinshua chang, 1959), 190.

10. Sigrid Schmalzer, *Red Revolution, Green Revolution: Scientific Farming in Socialist China* (Chicago: University of Chicago Press, 2016); see also Jia-Chen Fu, "Artemisinin and Chinese Medicine as Tu Science," *Endeavor* 41, no. 3 (2017): 127–135.

11. Marta E. Hanson, *Speaking of Epidemics in Chinese Medicine: Disease and the Geographic Imagination in Late Imperial China* (New York: Routledge, 2011), 1, 16, 171n3.

12. Wang Qizhong and Liu Shengming, "Sichuan Daxian dagao tunongyao de jingyan" [Experiences from Da County, Sichuan, in going big in native pesticides] *Kunchong zhishi* 1959, no. 1: 29.

13. Lü Youlan, "Zongjie tuiguang."

14. Yue Zong, "Zhongguo tunongyao lishi ziliao zhailu" [Excerpts from historical materials on Chinese native pesticides], *Kunchong zhishi* 1959, no. 1: 37.

15. Tunongyao kexue, "Dagao tunongyao."

16. Joseph Needham with Lu Gwei-djen and Huang Hsing-tsung, *Botany*, in *Science and Civilisation in China*, vol. 6: Biology and Biological Technology, pt. 1 (Cambridge: Cambridge University Press, 1986), 471–553.

17. Lü Youlan, "Zongjie tuiguang."

18. Sichuan Daxian shangyeju, "Daxian tunongyao zhizao, shiyong, chucun baoguan fangfa" [Methods of manufacturing, using, and safely storing native pesticides in Da County], *Nongye kexue tongxun* 1959, no. 12: 421–422.

19. Sichuan Daxian zhuanqu kexue yanjiusuo, *Tunongyao zhi*, 191.

20. Sichuan Daxian shangyeju, "Daxian tunongyao."

21. Lü Youlan, "Zongjie tuiguang."

22. Benkan weiyuanhui, "Kunchongxue."

23. Lü Youlan, "Zongjie tuiguang."

24. Zhongguo kexueyuan kunchong yanjiusuo huaxue fangzhi shi, "1958 nian quanguo tunongyao yaoxiao ceding chubu baogao" [Preliminary report on determining the efficacy of native pesticides throughout the nation in 1958], *Kunchong zhishi* 1959, no. 1: 4, 7–16; Zhongguo nongye kexueyuan zhiwu baohu yanjiusuo, "Tunongyao dui zhuyao nongzuowu bingchong de yaoxiao ceding" [Determining the efficacy of native pesticides in important crop diseases and insect pests], *Nongye kexue tongxun* 1959, no. 9: 297–300.

25. Tunongyao kexue, "Dagao tunongyao."

26. See, for example, Nongye ziliao bianji weiyuanhui, ed., *Tunongyao* [Native pesticides] (Beijing: Nongye chubanshe, 1958).

27. That said, an article on the native pesticide campaign from August 1958 in Jiangxi Daily made passing reference to the consultation of traditional doctors, suggesting that the practice was not limited to Sichuan. Reprinted in in Nongye ziliao, *Tunongyao*, 16.

28. Kaifengshi, *Tunongfei*, 54–58.

29. Lili Lai and Judith Farquhar, "Nationality Medicines in China: Institutional Rationality and Healing Charisma," *Comparative Studies in Society and History* 57.2 (2015): 381–406, 384.

30. Wang Saishi, "Gudai dui dasuan de yinjin yu liyong" [Introduction and use of garlic in ancient times], *Nongye kaogu* 1996, no. 1: 182–188.

31. Hai Binfu, "Yizhong baogui tunongyao dasuan de huaxue chengfen" [Chemical composition of garlic, a precious native pesticide], *Huaxue shijie* 1960, no. 2: 56–57.

32. Lü Youlan, "Zongjie tuiguang," 6.

33. Sichuan Daxian shangyeju, "Daxian tunongyao." See also Wang and Liu, "Sichuan Daxian," 31.

34. Gao Hua, "Food Augmentation Methods and Food Substitutes during the Great Famine," in *Eating Bitterness: New Perspectives on China's Great Leap Forward and Famine*, eds. Kimberly Ens Manning and Felix Wemheuer (Vancouver: University of British Columbia Press, 2011), 183–187.

35. Xun Zhou, *Great Famine in China, 1958–1962: A Documentary History* (New Haven, CT: Yale University Press, 2012), 51.

36. Sources promoting the use of these plants as insecticides include: Zhongguo renmin gongheguo nongyebu, ed., *Zhongguo tunongyao zhi* [Records of Chinese native pesticides] (Kexue chubanshe, 1959); Nongyebu, *Tunongyao zhi*; "Lixing jieyue, wei zuguo: wei zuguo chuangzao gengduo de caifu" [Be parsimonious to create more wealth for China], *Renmin ribao*, December 9, 1954, 2; and Yang Min, "Nongyao shihua," *Renmin ribao*, April 3, 1960, 8.

37. Xun Zhou, *Great Famine*, 5.

38. Gao Hua, "Food," 188–189; "Duo diao xie gongye yuanliao jincheng you shenme haochu" [Benefits of moving more industrial materials into the cities], *Renmin ribao*, October 27, 1961, 3.

39. See Xun Zhou, *Great Famine*, 8, among many other anecdotes.

40. See, for example, Zhang Lijian and Zhi Lizhi, "'Zhongyi nongye' fazhan zhanlüe ji qianjing" [Strategies and prospects for the development of "Chinese medicine agriculture"], *Nongyeshengchan zhanwang* 2018, no. 11: 72–76.

6 FOOD DELIVERY, THE PLATFORM ECONOMY, AND DIGITAL CULTURE IN CHINA: THE HUMAN-NONHUMAN ENTANGLEMENT OF URBAN CHINESE FOODWAYS

FAN YANG

In early 2019, a high-school friend in Shenzhen, the Special Economic Zone where I grew up, introduced me to a cooking robot (see figure 6.1). The machine was the product of a Shenzhen-based startup called Fanlai (饭来), which literally means "meals come." The word *fan* (饭) denotes "rice," the long-standing staple food for much of southern China, thus connoting "meals" in general in the Chinese language. To operate the machine, one had to download the Fanlai app on the mobile phone. Using the app, customers could order prearranged packets of ingredients, which would be delivered to households by human couriers. Once initiated, the robot would move an opened carton facing down from right to left, releasing the oil, the garlic/ginger, and other ingredients from its compartments into the pot below one by one, much like what a human cook would do.

After the meal, I offered to help with the dishes but was told that the *ayi* (阿姨 aunt, a common euphemism for female housekeepers) would usually do the work. We joked that a cooking robot would perhaps be even better if accompanied by a dish-washing robot. Then I recalled how much I'd become accustomed to the presence of a dishwasher in US households, an appliance that didn't seem very popular in the urban Chinese households that I visited, even in first-tier cities like Shanghai and Shenzhen. I suspected that

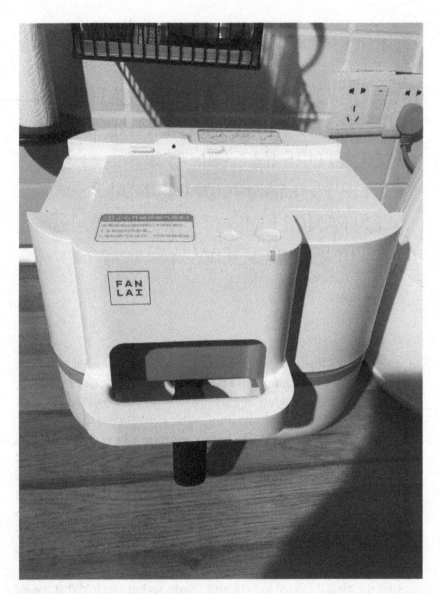

FIGURE 6.1
A Fanlai cooking robot. Photo by the author.

this unpopularity might have something to do with the limited space in city housing. But the abundance of rural-to-urban migrants willing to work as low-paid *ayi*s could just as well explain this absence.

My encounter with the cooking robot in Shenzhen, a city known for "Shenzhen speed," seemed entirely reasonable given the city's energy and image as a place of innovation.[1] During the same trip, another friend from high school, a former banker who recently joined a fintech startup, told me that she hadn't been to the grocery store in years thanks to widely available and highly efficient delivery services. During the pandemic in 2020, when I asked my friends in Shenzhen about how they had been getting food since the beginning of the quarantine orders, some mentioned that because of the well-established e-commerce network, not much had changed.

What struck me, however, was that the technical efficiency exemplified by the cooking robot—a nonhuman agent—appeared to mask the necessity of human agents in facilitating home cooking. The human couriers—often male, who deliver the prepacked ingredients—and the domestic workers—most often female, who wash the dishes—seem to be essential in bringing meals to the home, even though the robot's name, Fanlai, made it sound like the meals would just arrive by magic. The human-nonhuman entanglement is certainly not new in Chinese food cultures, given the long-standing and intertwined histories of tools, technologies, and craftsmanship involved in planting, growing, harvesting, cutting, preparing, preserving, cooking, and serving food (see, for example, DuBois's and Schmalzer's chapters [chapters 3 and 5] in this volume). Yet in the context of twenty-first-century urban China, particularly under the conditions of restricted mobility during the COVID-19 pandemic, recognizing this entanglement seems quite important for a number of reasons. For one thing, such a recognition helps to illuminate the myriad and uneven ways in which technologies, especially Information and Communication Technologies (ICTs), have come to shape urban Chinese food practices. Probing the entanglement of the human and the nonhuman also prompts a rethinking of such binaries as mobility and immobility,[2] private and public, and materiality and immateriality in the study of food and digital culture. The drastic rise of the food delivery industry in urban China in recent years, which I will turn to in the next section, provides a prime lens through which to examine these multifaceted and often unsettling impacts of ICTs on different humans, particularly along the lines of class (intertwined with rural or urban origins) and gender. These changing practices of eating also present broader implications for the nonhuman environment, from shifting notions of the domestic and urban space to the generation of long-lasting material waste often elided within an "immaterial" digital culture obsessed with speed and immediacy.

WAIMAI, OR FOOD DELIVERY

In 2010, my partner and I visited Taishan Mountain in Shandong Province. Being a Hangzhou native, I wasn't too keen on the local Shandong cuisine on offer at the restaurants we tried. And frankly, after traveling in China for over a week, my (somewhat Americanized) "Chinese stomach" was ready for something different.

I could tell that my partner, who was born and raised in New Jersey and never really complained about food, was probably missing salad and pizza. One night, as we scrolled through the various TV channels in the hotel and found a Hollywood movie to watch, we decided that perhaps we could get McDonald's for dinner. (I should mention that even though I rarely frequent McDonald's in the United States, I almost always go to one in China whenever I visit, just to be reminded of how *different* the experience usually is from the fast-food chain in the United States, which I invoke often in my Media and Globalization class.[3]) I found a branch online, but it was getting late, and we didn't feel like taking a cab there to pick up our order. I thought I'd just try calling them to see if they could deliver our food instead. Fifteen minutes later, a young man who was presumably an employee at McDonald's brought our burgers and fries to the hotel lobby. I handed him some cash and took the food, though I can't recall if I gave any extra delivery fees or tips, perhaps because the "extra cost" was too minimal to note when converted into US dollars. All I remember was the Big Mac I had that night was (shamefully) one of the most memorable meals I had on that trip back to China!

Of course, I am in no position to claim that I came up with the idea of *waimai* (外卖 food delivery), now a booming business sector in China that also has a longer history than typically acknowledged. One such story about a premodern, ultra-privileged form of food delivery immediately comes to mind. In the Tang dynasty (618–907), Empress Yang, a favorite concubine of Emperor Xuanzong, famously demanded *lizhi* (荔枝 lychee), a fruit that only grows well in Southern China, to be delivered to the capital Chang'an, or present-day Xi'an in western China. Quite often, the story is used to demonstrate the luxurious lifestyle of the regime that contributed to its eventual demise. The demand for express meal orders also could not possibly have been limited to a (privileged) diasporic Chinese like me who occasionally travels back to China and wants a Big Mac out of the blue, despite rarely eating at McDonald's while living in America. I also remember the novelty of pizza delivery when the Domino's opened in Shenzhen during my teenage years in the early 1990s, though the service was hardly ever as popular as it is in the United States. Meanwhile, in cities like New York, Chinese takeout is also delivered by mostly underpaid and sometimes undocumented Chinese immigrants on bicycles, as portrayed in numerous popular films and TV shows and more poignantly in the independent film *Take Out* (2004). Surely, food delivery has never been unique to urban China, nor is its boom under COVID distinctive to China.

What can be observed, however, is the voluminous growth of the food delivery sector in China in the twenty-first century. Even before the pandemic, the O2O (online to offline) food delivery service had become one of the fastest-growing sectors since 2015. Tech giants like Alibaba, Tencent, and Baidu have all participated in and contributed to this boom. A year after Baidu Deliveries' creation of "artificial intelligence logistics" (AIL) in 2015, Meituan and Eleme, two other major platforms, developed their versions of the "automated online food dispatching systems."[4] Since 2017, these two companies have emerged as a duopoly, the former backed by Alibaba and the latter by its rival Tencent. In 2018, it was estimated that three hundred million people used online food ordering in 1,300 cities.[5]

The quarantine orders of COVID in 2019 gave the O2O sector a major boost as restaurants moved to third-party delivery service as a means to continue their otherwise disrupted operations. The scale of delivery services surpassed 650 billion in 2019, and the number of customers exceeded 460 million, with more than half of the permanent residents of Chinese cities partaking in the service.[6] In the city of Wuhan, for example, there were five times more delivery orders than usual during the lockdown between January 23 and April 8, 2020.[7] Meituan received twice as many food orders, making up a good portion of its nearly four million total number of orders during this time. The upward trend for more O2O food services continued even as China returned to relative normalcy, as one survey indicated that "70 percent of restaurants planned to increase their spending on third-party delivery service after the pandemic."[8] According to an industry report released in 2021, the number of online food delivery users held steady at 456 million in 2020.[9]

URBAN FOODWAYS AND THE PLATFORM ECONOMY

The dramatic rise of food delivery as an integral part of urban Chinese foodways is inseparable from the rise of the platform economy, itself intricately intertwined with "the rapid penetration of smartphones and the widespread adoption of mobile payment applications" in China.[10] The informalization of the rural-to-urban migrant labor force, the density of residents and food offerings in many Chinese cities, and the changing demographics of urban consumers are also likely contributing factors. One report produced by Meituan attributes the steady growth of the delivery sector from 2015 to 2019 to the rise of the new consumer base consisting primarily of the so-called "post-90s" generation who are now in their twenties. Also invoked are such sociological shifts as "the miniaturization of the family structure" and the "urbanization process"; singles also make up the bulk of the clientele, and more female than male customers tend to participate in the service.[11]

Just like the Fanlai cooking robot whose operation relies on an assemblage of human-nonhuman actions, food delivery offers a distinctive opportunity to discern the uneven impacts of ICTs on different humans. The *waimai xiaoge* or *kuaidi xiaoge* (外卖小哥 or 快递小哥 delivery worker) emerged as a central figure in this assemblage. I distinctively recall President Xi Jinping invoking this group in his 2019 New Year speech to the national public. Literally translatable as "express delivery little bro," they were deemed exemplary of China's "*qianqian wanwan laodong renmin*" (千千万万劳动人民 tens of thousands of laboring people).[12]

In a special report about delivery services during Wuhan's COVID lockdown from late 2019 to early 2020, the platform Meituan describes the transformation of the delivery workers into the city's "*baidu ren*" (摆渡人 ferrymen). Caught between "the shortage in personal protective gear and the drastic rise in demand for delivery of all essential goods," these workers have rendered the company "a major force in the battle against the pandemic" by "delivering *aixin can* (爱心餐 loving meals) to hospitals, seamlessly connecting supermarkets and convenience stores, and ensuring the provisions of life-sustaining supplies."[13] Together, they processed 3,096,000 orders during the seventy-six-day lockdown, with ninety thousand delivered *pro bono* to the medical teams that offered aid to Hubei Province. Listing the "top five" food-related items at night, including "fast-food snacks, convenient stores, supermarkets, seafood and barbeque, and local cuisines," Meituan depicts its service as offering "late-night consolation for Wuhan people's heart."[14]

The report also highlights Meituan's role in the national campaign of poverty alleviation. In Hubei, 791,500 couriers were hired and compensated during this time between January 20 and April 30, 2019, lifting the province's employment rate from the fifteenth to the fifth position within China. Twenty-five percent of the new hires came from 832 poor counties. In 2019, 253,000 workers were "elevated from poverty" out of 257,000 who were registered to be living under the poverty line, allowing Meituan to claim a 98.4 percent rate in poverty alleviation during 2019. Plans are underway to expand the company's reach into other less developed provinces, with the hope of providing two hundred thousand "flexible" employment opportunities. Since April 27, 2019, the company has launched "New Beginnings Plan" to offer fifty thousand rider positions in fifty-two poor counties.[15]

The company naturally paints an uplifting picture that emphasizes the heroic role played by its delivery workers, and indeed its platform, during the COVID-19 crisis. The report highlights the delivery workers as the only people traveling in the streets of Wuhan, offering a kind of comfort to a deserted city.[16] One story features a rider who volunteered to purchase groceries by sending a WeChat message to his neighbors.[17] Often touted are the timely opportunities offered to workers who had faced employment obstacles during this time. One of the delivery workers, Yu Yimeng, for example, was a former bartender who lost his job due to COVID. Lured by an online recruitment ad for couriers and the prospect of working close to home, he signed up at a nearby station to become a rider. In an interview featured on the national channel China Central Television (CCTV), Yu says: "As long as you are willing to work hard, to struggle and run, you can have a stable income."[18] Other stories also heightened the company's projection of its "mission" to "bring the world into your hand"[19] while allowing many laid-off workers to become gainfully employed.

What these rosy representations often mask is the fact that the platform, operating through multiple human-nonhuman assemblages, works to deepen the unevenness between different humans, especially during the pandemic. To probe further into these uneven effects, it is important to recognize that companies like Meituan and Eleme are engaged in a kind of "mobility business," to borrow a term from anthropologist Biao Xiang. Focusing on "selling movements," the sector essentially provides "the service of having someone else move on your behalf."[20] In part due to the centrality of mobility in this line of work, the subject formation of the delivery workers who make up the human component of the urban food delivery business is intricately entangled with a web of nonhuman agents.

THE HUMAN, THE NONHUMAN, AND THE DEHUMANIZED

Key to the making of the delivery workers' subjectivity is what communication scholar Ping Sun calls "algorithmic governance." As Sun points out, algorithms are best understood as "a process that includes the assemblages of both human and non-human agents in social and technical contexts, where they meet, interact, and conflict with each other."[21] The configuration of algorithms in food delivery apps privileges the speed and efficiency of the delivery over the safety and security of the workers, thus exerting tremendous pressure on the laboring body. Not only does the platform's app-based interface monitor the riders' movements and their performance, which directly impacts their pay, penalty, and bonus, but it also allows customers, not riders, to access the maps, cancel orders, and rate the riders. This "customer supremacy" manifests a "kind of algorithmic control" by establishing "information asymmetries" between the riders and those they serve.[22] Ironically, due to their busy schedules, many riders often themselves partake in the delivery services as customers. While this temporarily inverts the power relations between the workers and the clients, it also arguably entrenches the workers into the system's logic even more, since the inversion further naturalizes the control of the algorithms.

The delivery app also limits the workers' ability to negotiate their relationship with the nonhuman surroundings. As rural-to-urban migrants, the riders often don't know their directions well in an urban environment and have to rely on the platform's imperfectly designed navigation system, which sometimes asks them to go "against the flow of traffic" or even "through walls."[23] The nonhuman infrastructure built to enhance the movement of goods and people, in this case, has made it difficult for human couriers to meet the platforms' demand to minimize their delivery time. Riders thus frequently resort to breaking traffic rules—what Sun calls the riders' "inverse algorithm," a labor practice that subjects the workers' bodies to the harm of increased accidents.[24] In major cities like Shanghai, Shenzhen, Chengdu, and Guangzhou, between 2017 and 2018, there were a rising number of traffic accidents, and many involved delivery riders. The trending of the hashtag, "#Delivery has become one of the most dangerous professions#" on the social media platform Weibo was a telling sign of this correlation.[25]

A widely circulated long-form report in *Renwu* (人物 *Portrait*, a state-run magazine) that was later translated as "Delivery Workers, Trapped in the System" by the Chuang blog includes a poignant accident experienced by a rider named Zhu Dahe. When he went off the road, and in the process spilled the *mala xiang guo* (麻辣香锅 spicy hotpot) on his bike, "the first thought that entered his head, even before the pain hit him, was 'Oh no, I'm going to be late!'" To avoid a negative review, Zhu contacted the customer to cancel the order but bought and ate the expensive 80-yuan hot pot dish himself. Thinking back, Zhu realized he could have given the money to the customer to reorder the dish, and at least kept the delivery fee of 6.5 yuan, a figure he still remembered well.[26]

Here, it is striking that Zhu's immediate reaction, which concerns the delivery time, *preceded* his bodily pain, suggesting the extent to which the nonhuman forces of the algorithms and infrastructure have seemingly come to *desensitize* the human body. Equally unsettling is the difference between the price of the spilled spicy hot pot and Zhu's delivery fee, offering a chilling glimpse into the income inequality in urban China. Also manifested is the mental pressure that the system has exerted on the workers. Zhu, a rider "from a rural area who was not familiar with roads in Beijing, let alone the massive flows of cars and people," experienced severe depression during the first months of becoming a rider, much of which came from the self-doubt that he "wasn't delivery rider material," since presumably "every deliver rider makes more than 10,000 yuan a month [sic]."[27]

Zhu's account brings to mind philosopher Yuk Hui's observation that the "food delivery industry and its online platforms provide a clear example of how human flesh is used to compensate for the imperfections of algorithms."[28] For Hui, this industry "is driven by a psychogeography dictated by hunger and desire," wherein the courier "endures more misery when his bike is damaged than when his organic body suffers."[29] Here, the transportation vehicle, another nonhuman entity, has come to shape and condition the agency of the couriers in profound ways. One key example of this is electronic bikes, the delivery workers' vehicle of choice due to their affordability. Communication scholars Julie Yujie Chen and Sun Ping have found that workers often have to carry an extra set of batteries throughout the day because each set can last no more than five hours. Riders are also denied access to elevators in offices and residential buildings as well as shopping malls, leaving them no choice but to leave their e-bikes in unattended outdoor spaces. The frequent thefts of bikes and batteries that result routinely exert mental and financial burdens on the delivery workers, who spend a significant portion of their own funds to purchase these means of transportation.[30]

What may be observed, then, is that the technological nonhuman—be it the e-bikes, the batteries, the platform, or its algorithms—has simultaneously enabled and delimited the workers' subjectivity. Sometimes, the technological nonhuman also encounters such ecological forces as the weather, thus implicating the riders' lives in even more contradictory ways. For the couriers, rainy days mean "more orders to run." Yet the sudden influx of orders can easily overwhelm the system. The algorithms' default goal to optimize delivery efficiency does not take these random natural phenomena into consideration. Therefore, severe weather not only causes delays and consequently deduction in the workers' pay but also increases the risks of traffic accidents among them. During Typhoon Lekima in Shanghai in August 2019, for example, an Eleme courier was electrocuted in an accident. Yet Eleme subsequently doubled the absentee fees for workers if they were to take time off in the following three days. The news of this caused an uproar on Chinese social media.[31]

Severe weather can also cause the system to fail at optimization, demanding a human to take over in manually scheduling the tasks. Nonetheless, even "this kind of manual intervention is not done on behalf of the workers, but rather in order to force every rider to operate at the limits of their speed and ability."[32] A former worker who quit Meituan later refused to run a delivery business for friends because "this industry suppresses any human sense of time."[33]

Recalling the days in which "there was an overwhelming number of orders," he told the reporter of *Renwu* "after finishing delivery everyone was numb. Everyone was running on instinct, without a human emotional response."[34] The system, in other words, also dehumanizes the workers by alienating their relation to time. It is perhaps no surprise that a female rider in a piece shared on WeChat, describes the platform as having "claws"—a nonhuman monster that threatens the well-being of the human-workers.[35]

Food delivery in urban China is thus entrenched in human-nonhuman entanglements, encompassing human interactions with apps, mobile phones, telecommunication networks, transportation vehicles, platform algorithms, and urban infrastructures, among other nonhuman entities. Indeed, it may be more accurate to say that the human courier's subjectivity is *co-constituted* with nonhuman forces. These forces in turn exert contradictory effects on the human body. Such an attention to the mutually constitutive making of the human and the nonhuman agents of food delivery helps to reveal the ways in which the techno-human assemblages that enable this urban food practice have worked to deepen the unevenness between different humans.

First and foremost, it is important to remember that even behind the nonhuman technologies, there are always the human decision makers that help shape the configuration of the platform. Platform capitalists, ultimately, are the people who stand to benefit the most from the dehumanization of the couriers. The emerging movement in a growing number of cities around the world known as "platform cooperativism" has come to challenge the power hierarchy embedded in this profit-driven model of the platform economy.[36] This movement aims to incorporate gig workers into the design of more humane algorithms. After all, in the words of a female courier, the "artificial intelligence" deployed by these platforms, which literally translates as *"rengong zhineng"* (人工智能 human-made intelligence), is more akin to *"shougong zhineng"* (兽工智能 beast-made intelligence),[37] a pun that uses the nonhuman *"shou"* (兽 beast) to replace the *ren* (人 human), inflecting the inhumane practices of the platforms. However, the movement has thus far had limited outreach and has yet to become a major force whether in China or globally to truly disrupt the hegemony of platform-based delivery systems.

Urban food delivery also manifests the ways in which the platform economy has helped shape, as anthropologist Yang Zhan points out, "a new type of class politics with regard to risks."[38] Key to this transformation is what Zhan calls "the outsourcing of risks—both in terms of health and of mobility—on the part of the privileged."[39] Such a reconfiguration is closely tied to the outsourcing of mobility discussed by Biao Xiang and is in many ways exacerbated by the pandemic. While mobility has typically been conceptualized as a form of capital, generative of "opportunities and resources," the COVID lockdown has "turned mobility from an asset into a liability." What it produced is a kind of "'immobility capital', i.e., the capacity of not moving," since "[t]o move is to expose oneself to danger and suspicion." Under quarantine orders, "some people's stasis necessitates others' hypermobility." One single rider, according to an Alibaba report, "enables 24 residents to stay at home."[40] The rural-to-urban migrants that make up the bulk of the delivery workforce in China, while on the surface have been "lifted" from poverty through a job that promises freedom, flexibility, and independence (from long hours of repetitive factory work), have come to bear the brunt of the redistributed risks due to the mobility required in their work.

GENDER, SPACE, AND SPEED

The outsourcing of mobility and risks reflects the *spatial* dimension of China's shifting urban foodways, which is here first and foremost manifested through deepened class inequality. It is important to note that the spatial impacts of food delivery are also intertwined with gender differences. On the one hand, scholars who examine female couriers' daily practices have found that they are negotiating their identities in ways that defy their "disadvantaged" status, often by forming "mediated and ritualized gender-based networks."[41] On the other hand, public discourses around outsourced food making also attribute its rise to the increased employment of women outside the household. Related to this are the skyrocketing real estate prices, which have led to the shrinking of domestic space in many cities. Within these smaller units, "the kitchen is increasingly marginalized" because of its lowered cost-benefit ratio, as singles and young couples are using it less often. The diminished demand for cooking space has prompted developers to reprioritize the design of housing.[42] This in turn contributes to the growing popularity of delivery services as well as speedy home cooking devices like the Fanlai machine my friend used in Shenzhen.

Meanwhile, a desire to "return to the kitchen" can be discerned among some female white collar workers and affluent housewives of the so-called *balinghou* (八零后 post-80s generation, referring to those born in the 1980s, the Chinese equivalent to millennials). A journalist in Xiangtan (Hunan Province), for example, observed that a cooking class scheduled at 10 a.m. to 12 noon attracted many executive-level working women. Some boyfriends and husbands also purchased the class as a gift to their (future) wives. The participants invoked the wish to "say goodbye to takeout" as a motivation for their enrollment, citing health and economic concerns as well as their male partners' excitement to enjoy homemade cuisine.[43] Other posts on WeChat and the popular online forum Zhihu also feature young people who have grown tired of takeout food and are now looking up recipes on the internet to experiment with home cooking. One contributor to a WeChat public account called *Sanmingzhi* (三明治 Sandwiches), for example, wrote about his experiences in 2020 learning to cook while living in a city where the only food delivery options were KFC and McDonald's. At the end of his essay documenting the experience, he reflects on his own pickiness when his mother tried to make new dishes, whose dexterity he now appreciates. Reminiscing about the past, he was learning to make familiar dishes from his hometown, adding more "home flavor" to a city he's been living in for three years.[44] Such an association between "mother" and "home" is not uncommon, pointing to a dominant gendered connection between cooking and women in modern China (see King's contribution [chapter 7] in this volume). Perhaps for this reason, in a 2021 report of the delivery industry, female customers' purchase of fresh ingredients surpasses the male customers' order of cooked meals.[45]

Indeed, a key contributing factor for this trend toward home cooking can very well be the rising health and safety concerns regarding takeout meals among middle-class consumers. Media and government investigations have revealed that the sanitary conditions of some restaurants specializing in food delivery were substandard. Even when the state has issued new legal measures to regulate the industry, profit-driven businesses have still found ways to take advantage of the online mode to conceal their lack of compliance.[46] Some restaurants, such as one in Hefei in Anhui Province, were disclosed to have heated up precooked, packaged dishes to pour on rice for faster delivery, even using long-expired ingredients.[47] Another extreme version of this abuse of the spatial separation between the production and consumption of meals is the emergence of so-called "ghost restaurants." Camouflaging as licensed businesses to become part of the delivery networks of Meituan and Eleme, these shops operate out of residential buildings with nothing more than a home kitchen.[48] It would appear that there may be a blurring of the domestic space of the kitchen and the public space of the "restaurant," to the dismay of many customers overly obsessed with delivery speed.

In close connection to the spatial impact of ICTs on urban Chinese foodways, then, is the *temporal* dimension of the delivery system. Communication scholars Chen and Sun have observed the emergence of a sort of "temporal arbitrage" whereby profits are generated from "stratifying the value of people's time." As "a regime that is inherent in capitalism but has been accentuated by the on-demand service economy," temporal arbitrage "pivots on structurally dismantling the collective time of a class or occupation to benefit another." It does so by normalizing two things: "the customer's cultural expectations for timed and closely monitored service fulfillment," on the one hand, and "the workers performing their duties in an increasingly frenetic and fragmented manner," on the other.[49] These expectations and practices, at times, appear incongruent with some of the long-standing practices of preparation and consumption in Chinese foodways. For example, some chefs would prefer to wait for the couriers to arrive before starting to make the order so as to maintain the quality of the taste. This then comes into conflict with the platforms' and customers' demand for immediacy. The consequence is that the delivery workers lose precious time waiting for the orders and even get penalized for being too slow.[50]

The preoccupation with speed in urban China's on-demand food delivery has an even more troubling layer of implication. In their *New York Times* article, "Food Delivery Apps Are Drowning China in Plastic," journalists Raymond Zhong and Carolyn Zhang use the following subtitle: "The noodles and barbecue arrive within 30 minutes. The containers they come in could be around for hundreds of years thereafter."[51] According to a study cited by the authors, "the online takeout business in China was responsible for 1.6 million tons of packaging waste in 2017, a ninefold jump from two years before. That includes 1.2 million tons of plastic containers, 175,000 tons of disposable chopsticks, 164,000 tons of plastic bags and 44,000 tons of plastic spoons." In 2018, the total packaging waste reached two million tons. Part of the contributing factors are the low cost of delivery and the frequent discounts that platforms offer, which makes it "possible to believe that ordering a single cup of coffee for delivery is a sane, reasonable thing to do." Additionally, the containers for delivery need to be cleaned before recycling. They are also extremely lightweight and therefore require a large amount to generate minuscule profit. These factors offer little incentive for informal recycling workers to collect them. The result is that most of this plastic waste "ends up discarded, buried or burned with the rest of the trash."[52]

There is, in other words, a latent tension between digital time and geological time, between the constant human desire to speed up delivery and disregard for the much slower pace at which nonhuman plastic objects disintegrate into the environment, be it air or sea. Given the state's recent promotion of "ecological civilization,"[53] some measures are underway to curb the polluting effects of the food delivery sector. As of 2020, for example, the central government banned the use of nondegradable plastic bags and utensils for food delivery services along with other items like plastic straws.[54]

Whether or not the implementation of this policy is effective on the ground has yet to be determined, as the complex history of recycling in a city like Beijing has shown that many tensions between the state and the nonstate environmental actors persist to this day.[55] When interviewed, a spokesperson for Meituan also told the *New York Times* reporters that the company "is deeply committed to reducing the environmental impact of food delivery," even "pointing to initiatives such as allowing users to choose not to receive disposable tableware."[56] Nonprofit organizations like Plastic Free China are also working to promote the use of recyclable containers among platforms so as to reduce the generation of waste.[57] Yet, it remains to be seen whether the speed at which these measures can take effect will catch up and keep pace with the rapid growth of the food delivery business, which is also entwined with the ongoing urbanization process that continues to intensify in the Chinese context.

CONCLUDING THOUGHTS: FOOD PRACTICES AND THE PLATFORM ECONOMY

The Fanlai cooking robot I encountered in Shenzhen first inspired me to turn critical attention to the interactions between digital cultures and food practices in urban China. It has led me to investigate the role of ICTs in deepening the divide between different humans. As the COVID-19 pandemic introduced a new normal of everyday practice in Chinese cities and beyond, this entanglement and its impact on the reconfiguration of class and gender (among other differences) is likely going to continue if not deepen.

Some platforms in China have indeed begun to experiment with the use of *wu ren ji* (无人机 humanless machines)—an apt Chinese name for drones—to deliver food so as to minimize human contact. Meituan, for example, launched "person-free delivery services," placing one thousand "smart meal cabinets" in Beijing, Shanghai, and Guangzhou, and tested self-driving delivery vehicles running at twenty kilometers per hour in Zunyi city outside of Beijing.[58] A comedic segment in the 2021 CCTV Chinese New Year Gala featured the use of drones to deliver food from one apartment building to another, as did the popular writer Peter Hessler's reportage in the *New Yorker* about university life under quarantine in Chengdu, Sichuan, in 2020.[59] Recently, a friend returning to China informed me that robot couriers delivered an order of fried dumplings to his hotel room in Shanghai (see

FIGURE 6.2
Robot couriers. Photo by Nianshen Song.

figure 6.2). Meanwhile, as media scholar Xiaowei Wu writes in *Blockchain Chicken Farm*, the perennial concern for food safety has given rise to the production of meticulously surveilled nonhuman animals, such as chickens and pigs, so that urban consumers can more conveniently track the source of their poultry and pork.[60]

Food for humans, except in extreme circumstances and imaginings, is almost always also about the nonhuman. Yet, one of the key lessons of the pandemic is precisely the importance of recognizing how intertwined the human and the nonhuman worlds have always been and increasingly have become. Turning attention to these entanglements, therefore, allows us to further inquire into the spatial and temporal aspects of ICTs' impacts on Chinese foodways. Digital tools such as mobile phone apps, their algorithms, and the platforms they deploy appear to be deepening the inequity among humans differently positioned in the social "food chain." Their varied relations to the practices of cooking, delivering, and eating food are intertwined with spatial mobility, the simultaneous separation and interpenetration of domestic and public realms, and the rising expectations for immediacy. The seeming immateriality of algorithms on which this culture depends, among other things, may obscure the enduring materiality of (plastic) waste generated by sped-up food delivery. Understanding the tensions and contradictions within modern Chinese foodways with a distinctive sensitivity to the embeddedness of food production, distribution, consumption, and recycling in China's digital cultures can perhaps aid us in exploring new horizons for tackling the widening injustices that permeate a food system so fully integrated with the global platform economy.

NOTES

1. See, for example, Silvia M. Lindtner, *Prototype Nation: China and the Contested Promise of Innovation* (Princeton, NJ: Princeton University Press, 2020).

2. For a more extended reflection on (im)mobility in contemporary China that also draws on the example of food delivery, especially under COVID-19, see Fan Yang and Cara Wallis, "Mobile Technologies and the Unevenness of (Im)Mobility," in *Oxford Handbook of Chinese Digital Media*, ed. Carlos Rojas, Jinying Li, and Yomi Braester (Oxford: Oxford University Press, forthcoming).

3. James L. Watson, *Golden Arches East: McDonald's in East Asia*, 2nd ed. (Stanford, CA: Stanford University Press, 2006).

4. Ping Sun, "Your Order, Their Labor: An Exploration of Algorithms and Laboring on Food Delivery Platforms in China," *Chinese Journal of Communication* 12, no. 3 (September 2019): 312, https://doi.org/10.1080/17544750.2019.1583676.

5. Sun, "Your Order, Their Labor," 308.

6. Meituan Research Institute and Special Committee on Delivery, the China Hospitality Association, "Report on the Development of China's Delivery Sector from 2019 to the First Half of 2020" (China: Meituan Research Institute, June 2020), 3, https://s3plus.meituan.net/v1/mss_531b5a3906864f438395a28a5baec011/official-website/1898d746-f44c-431f-98b3-3c41a39f3421. It is worth noting that China's *hukou*, or household registration system, has long restricted the movement of rural residents to urban centers and denied migrant workers access to social welfare benefits and other opportunities, despite the government's reforms in recent years.

7. Minghe Hu et al., "How China's Delivery Services Platforms Are Evolving, from Smart Lockers to 'Semi-Finished' Meals," *South China Morning Post*, April 24, 2020, sec. Tech, https://www.scmp.com/tech/big-tech/article/3081251/smart-lockers-semi-finished-meals-howchinas-delivery-services.

8. Aimei shenghuo yu chuxing yanjiu zhongxin, "Aimei baogao - 2020 yiqing qijian Zhongguo canying waimai shichang shanghu zhuanti yanjiu baogao" [Ai Media

report - Special research report on businesses in China's takeout market during the pandemic in 2020], Aimei wang, April 12, 2020, https://www.iimedia.cn/c400/70742.html.

9. "2020–2021 Nian Zhongguo waimai hangye fazhan yanjiu baogao" [Report on the development of China's food delivery industry in 2020–2021], Sohu.com, June 1, 2021, https://www.sohu.com/a/469748860_407401.

10. Julie Yujie Chen and Ping Sun, "Temporal Arbitrage, Fragmented Rush, and Opportunistic Behaviors: The Labor Politics of Time in the Platform Economy," *New Media & Society* 22, no. 9 (September 1, 2020): 1562, https://doi.org/10.1177/1461444820913567.

11. Meituan waimai yonghu diaoyan, "Zhongbang fabu! 2020 waimai hangye baogao" [Published! 2020 report on the delivery sector], *Zhihu zhuanlan*, June 28, 2020, https://zhuanlan.zhihu.com/p/151608872.

12. Zhongguo xinwen, Guojia zhuxi Xi Jinping fabiao 2019 xinnian heci [China News: Nation's President Xi Jinping presents 2019 new year's speech], 2018, see full text at http://cpc.people.com.cn/n1/2019/0101/c64094-30497657.html.

13. Meituan Peisong, "Yiqingzhong de jishi peisong - 2020 Meituan Peisong kangji xinguan yiqing xingdong baogao" [Instant delivery during the pandemic - Meituan Delivery combatting the coronavirus], Zhongguo Wuliu yu Caigou Lianhehui Tongcheng Jishi Wuliu Fenhui, May 2020, 2, http://www.chinawuliu.com.cn/upload/resources/file/2020/05/19/46557.pdf.

14. Meituan Peisong, "Yiqingzhong," 6–8.

15. Meituan Peisong, "Yiqingzhong," 13–31.

16. Notably, not only were humans the recipients of food but also nonhuman pets. For example, one order asked the rider to feed the three cats stuck at the home of a frontline worker who couldn't be home for days. Meituan Peisong, "Yiqingzhong," 27.

17. Meituan Peisong, "Yiqingzhong," 40.

18. Meituan Peisong, "Yiqingzhong," 30.

19. Meituan Peisong, "About Us," Meituan guanwang, 2020, https://peisong.meituan.com/about.

20. Biao Xiang, "The Emerging Business of Mobility," accessed January 31, 2021, https://www.eth.mpg.de/5517502/blog_2020_07_17_01.

21. Sun, "Your Order, Their Labor," 310.

22. Sun, "Your Order, Their Labor," 317.

23. Chuang, "Delivery Workers, Trapped in the System," *Chuang* (blog), November 12, 2020, https://chuangcn.org/2020/11/delivery-renwu-translation/.

24. Chuang, "Delivery Workers."

25. Chuang, "Delivery Workers."

26. Chuang, "Delivery Workers."
27. Chuang, "Delivery Workers."
28. Yuk Hui, "For a Planetary Thinking," *E-Flux* 114 (December 2020), https://www.eflux.com/journal/114/366703/for-a-planetary-thinking/.
29. Hui, "Planetary Thinking."
30. Chen and Sun, "Temporal Arbitrage," 1569–1573.
31. Chuang, "Delivery Workers."
32. Chuang, "Delivery Workers."
33. Chuang, "Delivery Workers."
34. Chuang, "Delivery Workers."
35. Sisi Tang, "Waimai nüqishou: Redian guohou, yulun shengle, er women yiran kuibai" [Female delivery rider: After the hotspot, the public opinion won, but we remain defeated], *Zhihu*, January 20, 2021, https://zhuanlan.zhihu.com/p/345764992.
36. Trebor Scholz, "Platform Cooperativism vs. the Sharing Economy," *Medium*, July 10, 2015, https://medium.com/@trebors/platform-cooperativism-vs-the-sharing-economy-2ea737f1b5ad.
37. Tang, "Waimai Nüqishou."
38. Yang Zhan, "COVID-19, Mobility, and Care Work," *Pathologies of Governance: Lockdowns, Shut-Offs, and Chokeholds in the Time of COVID-19*, online, January 30, 2021.
39. Zhan, "COVID-19."
40. Xiang, "Mobility."
41. Ping Sun, Yuchao Zhao, and Qianyu Zhang, "Pingtai, xingbie yu laodong: 'Nüqishoude' xingbie zhanyan" [Platforms, gender, and labor: Female food delivery couriers' gender performativity], *Funü yanjiu luncong* 6 (2021): 5–16.
42. Cheng Zhang, "Yinshi kuaicanhua shehui" [Fast food society], *Jiancha fengyun* 17 (2021): 68–69.
43. "Shidaixinchaoliu, bailing ai zhufan" [New trend, white collars love to cook], *Sina*, December 06, 2007, https://news.sina.com.cn/c/2007-12-06/013113026784s.shtml.
44. Xiaobin, "Zai meiyou waimai de rizi li, wo chengwei le yige chuyi xinshou" [I have become a novice in cooking in the days without takeout], *The Paper*, April 12, 2020, https://www.thepaper.cn/newsDetail_forward_6916477.
45. "2020–2021 nian Zhongguo waimai hangye fazhan yanjiu baogao" [Report on the development of China's food delivery industry in 2020–2021].
46. Yao Liu, "Toushi 'waimaire': Waimai weihe nanguan?" [Perspective on 'food delivery fever': Why is delivery hard to regulate?], *Renmin wang*, December 5, 2016, http://health.people.cn/n1/2016/1205/c14739-28924835.html.

47. "Lianjia waimai sushibao de mimi: Shengchan guocheng lingren zuo'ou, rixiao sishiwan fen" [The secret of instant cheap food delivery: The production process is disgusting, and 400,000 orders are sold per day]," Sohu.com, November 16, 2018, https://www.sohu.com/a/www.sohu.com/a/275905646_99964807.

48. "Meituan, Eleme naxie 'youling canting,' ni chiguo ma?" [Have you been to those 'ghost restaurants' on Meituan and Eleme?], Sohu.com, July 21, 2021, https://www.sohu.com/a/478937912_120824046.

49. Chen and Sun, "Temporal Arbitrage," 1563.

50. Chuang, "Delivery Workers."

51. Raymond Zhong and Carolyn Zhang, "Food Delivery Apps Are Drowning China in Plastic," *New York Times*, May 29, 2019, https://www.nytimes.com/2019/05/28/technology/china-food-delivery-trash.html.

52. Zhong and Zhang, "Food Delivery Apps."

53. Yifei Li and Judith Shapiro, *China Goes Green: Coercive Environmentalism for a Troubled Planet* (Cambridge, UK: Polity, 2020).

54. Guojia fazhan gaiwei - shengtai huanjingbu, "Guojia fazhan gaige wei shengtai huanjing bu guanyu jin yibu jiaqiang suliao wuran zhili de yijian" [Suggestions of the National Development and Reform Commission's Ministry of Ecological Environment on strengthening the regulation of plastic pollution], Zhonghua Renmin Gongheguo zhongyang Renmin Zhengfu, January 16, 2020, http://www.gov.cn/zhengce/zhengceku/2020-01/20/content_5470895.htm.

55. Joshua Goldstein, *Remains of the Everyday: A Century of Recycling in Beijing* (Oakland: University of California Press, 2020).

56. Quoted in Zhong and Zhang, "Food Delivery Apps."

57. Sherry Lu, "Plastic Policy," Webcast, The Plastic First Mile - Closing the Loop on Plastic Waste in Asia, (Wilson Center, December 7, 2020), https://www.wilsoncenter.org/event/webcast-plastic-first-mile-closing-loop-plastic-waste-asia.

58. Meituan Peisong, "Yiqingzhong," 10.

59. Peter Hessler, "How China Controlled the Coronavirus," *New Yorker*, accessed September 9, 2020, https://www.newyorker.com/magazine/2020/08/17/how-china-controlled-the-coronavirus.

60. Xiaowei Wang, *Blockchain Chicken Farm: And Other Stories of Tech in China's Countryside* (New York: FSG Originals, 2020).

III CONSTRUCTING CULINARY IDENTITIES: GENDER, NATION, AND ETHNICITY

7 DOMESTIC COOKBOOKS AND THE EMERGENCE OF FEMALE CULINARY AUTHORITY IN TWENTIETH-CENTURY CHINA

MICHELLE T. KING

Seen from the vantage point of our twenty-first century era of celebrity chefs and hyper-personal branding, the mid-twentieth-century appearance of almost a dozen Chinese cookbooks in Taiwan titled with the names of their female authors does not seem terribly unusual. Today's readers may not recognize the names of Huang Yuanshan or Fu Pei-mei, but the impulse of these women to lend their own names to their *shipu* (食譜 cookbooks)—*Yuanshan shipu* (3 vols., 1954–1964), *Peimei shipu* (3 vols., 1969–1979)—just seems like a familiar form of savvy marketing. Another popular Chinese cookbook of the time, *The Joyce Chen Cookbook* (1962), written by Joyce Chen and published in English for an audience of American housewives, shares the same authorial naming convention.

Yet, when we consider these cookbooks from the other end of history, from the vantage point of China's late imperial period or the early twentieth century, the monumental shift that they represent in the gendered conventions of Chinese culinary authority becomes much more obvious, as does the sheer audacity of these Chinese women in naming their cookbooks after themselves. These postwar cookbooks upend the traditional Chinese vision of cooking and eating as a hierarchy of the male gourmand over his male cook, replacing it instead with a female cooking instructor speaking to her fellow housewives, who were focused on feeding their families nutritious and tasty meals. Once the modern home cook was explicitly gendered as female, a housewife's decision-making and her skills—where to shop, what to buy, and how to cook—became of central importance in the whole enterprise of nourishment within the family home.

The twentieth-century emergence of the Chinese domestic cookbook, written by Chinese women for audiences of other women, must certainly number as one of the most significant transformations of modern Chinese foodways, in light of the overwhelming dominance of men as both authors and subjects in the long history of Chinese food writing. Although Chinese women labored in food preparation for millennia prior to the twentieth century, for the most part they did so silently, with few of their voices or perspectives on the subject captured in the historical record. Gradually, over the course of the twentieth century, Chinese women vocally claimed the kitchen, food, and cooking as domains of their expertise through diverse media, including cookbooks, newspaper cooking columns, essays, home economics textbooks, lectures, radio broadcasts, and television appearances. In so doing, they validated the worth of female domestic labor as a means of sustaining both individual families and the nation as a whole.

THE MALE GOURMET

Almost all Chinese writings on food prior to the twentieth century were authored by male literati, and a great deal of it depicts eating and drinking as a male homosocial activity. The eighteenth-century epicure Yuan Mei's (1715–1797) well-known culinary treatise, *Suiyuan shidan* (隨園食單 *Recipes from the garden of contentment,* 1792), for example, privileges the eating experience of the male literatus, providing a series of buying and tasting notes with a plethora of dos and don'ts for like-minded gentlemen connoisseurs. Even the preparation of food was detailed as the domain of men. Yuan Mei's homage to his personal chef, Wang Xiaoyu, serves as an illuminating example. Yuan Mei admired Wang not only for his exceptional culinary skills but also for his sense of economy. Wang did not need fancy, rare, or expensive ingredients to make something taste delicious. When Yuan Mei once asked Wang why he stayed with him, when he could have commanded a much higher salary in a wealthier household, Wang was clear:

> To find an employer who appreciates one is not easy, but to find one who understands anything about cookery is harder still. So much imagination and hard thinking go into the making of every dish that one may well say I serve up along with it my whole mind and heart. The ordinary hard-drinking revelers at

a fashionable dinner-party would be equally happy to gulp down any stinking mess. They may say what a wonderful cook I am, but in the service of such people my art can only decline. True appreciation consists as much in detecting faults as discovering merits. You, on the contrary, continually criticize me, abuse me, fly into a rage with me, but on every such occasion make me aware of some real defect, so that I would a thousand times rather listen to your bitter admonitions than to the sweetest praise. In your service, my art progresses day by day. Say no more! I mean to stay on here![1]

What is most significant in this encounter is that Yuan Mei's encomium to the consummate skill of his chef as culinary artist ends up as praise for his own refined taste as a gourmet. Within this dynamic, there is no question of the motor ultimately driving the culinary enterprise: it is the subtle sensibilities of the gourmet's tongue (and his constant haranguing) that determine its success, not the hands or skill of the cook.

This appreciation of the male gourmet as the most important factor in good Chinese cooking was still evident into the mid-twentieth century. F. T. Cheng (1884–1970) served as the Republic of China's last ambassador in London from 1946 to 1950, and at the urging of friends wrote a book titled *Musings of a Chinese Gourmet* (1954). With regard to Chinese cooking, Cheng asserted, "It may be said that the man of taste, the gourmet, has played a greater part in the development of good cooking than the person called the cook; for the latter is in general merely one who is able to prepare a meal of some sort, whereas the former, though he may not have done more cooking himself in his life than boiling an egg, knows what is good or bad, and it is this knowledge, when imparted to the cook, that leads to the improvement of cooking." Note, again, that Cheng assumes the gourmet to be a "*man* of taste," while his cook is also presumably male. Cheng claimed that he himself had "trained" two cooks, one in Peking and one in Nanking, who later went on to work for the foreign minister and prime minister, and that he had "polished" another cook in London.[2]

In most imperial Chinese writings on food, women are rarely seen, either as subjects or authors.[3] The single notable exception to this paucity of premodern female authorship of Chinese historical sources on food is the Southern Song dynasty (1127–1279) cookbook, *Wushi zhongkui lu* (吳氏中饋錄 *Madame Wu's kitchen records*), which appears to have been written by a woman of the Wu family, detailing food preservation methods, as well as recipes for individual dishes. According to cookbook author and culinary historian

Hsiang Ju Lin, Madame Wu's cookbook is "straightforward with no scholarly asides . . . She had not much in common with gentleman cooks."[4] Little more can be said about Madame Wu herself, except to posit her existence in a well-to-do family in Pujiang County near Jinhua, Zhejiang (famous for its dry-cured hams). The text itself was included in a Yuan dynasty (1271–1368) compilation, and later copied without attribution in the fourteenth and sixteenth centuries.[5]

Chinese historians have not yet found, to the best of my knowledge, any early Chinese cooking manuscripts handwritten by women, to be copied by other women in the family, as in the case of two Korean manuscripts from the late Chosŏn period. Chang Kyehyang (1598–1680) wrote *Umsik timibang* (*Recipes of tasty foods*) explicitly for her daughters, without the intention of wider dissemination. More broadly circulated was the later *Kyuhap ch'ongsŏ* (*Home encyclopedia for women in the inner chamber*) by Yi Pinghŏgak (1759–1824), a general compendium of domestic knowledge for women. Yi's manuscript was originally hand-copied by relatives, but it was later eventually published in a woodblock print in 1869. Not surprisingly, the emphasis in these female-authored culinary texts, written for other women, centers on recipes and methods for actually making food, not on guidelines for the discerning gentleman's dictates of taste. Yi wrote, "My purpose is to make one book by which [female] readers can easily learn and practice methods in their real life."[6]

In China, it is not until the twentieth century that we begin to see significant numbers of Chinese food texts written for, and more notably, by Chinese women. Jin Feng describes the earliest known verifiable example, Zeng Yi's (1852–1927) *Zhongkui lu* (中饋錄 *Kitchen records*, 1907), published by her son along with her other writings on the occasion of her fifty-fifth birthday "as an act of filial tribute."[7] The term *zhongkui* (中饋), Feng explains, was a classical reference to women's ascribed household duties in the ancient divination text, the *Book of Changes*: "Only by making food at home can women ensure the harmony and prosperity of the whole clan."[8] Like Madame Wu's much earlier cooking guide, Zeng Yi focuses primarily on food preservation methods, such as drying, pickling, salting, and fermenting. The cookbook section itself is slender, containing only twenty recipes, yet Zeng clearly saw cooking as a central task for women: "Every girl must learn how to cook before she gets married," she wrote.[9] Feng underscores the fact that Zeng's admirers praised her most for her ability to write excellent poetry *in addition* to ably managing her household. Mirroring the moral hierarchies of male literati writing, certain forms, such as poetry, were seen as superior to "marginalized" dilletante writing on foodways.

THE EMERGENCE OF THE CHINESE HOUSEWIFE

Starting in the early twentieth century in China, all of the basic conditions needed for the emergence of the modern domestic cookbook—growing interest in cooking, widespread basic literacy, and sufficient economic ability—would arise at the same time as the figure of the modern middle-class Chinese housewife. Although Chinese women had for centuries carried out the domestic tasks of cooking, cleaning, and child-rearing (or overseen female servants who performed these same duties), there was no specific Chinese term to identify the "housewife" as such in earlier periods. Prior to the twentieth century, Chinese women were instead identified almost exclusively by their familial role—daughter, sister, wife, mother, mother-in-law, sister-in-law, aunt, grandmother, concubine, servant—not by the site of their labor. Instead, the concept of the "housewife" appears to have made its way into modern Chinese from both English and Japanese sources in the early twentieth century.

As a point of comparison, in the Anglophone world, where the term "housewife" first originated, the housewife's purview had long included cooking (or at least overseeing it when performed by servants). Early examples of English cookbooks written explicitly for or even by women for use in a domestic setting include Thomas Dawson's *Good Huswives Jewell* (1585), Gervase Markham's *The English Huswife* (1615), Eliza Smith's *The Compleat Housewife* (1727), and Susannah Carter's *The Frugal Housewife* (1765). (By contrast, according to Stephan Mennell, early modern French cookbooks were largely written by and for male chefs in professional kitchens.[10]) Later nineteenth-century American cookbooks followed this English model, focusing on the role of the housewife in the management of the household, particularly in the kitchen. Cookbooks such as Mary Randolph's *The Virginia Housewife* (1824), Lettice Pierce Bryan's *The Kentucky Housewife* (1839), and Sarah Rutledge's *The Carolina Housewife* (1847), for example, were popular in their mid-nineteenth century moment, a period Glenna Matthews has called the "golden age of domesticity" in the United States, when the status and role of the housewife were held in their highest esteem.[11]

A closer analogue to the Chinese historical example can be found in the Meiji-era (1868–1912) emergence of the Japanese housewife and Japanese domestic cookbooks. The first Japanese cookbooks written explicitly for female audiences appeared in 1820 and 1849 and featured Japanese cuisine.[12] By the 1890s, "women became the central audience for cookbooks in general," and cookbooks encompassed both Japanese and Western cuisines.[13] Meanwhile, the Japanese term for "housewife" (*shufu* 主婦) first appeared in a 1874 translation of a Scottish text on home economics.[14] The Meiji *shufu*'s most significant role was "as the economic manager of the household and supervisor of domestic affairs," which was distinct from other, more common Japanese terms for women based on their familial relations and status alone.[15] During the Taishō period (1912–1926), the concept of *shufu* expanded beyond upper-class women to include growing ranks of middle-class women, who devoured popular magazines, the most notable of which was *Shufu no tomo* (主婦の友 *The housewife's companion*), which was founded in 1917 and remained in print until 2008.[16]

The Chinese housewife, *jiating funü* (家庭婦女 woman of the house), *jiating zhufu* (家庭主婦 woman in charge of the house), or just *zhufu* (主婦 housewife or hostess), was born slightly later, in the Republican era (1911–1949), amid the efflorescence of new magazines and newspapers targeting female audiences. The precise etymology, first usage, and interrelationship of all these Chinese terms for "housewife" await further detailed study. Yet certainly by the 1930s and 1940s, the figure of the Chinese housewife and her assigned household tasks, including cooking, were embedded in popular consciousness, as evidenced by the myriad Republican magazines devoted to homemaking and family life. The entire field of modern home economics in China, which also rose to prominence in the Republican era, precisely targeted the perceived need for training Chinese women how to manage their domestic affairs.[17]

Li Gong'er's 1917 publication of *Jiating shipu* (家庭食譜 *Home cookbook*) likely constitutes the first stand-alone domestic Chinese cookbook written expressly for a female audience. (Though Zeng Yi's recipes were published earlier in 1907, they only appeared as part of her collected writings, alongside her poetry and other essays.) In his preface, Li emphasizes that he wrote the cookbook "*wei wuguo funü er zuo*" (為吾國婦女而作 for our country's women). Though he does not explain how he himself learned how to cook, he states that the contents of the cookbook are "entirely from the author's knowledge, based on everyday experience." In Li's estimation, nothing is more important than cooking in managing and bringing order to a household. "If this [task] does not belong to women, then who?" Li asks.[18]

Yet Li was discouraged that early Republican Chinese women did not seem eager to take on this responsibility, and his overall aim was to overturn the disdainful attitude women in China had against cooking. In his opinion, women either had high ambitions and wanted to continue with their formal school studies, or they had low ambitions and were self-indulgent and extravagant. In both cases, they were uninterested in learning how to cook. Even for those few, exceptional women who might have a real desire to learn how to cook, there were few good books on cooking to guide them. The ones that did exist were too complicated or unclear for beginners to use as practical kitchen guides. He purposely wrote his text in simple language, so that women young and old could understand it. One gets more than a whiff of what we might today call mansplaining in Li's preface: Li seems most motivated to write a cookbook because *he* sees a critical need for it, not necessarily because any women have urged him to do so—and he is prepared to teach women what he knows.[19]

In a preliminary assessment, it appears that more Chinese cookbooks in the Republican era were written by men than women.[20] Tao Xiaotao published his cookbook, *Taomu pengrenfa* (陶母烹飪法 *Grandma Tao's cooking methods*), in 1936 when he was only eighteen. Tao was the son of the mass education reformer Tao Xingzhi, who encouraged his children not to rely on the labor of others but to wash and mend their clothes, cook, and clean by themselves.[21] Tao's father wanted him to learn how to cook from his paternal grandmother, whose specialty was Huizhou dishes from their home province of Anhui. In his preface, Tao makes no specific mention of writing the cookbook for women readers, as did Li Gong'er before him, but instead scrupulously uses generic, non-gendered terms regarding his audience: "Cooking is something that every living person (*gege huozhe de ren* 各個活著的人) should learn," or "The average beginning cook (*yiban chuxue pengren de ren* 一般初學烹飪的人) wants to find the most appropriate ways of doing things."[22] What is perhaps most interesting, however, is that despite envisioning a potential audience of both male and female readers, it is solely Tao's own male authorial voice that emerges from the cookbook, despite the titular nod to Grandma Tao's foundational culinary expertise.

FEMALE COOKBOOK AUTHORSHIP

We do not see a real flourishing of Chinese cookbooks written by and expressly for other women until the postwar era. The vast majority of these postwar Chinese cookbooks were published *outside* of mainland China, primarily in Taiwan, Hong Kong, Singapore, and to a lesser degree in the United States or the United Kingdom. It was outside of the PRC that the bourgeois concern for the pleasures of the Chinese table (and the food to put on it) would be most clearly articulated in the postwar years, by a new generation of female Chinese cookbook authors, in both English and Chinese. These women did not hide behind pen names and flowery literary titles but boldly declared themselves under their own real names as cooking instructors, cookbook authors, food column writers, television cooking personalities, radio cooking program hosts, restaurateurs, and entrepreneurs. Their publications, moreover, reached global audiences, Chinese and English readers alike. Although a few studies have been written about these women as individual cookbook authors, it seems critical to consider them together in the same discursive field, in order to discern a broader generational pattern of authorship and gendered culinary authority.[23]

Arjun Appadurai has described the social conditions in postwar India that allowed for the flourishing of cookbooks written by and for Indian women between the 1960s and 1980s; many similar circumstances can be identified in postwar Taiwan as well.[24] Postwar Taiwan boasted an extraordinarily rich range of Chinese regional and ethnic cuisines, as migrants from every corner of each of China's provinces had moved to the island after 1949. Seeing how one's neighbors from a different region of China ate (a process Yujen Chen describes at more length in chapter 8 of this volume) heightened people's curiosity at trying different regional dishes, reflecting a kind of "culinary cosmopolitanism."[25] Moreover, postwar peace meant more leisure time and affluence to cook new types of dishes, demonstrating an urban middle-class woman's ability to both impress her husband's guests and please her children's demands for culinary novelty.

In considering the explosion of postwar Chinese cookbooks, what is most compelling is the dynamic twentieth-century transformation of culinary authority, from the male-centered, hierarchical relationship between a culinary connoisseur and his cook (à la Yuan Mei) to the female-centered, instructional relationship between a middle-class female cookbook author and her audience of fellow housewives. Chinese female cookbook authors of the postwar period fall into two distinct categories: one group represents a kind of intermediary, liminal stage between the traditional Chinese model of male culinary authority and the new model of female culinary authority (Huang Yuanshan, Lin Tsuifeng, Lin Hsiang Ju), and another group represents a new model of female culinary authority standing alone, without apparent male influence (Joyce Chen, Fu Pei-mei). For the former group of women, their husbands, fathers, or fathers-in-law had an outsized impact on the publication of their cookbooks. The latter group of women, on the other hand, developed their cookbooks independently of male involvement. Both groups of cookbook authors broke new ground by speaking to broad audiences of other women, primarily housewives. (One must note that both in Taiwan and the United States, almost all of these published mid-century cookbook authors were mainlanders, teaching women how to cook various regional Chinese cuisines, not native Taiwanese, an ethnic distinction that would gain more political recognition in the 1980s, as Chen describes in chapter 8.)

In postwar Taiwan, the first Chinese cookbooks to achieve commercial success were those of Huang Yuanshan (1920–2017), a Hong Kong native, who established herself as the country's most accomplished cookbook author in the 1950s. Huang was a professor in the new department of home economics at National Taiwan Normal University and had previously published recipes in the *Central Daily News* and the *Funü zhoukan* (婦女週刊 *Women's weekly*).[26] Huang published her first eponymous cookbook, the *Yuanshan shipu* (媛珊食譜 *Yuanshan cookbook*), in Taipei in 1954. This was followed by four other cookbooks, all bearing her own name in the title: the *Yuanshan dianxinpu* (媛珊點心譜 *Yuanshan snacks and desserts cookbook*, 1956), the second volume of the *Yuanshan cookbook* (1957), the *Yuanshan xicanpu* (媛珊西餐譜 *Yuanshan Western cuisine cookbook*, 1960), and finally, the third volume of the *Yuanshan cookbook* (1964).[27]

Huang readily acknowledged that her father-in-law, Qi Rushan (1877–1962), a noted scholar of Chinese opera and self-professed gourmand, was more responsible than anyone else for launching her culinary career. After the entire family moved to Taiwan in 1949, Qi Rushan lived with Huang and her husband, Qi Ying (the couple had no children). Both her father-in-law and her husband would frequently entertain guests, and Huang, who did not work outside of the home, was "personally responsible for going into the kitchen to make all the food."[28] Their housemaid offered no help, since she was only a young girl, who knew nothing about cooking herself and had to be trained in the kitchen by Huang. Since Qi Rushan was well traveled, had eaten all kinds of dishes at famous restaurants, and dabbled in food history himself, Huang always asked for his advice. "He would tell me in general how to make something, then I would go and experiment. My father-in-law would then taste it, and if he said it was good, then I would make a record of it."[29] Eventually she began to teach other women how to cook at women's association meetings, and the editor of the women's weekly at the *Central Daily News* approached her to write a regular cooking column.

In one sense, the dynamic Huang describes closely mirrors that recorded by Yuan Mei about his personal chef Wang Xiaoyu more than a century and a half earlier: the tongue and memory of the gourmet guides the hands of the cook. It is also not difficult to imagine that Huang's association with her famous father-in-law, who wrote a preface for her first cookbook, smoothed the way for its publication. Yet here the major difference is of course that Huang, a woman, is the recipient of Qi's cooking tips, and instead of just cooking to please her father-in-law's palate, she shares what she has learned by becoming a cooking teacher for other women. Eventually Huang would become a respected culinary professional in her own right, but she never stopped acknowledging her father-in-law's culinary influence on her development.

We can see another example of this transitional model of culinary authority in the English-language Chinese cookbooks of Lin (neé Liao) Tsuifeng (dates unknown) and Lin Hsiang Ju (1931–present), the wife and daughter of noted bon vivant, philosopher, and gastronome Lin Yutang (1895–1976). Lin Yutang wrote many well-received volumes in English about Chinese habits and customs, the most notable of which were *My Country and My People* (1935) and *The Importance of Living* (1937). The Lins lived mostly in the United States after 1935, and the mother-daughter pair wrote two English-language Chinese cookbooks, *Cooking with the Chinese Flavor* (1956) and *Chinese Gastronomy* (1969).[30] Lin Yutang recognized his daughter Hsiang Ju for her extraordinary sense of taste, calling her a "born gastronome" with a "prodigious gustatory memory," while his wife Tsuifeng "provides the expert knowledge and skill and guidance."[31] Hsiang Ju reports that she wrote the cookbooks directly in English by herself.[32] This was after lengthy discussions with her mother about proper cooking techniques and mistakes to avoid, which Lin Yutang witnessed. "I heard the conversations going on in the next room," he writes. "It was like a party recalling a hunting or fishing trip, or a trip to the Arctic."[33]

While Lin Yutang gives ample credit to his wife and daughter for their practical culinary expertise, he still assigns himself the very same role occupied by Yuan Mei, that of the gentleman connoisseur with exquisite taste. Despite living with two exceptional home cooks, Lin claims that "acting as a fastidious and discerning critic" is his own "contribution to the excellent level of my household dinners . . . The wine-taster still serves a valuable function, even though he does not grow the vine himself."[34] Indeed, it seems that for Lin Yutang, eating and having no hand in the cooking is the man's rightful place: "When a man has driven himself to work all day, he has the right to look forward to a delightful supper at home, cooked with respect for the food and for its eater. Admittedly, this is a man's point of view; he does not have to do the cooking himself."[35] In this sense, it is no accident that Lin compares his daughter and wife's culinary conversations to the masculine activities of hunting, fishing, or Arctic exploring: food preparation is *their* job, one which they undertake with all the seriousness of (male) scientists and scholars, and with the gusto of adventurers.

One revealing anecdote, however, cuts through Lin Yutang's chauvinist bluster (which, to be fair, is written with a certain degree of self-deprecating humor). More than once, he writes, when "telling my friends about the secret of preparing certain dishes, I often have to draw up sharp." His wife Tsuifeng quickly interjects, and tells him, "No, no, Y.T. That is not the way."[36] Quite literally, the female home cook has found her voice and puts the male eater in his rightful place, as the passive recipient of the meal, not its inventive and knowledgeable creator. Elsewhere in their own introduction to the cookbook, Tsuifeng and Hsiang Ju acknowledge that in Chinese culinary history, "the best team is made up of the gastronome and his cook . . . Cuisine did not come into its own until the critics became articulate. They found fault with the food. They developed ideas and harassed their cooks."[37] While Tsuifeng and Hsiang Ju use male pronouns to describe the Chinese gastronome and his cook, what is most interesting is the way in which their own mother-daughter partnership upends these traditional Chinese gendered norms of culinary authority: both the gastronome (Hsiang Ju) and cook (Tsuifeng) are now female, and equal partners in the culinary enterprise.

Seeing these earlier transitional models of postwar cookbooks authored by Chinese women allows us to better understand how truly remarkable it was for a second group of postwar Chinese women to write their cookbooks and build their culinary careers entirely on their own terms.[38] Joyce Chen (née Liao Jia'ai) (1917–1994) gained fame in the postwar period as chef-owner of several Boston-area restaurants specializing in northern Chinese cuisine, opening her first Joyce Chen Restaurant in 1958. Her eponymous *The Joyce Chen Cookbook* was published in English for American audiences in 1962. Originally self-published in a small run of six thousand, *The Joyce Chen Cookbook* eventually sold more than seventy thousand copies nationwide after it was republished by J. B. Lippincott.[39] Chen also appeared on a short-lived public television program on Chinese cooking from 1966–1967, *Joyce Chen Cooks*. Chen later patented her own line of flat-bottomed woks suitable for American electric ranges and developed a line of readymade bottled Chinese sauces under the Joyce Chen Foods brand. Even today, Chen's name still retains some of its former culinary cachet. The family still produces her branded line of sauces and frozen food products, while another company continues to produce and market Joyce Chen–branded cookware, including woks, kitchen scissors, steamers, and bamboo cutting boards.[40]

Chen described the circumstances of how she came to write her cookbook, a story intimately intertwined with her life as a housewife. She was born into a wealthy family (with a skilled family chef) in Beijing, married her husband there in 1943, and eventually moved with him and their two young children to the United States in 1949. She describes how she made "two dozen egg rolls and a few dozen cookies for the mothers' food table" to sell at her children's school bazaar in 1957.[41] Later, when she asked her daughter Helen about it, her daughter said that she only saw the cookies for sale. "I thought they did not like [the eggrolls] so they had not put them up for sale," recalled Chen.[42] It was thus a great surprise when later that evening she ran into the mother in charge of the sale, who asked if Chen would make more eggrolls—they were so popular they had sold out in five minutes. This surprise success led to cooking classes for friends interested in Chinese cooking, and then eventually, the opening of her own restaurant. While Chen thanks more than two dozen people who helped make her cookbook possible, she appears to have written it more or less on her own. She told her daughter that it was "written with blood, sweat and love. Blood and sweat for the hard work in testing, researching, and struggling with my poor English."[43] Indeed, her daughter, Helen, remembers her mother as having accomplished everything through "sheer intelligence, perseverance and sweat," as Chen had had no college education or other formal culinary training.[44] The fact that Chen's later culinary career was fashioned on her own terms as an independent woman was quite literally the case: Chen got divorced from her husband in 1966 and never remarried.

Yet, as a Chinese woman in the United States, Chen's success was relatively circumscribed, and limited primarily to her restaurants in the Boston area. Though her short-lived public television program on Chinese cooking had been filmed on the same set as Julia Child's, she was never embraced as America's culinary darling in the same way, due to a range of racial, linguistic, and cultural issues. Mainstream white American television audiences (and corporate sponsors) apparently did not a consider a cooking program with a Chinese woman speaking accented English and cooking only Chinese food as anything but a minor novelty; the program was canceled after only one season of twenty-six episodes.[45]

In comparison to other postwar female Chinese cookbook authors, the one who achieved the most lasting and widespread professional success was Fu Pei-mei (1931–2004), who published her own eponymous three-volume bilingual cookbook series, *Peimei shipu* (培梅食譜 *Pei Mei's Chinese Cook Book*), from 1969 to 1979.[46] Domestically in Taiwan, Fu managed to achieve much wider name recognition and influence than other women through the new medium of television, which began broadcasting in Taiwan in 1962. Fu began her career as a television cooking program host and instructor that same year, eventually parlaying her popularity into four continuous decades on Taiwan Television. In this sense, she literally *was* the "Julia Child of Chinese cooking," as she was dubbed by New York Times food critic Raymond Sokolov in 1971, in a way that Joyce Chen could never be in the United States.[47] One might even argue that Fu's success transcended Child's, since her bilingual cookbooks (the first of their kind in Taiwan), with recipes published in both Chinese and English on facing pages, gave her access to a much broader Chinese and English-speaking audiences around the world, beyond the confines of her own country.

Fu was always very conscious of her primary audience of housewives in Taiwan, in large part because she was a housewife herself, and never abandoned that identity. The popular press in Taiwan, for example, praised Fu for having achieved fulfillment in both her career *and* her family life. Writing in 1978, one reporter enthused that Fu had not only "received the respect and love of countless people" through her work but had also demonstrated "filial devotion to her in-laws at home," "love and respect" with her husband, and a maternal love toward her three children.[48] Ironically, when Fu first got married in Taiwan in 1951, she had no idea how to cook. Unlike Huang Yuanshan, who had an encouraging culinary coach in her epicurean father-in-law Qi Rushan, Fu's primary eater was her picky husband, Cheng Shaoqing, who only ever complained about the terrible food she made for his mahjong buddies. Embarrassed, Fu personally hired a series of regional restaurant chefs to teach her how to cook. She went on to start her own cooking school in the late 1950s, before she was discovered by Taiwan Television in 1962.

Fu's English was rudimentary, but she was determined to publish her cookbook in a bilingual format, mostly to satisfy the demands of many foreign women who joined her cooking classes in Taipei, and overseas Chinese who did not read Chinese.[49] Fu solved the problem of English translation, just as she had solved the problem of not knowing how to cook in the first place, not through her husband's knowledge or networks but by relying on her own. When writing the first volume, she took the entire English portion of the manuscript, which she had first tried to translate herself, and asked an American student from one her cooking classes, Yvonne Zeck, the wife of a US Air Force colonel stationed in Taipei, to help correct it. Zeck, who knew no Chinese, read through the manuscript and rendered it into readable English.[50] Fu seems to have used similar tactics to translate the second and third volumes of her bilingual cookbooks as well, as she thanks a different American woman in each one.[51]

Apart from the translation of the cookbook, the research and testing on all of the recipes was also the result of Fu's own labor. In her autobiography, she recalled, "The cookbook was the result of so much hard work, blood, sweat and tears. Every word and every sentence had been corrected over and over; the measurement of every ingredient and seasoning, cooking time and fire power, had all been the results of actual experiments . . . so that I could personally guarantee to readers that they too could make every dish with their own hands."[52] Fu had not known how to cook when she first married, and had not exhibited any particular interest in cooking or food as a young girl. Moreover, her upper middle-class family had no links to any notable culinary or cultural luminaries. (As one sign of Fu's entrepreneurial savvy, Fu self-published all of her cookbooks, as did Joyce Chen.) Yet in the repeated act of cooking and tasting, again and again, Fu had trained the sensitivity of her own palate, as well as the skill of her hands. She did have the help of others on her culinary journey, to be sure—the male restaurant chefs who taught her, the American military wives who helped translate her cookbooks—but the overall shape of her career and the publication of her cookbooks could only be credited to her own gumption, and hers alone. In the evolution of twentieth-century Chinese domestic cookbooks, the gastronome and the cook had become one: Chinese culinary authority could now be claimed by this entrepreneurial Chinese housewife, speaking to an international audience of fellow housewives.

Today's home cooks have a dizzying array of ways to learn how to make a new dish. Printed cookbooks and television programs have been replaced by social media and online content—blogs, vlogs, websites, cooking apps, videos, crowd-sourced recipe platforms, virtual cooking classes, Instagram (see, for example, Sabban's discussion of food vlogger Li Ziqi in chapter 10). With each new technological twist, culinary authority has continued to devolve, away from established mid-century female cooking teachers, cookbook authors, and television stars to anyone with a smartphone or computer and internet access. The reach of these twenty-first century culinary content creators, moreover, is seamlessly global. While this latest evolution of modern Chinese cooking instruction may seem inevitable, ubiquitous, and universal, it is not. As with any taken-for-granted historical trajectory, looking back across the twentieth century helps us understand how a mid-century generation of entrepreneurial Chinese women helped to dramatically shift the long-standing pattern of male dominance over centuries of Chinese culinary writing. Although their contributions to the Chinese culinary world have since by and large been neglected, ignored, forgotten, or minimized, these pioneering women deserve full recognition for asserting their own culinary authority as both tastemakers and skilled home cooks.

NOTES

1. Translation of Yuan Mei's "Chuzhe Wang Xiaoyu zhuan" [Biography of Chef Wang Xiaoyu] from Arthur Waley, *Yuan Mei: Eighteenth Century Chinese Poet* (New York: Grove Press, 1956), 52–53.

2. F. T. Cheng, *Musings of a Chinese Gourmet* (London: Hutchinson, 1954), 29–30.

3. Until recently, Chinese food historians paid little attention to gender. Out of the four hundred pages in K. C. Chang's edited volume *Food in Chinese Culture: Anthropological and Historical Perspectives* (New Haven, CT: Yale University Press, 1977), women are briefly mentioned on ten pages. More recent scholarly work has begun to address this lacuna, particularly through analysis of classic Chinese literature. See,

for example, Issac Yue, "Tasting the Lotus: Food, Drink, and the Objectification of the Female Body in *Gold, Vase, and Plum Blossom*," in *Scribes of Gastronomy: Representations of Food and Drink in Imperial Chinese Literature*, ed. Issac Yue and Siufu Tang (Hong Kong: Hong Kong University Press, 2013), 97–112.

4. Hsiang Ju Lin, "Madame Wu's Home Cooking," *Slippery Noodles: A Culinary History of China* (London: Prospect Books, 2015), 178.

5. Lin, "Madame Wu's," 183.

6. Ro Sang-ho, "Cookbooks and Female Writers in Late *Choson* Korea," *Seoul Journal of Korean Studies* 29, no. 1 (June 2016): 133–157 (148).

7. Jin Feng, "The Female Chef and the Nation: Zeng Yi's *Zhongkui lu* (*Records from the kitchen*)," *Modern Chinese Literature and Culture* 28, no. 1 (2016): 1.

8. Feng, "Female Chef."

9. Hsiang Ju Lin, "Domestic Duties," *Slippery Noodles*, 317.

10. Stephen Mennell, *All Manners of Food: Eating and Taste in England and France from the Middle Ages to the Present* (Oxford: Basil Blackwell, 1985), 135.

11. Glenna Matthews, *"Just a Housewife": The Rise and Fall of Domesticity in the United States* (Oxford: Oxford University Press, 1989).

12. Shoko Higashiyotsuyanagi, "The History of Domestic Cookbooks in Modern Japan," in *Japanese Foodways, Past and Present*, ed. Eric C. Rath and Stephanie Assmann (Urbana: University of Illinois Press, 2010), 130–131.

13. Higashiyotsuyanagi, "Domestic Cookbooks," 132.

14. Kazumi Ishii and Nerida Jarkey, "The Housewife Is Born: The Establishment of the Notion and Identity of the Shufu in Modern Japan," *Japanese Studies* 22, no. 1 (2002): 36.

15. Ishii and Jarkey, "Housewife," 37.

16. Ishii and Jarkey, "Housewife," 40.

17. Helen Schneider, *Keeping the Nation's House: Domestic Management and the Making of Modern China* (Vancouver: UBC Press, 2011).

18. Li Gong'er, "Zixu," *Jiating shipu* [Home cookbook] (Shanghai: Zhonghua shuju, 1917), 1. In the same volume see also his "Bianji dayi," 1–2.

19. I have no other biographical information on Li Gong'er, apart from the titles of his other books. These titles, several of which focus on letter writing for women, suggest that he may have been involved in women's education himself. See Li Gong'er, *Yan wen duizhao nüzi chidu* [A comparison of speech and writing for women's correspondence] (Shanghai: Huiwentangxinji shuju, 1930) and *Xinfunü shuxin* [New women's correspondence] (Shanghai: Chunming, 1940).

20. This preliminary conclusion is based on Ren Baizun's list of Republican era cookbooks. Of the nineteen pre-1949 cookbooks listed, two appear to be authored

by women. Table 8, "Zhongguo xiandai shipushu minglu" [List of modern Chinese cookbooks] in Ren Baizun, ed., *Zhongguo shijing* [Classic of Chinese food] (Shanghai: Shanghai wenhua chubanshe, 1999), 939–940.

21. Tao Xiaoguang, "Zhuiqiu zhenli zuo zhenren: Huainian wo de fuqin Tao Xingzhi" [Pursuing the truth to become a real person: Remembering my father Tao Xingzhi], *Wenshi ziliao xuanji* 72 (Wenshiziliao chubanshe, 1980), 201.

22. Tao Xiaotao, "Xu," in *Taomu pengrenfa* [Grandma Tao's cooking methods] (Shanghai: Shangwu yinshuguan, 1936), 1.

23. Buwei Yang Chao's cookbook has been discussed at length by scholars, but only in the context of her impact in America. See Charles W. Hayford, "Open Recipes Openly Arrived At: Mrs. Chao's *How to Cook and Eat in Chinese* (1945) and the Translation of Chinese Food," *Journal of Oriental Studies* 45, nos. 1–2 (December 2012): 67–87; Anne Mendelson, "Change, Interchange, and the First Successful 'Translators,'" chapter 6 in *Chow Chop Suey: Food and the Chinese American Journey* (New York: Columbia University Press, 2016).

24. Arjun Appadurai, "How to Make a National Cuisine: Cookbooks in Contemporary India," *Comparative Studies in Society and History* 30, no. 1 (January 1988): 3–24.

25. Appadurai, "How to," 7.

26. "Huang Yuanshan," *Baike zhishi*, accessed November 17, 2021, https://www.baike.com/wikiid/1904235602217580681. https://www.easyatm.com.tw/wiki/E9%BB%83%E5%AA%9B%E7%8F%8A.

27. Huang Yuanshan, *Yuanshan shipu*, vol. 1 [Yuanshan cookbook, vol. 1] (Taipei: Sanmin shuju, 1954); *Yuanshan shipu*, vol. 2 [Yuanshan cookbook, vol. 2] (Taipei: Sanmin shuju, 1957); *Yuanshan dianxinpu* [Yuanshan dim sum cookbook] (Taipei: Sanmin shuju, 1956); *Yuanshan xicanpu* [Yuanshan Western cuisine cookbook], (Taipei: Sanmin shuju, 1960); *Yuanshan shipu*, vol. 3 [Yuanshan cookbook, vol. 3] (Taipei: Sanmin shuju, 1964).

28. "Huang," Baike.

29. "Huang," Baike.

30. Lin Tsuifeng and Lin Hsiangju, *Cooking with the Chinese Flavour* (London: William Heinemann, 1957) [orig. pub. 1956]; Hsiang Ju Lin and Tsuifeng Lin, *Chinese Gastronomy* (New York: Pyramid Publications, 1972) [orig. pub. 1969].

31. Lin Yutang, "Foreword," in *Chinese Gastronomy*, 8.

32. Email correspondence from Hsiang Ju Lin to author, November 28, 2018.

33. Lin Yutang, "The Art of Cooking," in *Cooking with the Chinese Flavour*, xii.

34. Lin, "Art," xii.

35. Lin, "Art," xiii.

36. Lin, "Art," xi–xii.

37. Lin and Lin, *Chinese Gastronomy*, 9.

38. Pan Peizhi and Hu Peiqiang also named their cookbooks after themselves and built their careers independently of the men in their lives. See Pan Peizhi, *Pan Peizhi shipu*, vol. 1 [Pan Peizhi cookbook, vol. 1] (Taipei: Jiwen shuju, 1965) and *Pan Peizhi shipu*, vol. 2 [Pan Peizhi cookbook, vol. 2] (Taipei: Jiwen shuju, 1966); Hu Peiqiang, *Hu Peiqiang shipu* [The Hu Peiqiang cookbook] (Taipei: Zhonghua jiating zazhishe, 1971).

39. Niu Yue, "Carrying on a Chinese food legacy," *China Daily USA*, uploaded April 2, 2015, http://usa.chinadaily.com.cn/world/2015-04/02/content_19985670.htm.

40. "Joyce Chen Collection," Honey Can Do, accessed February 25, 2022, https://honeycando.com/collections/joyce-chen. See also https://kitchensupply.com/pages/joycechen.

41. Joyce Chen, "How I Came to Write This Book," in Joyce Chen, *The Joyce Chen Cookbook* (Philadelphia: J. B. Lippincott, 1962), 2. See also Stephen Chen, "Savoring the Legacy of Joyce Chen: Chef. Restaurateur. Entrepreneur," Joyce Chen Foods, accessed February 25, 2022, https://joycechenfoods.com/legacy/.

42. Chen, "How I Came to Write This Book," 2.

43. Chen, "Words from the Author," in *The Joyce Chen Cookbook*, 221.

44. Anne-Marie Seltzer, "Helen Chen Remembers Her Mother," Patch.com, uploaded Sept 3, 2010, https://patch.com/massachusetts/lexington/helen-chen-remembers-her-mother.

45. Dana Polan, "Joyce Chen Cooks and the Upscaling of Chinese Food in American in the 1960s," Open Vault from GBH, accessed February 25, 2022, https://openvault.wgbh.org/exhibits/art_of_asian_cooking/media.

46. Fu Pei-mei, *Peimei shipu diyice* [Pei Mei's Chinese Cook Book Vol. I] (Taipei: Self-published, 1969); *Peimei shipu di'erce* [Pei Mei's Chinese Cook Book Vol. II] (Taipei: Self-published, 1974); *Peimei shipu disance* [Pei Mei's Chinese Cook Book Vol. III] (Taipei: Self-published, 1979).

47. Raymond Sokolov, "Pei-Mei's cold (and hot) salads," *New York Times Magazine*, July 25, 1971.

48. Da Fang, "Pengtiao dashi Fu Peimei" [Cooking ambassador Fu Pei-mei], *Jiating Yuekan* (May 1978).

49. Fu, *Peimei shipu diyice*, 2.

50. Fu Pei-mei, *Wuwei bazhen de suiyue* [Years of five flavors and eight treasures] (Taipei: Juzi chuban youxian gongsi, 2000), 126.

51. Fu thanks Monica Croghan in vol. 2 and Nancy Murphy in vol. 3 for their translation help. See Fu, *Peimei shipu di'erce*, 8 and *Peimei shipu disance*, 6.

52. Fu, *Wuwei*, 127.

8 TAIWANESE CUISINE AND NATIONHOOD IN THE TWENTIETH CENTURY

YUJEN CHEN

Taiwanese cuisine has been a major attraction for international tourists and a national symbol of Taiwan since the early 1990s.[1] Vibrant discussions on how to define authentic Taiwanese cuisine and tastes have been held among chefs, gourmets, food writers, and government authorities. Gourmets and food writers tend to argue that Taiwanese cuisine, despite having originated from Chinese cuisine, has become a distinct tradition because of years of adaptation and indigenization. Along with the discussion of "national cuisine," beef noodles and pearl milk tea/bubble tea have been respectively characterized as the national food and drink of Taiwan, demonstrating the social needs of Taiwanese people to be recognized as a political entity, a nation, and part of international society.

What is Taiwanese cuisine? Taiwanese cuisine is a unique combination of the flavors and ingredients of aboriginal people, Austronesians, and different waves of immigrants to Taiwan. The definition of Taiwanese cuisine has shifted through each political regime, or more precisely, with the changing status of Taiwan as a colony, province, or nation. These changes are reflected in the changes in words used to describe Taiwanese cuisine: *Taiwan ryôri* in Japanese (台灣料理), *Tai cai* (台菜), and *Taiwan cai* (台灣菜).

The Qing dynasty governed Taiwan from 1684 until it was succeeded by imperial Japan, which colonized Taiwan from 1895 to 1945. In 1949, the Chinese Nationalist Party fled the Chinese mainland, where the Chinese Communist Party (CCP) had established power, for Taiwan and proclaimed themselves to be the only legitimate power representing all of China. Even today, the sovereignty over Taiwan is still hotly contested. The uncertainty of nationhood throughout much of modern Taiwanese history has resulted in changing ideas of Taiwanese cuisine, reflecting the parallelism of political ideology and dietary culture. This chapter traces the emergence and transformation of Taiwanese cuisine during the twentieth century, analyzing how Taiwanese cuisine has been shaped and transformed into a national symbol and how a national cuisine can be framed in relational contexts. The contemporary search for a discourse of national cuisine reveals the desire of the Taiwanese people to form a distinct national and cultural identity.

EMERGENCE OF TAIWAN *RYÔRI* AS ELITE FOOD IN THE JAPANESE COLONIAL ERA

Although the written history of Taiwan can be traced to the early seventeenth century, terms exclusively referring to "the cuisine of Taiwan" cannot be identified in the literature during the Qing dynasty (1644–1911).[2] Under Qing rule, dining out in Taiwan was limited to simple eateries offering rice, noodles, and snack foods. Wealthy households had private cooks preparing daily family meals, with chefs hired for banquets on special occasions, such as weddings and birthdays. However, food served at these banquets was not specifically termed "Taiwanese cuisine."[3] Taiwanese cuisine as a new culinary category began to develop during the Japanese colonial era.

The Japanese term *Taiwan ryôri* first appeared in print media in January 1898, shortly after the initiation of Japanese colonial rule in Taiwan in 1895. An article on an official New Year banquet held by the local administration in Tainan and attended by many Taiwanese officials of junior rank reported that some local Taiwanese dishes were served to great praise.[4] In addition to feasts, private parties of Japanese colonial officials at which Taiwanese food was served were recorded in the early 1900s publication of *Taiwan kanshû kiji* (臺灣慣習記事 *Records of Taiwanese customs*), providing insights into the interests of Japanese officers and folk scholars in Taiwanese dishes.[5]

This haute cuisine served at banquets, however, was not specifically identified as Taiwanese cuisine; rather, it was also called "Chinese cuisine." Although the term *Taiwan ryôri* had been in use since the end of the nineteenth century, it was often used to mean "Taiwanese or Chinese cuisine." This overlap in meaning did not mean that Japanese colonial officials were more familiar with or had a deeper understanding of Chinese cuisine. Although

Chinese food had long been an essential element of Japanese cuisine, after Japan's victory in the Russo–Japanese War in 1905, Japanese people began to change their attitude toward Chinese cuisine, shifting from a generally negative perception to curiosity. In the late nineteenth century, Chinese cuisine in Japan manifested in two extremes. Chinese cuisine was either esteemed as exclusive and exquisite or disdained as unclean and foul tasting, which was consistent with the negative impression that Japanese people had of Chinese people at the time. By the early 1900s, Japanese people had partially shifted their interest from Western food to Chinese food.[6] In the early twentieth century, sampling Chinese cuisine was a novel experience for most Japanese people. However, Japanese officials in Taiwan did not perceive a clear distinction between Chinese and Taiwanese cuisine.

Books and articles from the early twentieth century that document the introduction of Taiwanese cuisine have revealed that familiarity with the local cuisine was considered worthy knowledge for Japanese people holding and attending banquets or parties, with Taiwanese cuisine representing a new culinary category for Japanese people in Taiwan (figure 8.1). On these occasions, Taiwanese cuisine was regarded as exotic local fare, and Taiwanese banquets provided an opportunity for Japanese people to deepen their knowledge of local customs.

The distinction of Taiwanese cuisine was signified by a "Taiwanese lunch banquet" held by the Crown Prince of Japan, Hirohito, on April 24, 1923, during his visit to Taiwan.[7] All dishes served at the banquet were prepared by the chefs of two iconic restaurants, Jiangshan Lou (江山樓) and Donghuifang (東薈芳). The banquet menu for Hirohito included the following delicacies: snow-white bird's nest, turkey and pork medallions,[8] crystal pigeon eggs, braised shark fin, grilled eight-treasure crab,[9] snow-white tree fungus, fried spring rolls, braised soft-shelled turtle, sea cucumber with fungus, steamed fish fillet with ham, ham and white gourd soup, eight-treasure rice, and almond tea. The banquet included expensive ingredients such as bird's nest, shark fin, pigeon, crab, and duck, which are all traditional Chinese delicacies.[10]

The banquet considerably elevated the reputation of Taiwanese cuisine as well as that of the restaurant Jiangshan Lou. Jiangshan Lou's owner Wu Jiang-shan wrote a series of twenty-three articles on Taiwanese cuisine that were published in the official newspaper *Taiwan nichinichi shinpô* (臺灣日日新報

FIGURE 8.1
The first cookbook on "Taiwanese cuisine" was published in 1912. Photo from the Institute of Taiwan History, Academia Sinica, Taiwan.

Taiwan daily news) in 1927, as if Taiwanese cuisine had been an authentic and long-standing traditional cuisine.

Wu described a complete Taiwanese banquet. A Taiwanese banquet consists of thirteen courses, the seventh of which must be a snack that marks the middle of the banquet. Dry dishes must be served after wet dishes (i.e., soup or stew). A break period is provided once the snack for the seventh course is served, during which time guests may rest, smoke, explore the restaurant, or have a short nap. Sometimes, performances are held before the remaining six courses are served. In the second half, guests continue to enjoy elaborate dishes and conversation until a snack is served again, marking the end of the banquet.[11]

The significance of such large restaurants lay not only in the haute cuisine they served; these restaurants preserved cultural elements of the Han people and provided a space for Taiwanese intellectuals to connect (figure 8.2). As Pintsang Tseng has suggested, such restaurants constituted a potential public sphere in Taiwan within the colonial regime.[12] Interestingly, as a colony of Japan, although both Jiangshan Lou and Penglai Ge (蓬萊閣), another high-end restaurant established in 1927, were initially Chinese restaurants providing various regional Chinese cuisines, by the 1930s, they self-identified as Taiwanese restaurants.[13] The owners of each restaurant argued that although Taiwanese cuisine primarily used ingredients from four categories of Chinese regional cuisine, namely Sichuan, Beijing, Guangdong, and Fujian, these ingredients had been changed and adapted to suit the tastes of people living in Taiwan, thereby differentiating Taiwanese cuisine from Chinese cuisine. Thus, the boundary of Taiwanese cuisine was set clearly for the first time in the context of colonialization to distinguish it from Japanese cuisine and Chinese cuisine.

From the restaurateurs' assertions that they operated Chinese restaurants to the claims that Taiwanese cuisine was superior, these two elite restaurants transformed from Chinese to Taiwanese restaurants. The articles introducing Taiwanese cuisine published in the official newspaper *Taiwan nichinichi shinpô* penned by the owner of Jiangshan Lou helped to establish Taiwanese cuisine as a culinary category. Although the patrons of such Taiwanese restaurants were limited to government officials, merchants, intellectuals, and urban dwellers, knowledge and recognition of Taiwanese cuisine gradually trickled down from the top of the social ladder through the dissemination of such articles in print media.

FIGURE 8.2
The interior decoration of Jiangshan Lou, one high-end restaurant during colonial period, presented cultural elements of the Han people. Photo from National Museum of Taiwan History.

EVERYDAY FOOD OF THE TAIWANESE PEOPLE

The term *Taiwan ryôri* referred primarily to banquet dishes largely consumed by the Japanese and Taiwanese gentries. Everyday food of the general population was referred to as "food of the Taiwanese" by folklorists during the colonial period.[14] Despite not appearing publicly as a facet of Taiwanese food culture until the second half of the 1990s, these key features in the dietary structure of the everyday populace served as the basis of the diet of most Taiwanese people. More interestingly, the foodways of the common people were eventually highlighted as symbolic of Taiwanese cuisine after the rise of Taiwanese identity and culinary nationalism following the end of the authoritarian regime in the 1990s.

DAILY MEALS

Sweet potatoes and rice were both staple foods in Taiwan. Sweet potatoes were incorporated into the diets of many Taiwanese Indigenous peoples in the seventeenth century along with traditional millet and taro. Later, although rice was grown in large quantities, its high economic value limited its consumption among common people, and only wealthy people consumed rice on a daily basis.

Most households relied on homegrown vegetables and their own fishing hauls for side dishes. Bamboo shoots, ginger, black beans, and soybeans could be preserved through pickling or drying. Each region developed its own pickled products, which played a key role in the daily meals of Taiwanese people. Although pickles play a much less prominent role today and are no longer a daily side dish, they are still often included on family tables in dishes such as eggs with preserved radish, chicken soup with pickled pineapple and bitter gourd, and oysters with black bean sauce, which are typical Taiwanese family dishes.

The most common cooking methods used in everyday households prior to the mid-twentieth century included boiling, stir-frying, pan-frying, braising, and stewing. Only after the establishment of the cooking oil industry following World War II (WWII) and the importation of cooking oil in large quantities did oil-based cooking methods, such as deep-frying, become popular. During this period, seasoning methods were relatively simple. Most households used salt as the main seasoning in addition to natural plants such as spring onion, garlic, and ginger.

Soy sauce, although expensive, was used alongside other sauces such as homemade or purchased bean paste. In offshore areas, homemade shrimp oil or sauces made from small fish and shrimp were used.

Pork, in addition to vegetables and aquatic products, was one of the most highly valued side dishes in Taiwan. Pork was not always available. Raising pigs was costly, and the slaughter of pigs, cows, and sheep was taxable under Japanese rule. Fresh pork dishes, including pork liver, fried pork chop, and meatballs, were often the main foods offered at festivals.[15]

Although Taiwan is surrounded by the sea, fresh fish and shellfish were not easily accessible before refrigeration facilities became widespread, and pickled aquatic products were the mainstay. The coastal areas of southwestern Taiwan, including Chiayi, Tainan, and Kaohsiung, had similar diets. Pickled aquatic products were the main side dish in these areas. By contrast, mountainous areas relied on dry farming, reflecting the effect of the environment on diet. This regionally specific diet continued in rural areas until the 1960s.

SNACKS AND FESTIVAL FOODS

Night market food, especially *xiaochi* (小吃 snacks, a contemporary term that literally means small-eating), are popular among international tourists in Taiwan. Night market snacks can be traced to the *dianxin* (點心 snacks) and festive dishes of the Qing dynasty. Prior to the mid-twentieth century, snacks were often eaten between meals to replenish physical energy during busy farming periods in rural communities. Urban residents and wealthy people had a wide variety of snacks available for entertaining guests or for leisure.

Taiwanese festive food was also closely related to the rhythms and routines of farming life. An iconic festive occasion that Taiwanese people frequently enjoyed was the outdoor feasts known as *pān-toh*, the Hokkien[16] pronunciation of *banzhuo* (辦桌). Literally meaning "managing the tables," *pān-toh* refers to the outdoor feasts held on occasions such as weddings, religious feasts, house warmings, birthdays of older adults in the community, and the first-month birthday of newborns. Today, *pān-toh* is viewed as folk culture, with the dishes served at *pān-toh* often regarded as authentic Taiwanese cuisine (figure 8.3).

Pān-toh in Taiwan began in the Qing dynasty. The characteristics of the society at that time played a crucial role in shaping *pān-toh* as a major social activity. When groups of Han Chinese first migrated to Taiwan from the southern coast of China, constructing a social network in their new home was essential for survival; feasting was a key means of achieving this purpose. By treating guests to meals at feasts, hosts forged and strengthened connections with kin, countryfolk, and newcomers, establishing collaborative community systems. For the newly arrived immigrants, Taiwan afforded greater possibilities for social mobility than did mainland China, and feasts were an opportunity for individuals to demonstrate their generosity and enhance their social status.[17] *Pān-toh* was a means of competing for fame and bolstering reputations, and this custom persisted into the Japanese colonial era.

In rural societies, *pān-toh* was regarded as an inclusive local event involving the host family and all of their neighbors. Preparations by the host would last almost half a year. By the 1960s, professional *pān-toh* chefs had become

FIGURE 8.3
Pān-toh scene. Photo from the author's own collection.

widespread; however, before then, host families would invite talented neighbors to shoulder the task of cooking. A key characteristic of *pān-toh* was that the dishes were cooked outdoors rather than in kitchens. Cooks responsible for *pān-toh* constructed stoves under large tents, which served as makeshift kitchens. Dining tables were set up in open squares or along the street, with the tables, chairs, cooking utensils, and tableware all borrowed from neighbors. Neighbors also assisted with the cleaning and cutting of foodstuffs and the serving of dishes. A *pān-toh* was essentially a communal neighborhood event, and its success relied on a well-organized neighbor network and strong social bonds. An unsuccessful banquet would lead to a loss of face and friendship and might generate difficulties in the communal life of host families.

Another feature of *pān-toh* was *dabao* (打包 taking food home), which remains common at Taiwanese banquets today. Because many guests traveled from faraway places and had to spend half a day returning home, the host was expected to prepare enough food for guests to take home as supplies for their homeward journey. To meet such needs, dishes served during the second half of *pān-toh* were typically dry food items such as *zha huazhi wan* (炸花枝丸 deep-fried squid balls) and *zha yutou* (炸芋頭 deep-fried taro), which are easy to pack and carry when traveling.

CULINARY REMAPPING BY THE STATE AND MIGRATION AFTER 1945

Although Taiwanese cuisine emerged as an elite food and haute cuisine during the Japanese colonial era, its status and meaning changed drastically during the postwar period. After the 1970s, the term *Tai cai* (台菜) was widely employed in popular media discourse and official food guides to indicate Taiwanese dishes. However, dishes described using this term were regarded as unsophisticated, simple, and plain, marking a stark departure from its previous status as elite food.

This transition did not occur by chance, nor was it an isolated phenomenon. It was a part of the culinary remapping occurring alongside the political and social transformations soon after the end of the colonial era. With the surrender of Japan at the end of WWII and the defeat of the Nationalist Party by the CCP, the Nationalists lost control in mainland China and retreated to Taiwan in 1949. The Nationalist Party brought over one million soldiers and civilians to the island. For most of them, the relocation to Taiwan was initially nothing more than a temporary retreat. However, this temporary posting ultimately became an accidental migration, with most people unable to return to their hometowns in mainland China until the late 1980s.

This resettlement in Taiwan constituted a social reorganization that involved the transplantation of the Nationalist government and a large Chinese population from mainland China to Taiwan. The new political regime and migrant group were the two major forces driving the culinary remapping in Taiwan, which included the substantial transformation of restaurants and discursive changes in cookbooks and culinary literature. The new political regime implemented regulations changing the operations of restaurant businesses, and migrants introduced various regional Chinese foods into restaurants, markets, and families.

Japan formally ceded control of Taiwan to the Nationalist government on October 25, 1945, two months after surrendering. In the new political system, the participation of local citizens and other government ministries was undermined, thereby centralizing all political power in Nationalist Party authority. In addition, the growing conflict between the Nationalists and Communists on the mainland generated economic chaos that affected the Taiwanese economy. Prices of goods rose steeply, and food was in short supply.[18] The price of rice was sixty times higher in February 1947 than it was in 1945 and was higher than anywhere else in China at the time.[19]

The Nationalist Party proceeded to import Nationalist troops, officials, and civilians from mainland China to Taiwan in a short period. Although these forced migrants were a minority in Taiwan, accounting for 14 percent of the population, they were given the majority of senior government positions. In addition to strengthening political supervision and social control, the Nationalist government made efforts to reestablish Chinese culture in Taiwan. For the Nationalists to maintain their status as the legitimate government of China, policies aimed at resinicization were implemented across cultural and educational domains.

In addition to the overwhelming emphasis on Chinese culture, Chinese culinary culture played a key role in culinary remapping. Culinary remapping included both substantial and symbolic changes. Practical and symbolic boundaries were redrawn. Practical boundaries in cuisine are constituted by cooking methods, ingredients, and seasonings that create different tastes, smells, and appearances. Symbolic boundaries are shaped using descriptive classifications, representations, and narratives in cookbooks and literature.

By the 1950s, Chinese restaurants established by Mainlanders had become widespread, including restaurants providing various regional dishes. These new restaurants often assumed the names of famous restaurants on the mainland. A large portion of their clients were Mainlanders holding government positions. Some government institutions even had their own restaurants serving specific regional Chinese cuisines. For instance, Peng Chang-kuei, the inventor of the dish General Tso's chicken, was regarded as a key person in the development of Hunan cuisine in Taiwan. In addition to those restaurants established by Mainlanders, some exclusive government-operated restaurants were critical to the culinary map in Taiwan during the 1950s and 1960s. For example, the Yuanshan Dafandian (圓山大飯店 Grand Hotel) is a landmark building in Taiwan that was established with strong support from the state.[20]

Markets and villages for *juancun* (眷村 military dependents' villages) provided opportunities for the hybridization of various regional Chinese cuisines. In 1949, the Nationalist government began to build villages in which military dependents could settle. These villages also introduced Chinese foods into Taiwan and had a twofold influence on the culinary scene. First, those living in the villages brought and preserved a variety of Chinese foods from various provinces, which led to an exchange of regional cuisines. Villagers could taste, learn about, and modify dishes from other provinces, resulting in the hybridization of dishes. Second, in the villages, Chinese regional foods were more widely dispersed to non-Mainlander families in Taiwan. To expand their markets, Mainlander vendors sold their hometown specialties outside their villages, with these regional specialties from the mainland attracting both Mainlanders and local consumers.

These migrant chefs, cooks, and vendors altered the culinary map in Taiwan by transplanting various regional Chinese cuisines, specialties, and snacks from mainland China to Taiwan. Particularly in Taipei, the culinary map was redrawn as a condensed Chinese culinary map, a development that can be traced to the cookbooks written by Fu Pei-mei.

Fu Pei-mei (1931–2004) published more than fifty cookbooks, comprising more than four thousand recipes, during her lifetime. Her first bilingual (Chinese and English) cookbook, *Pei Mei's Chinese Cook Book*, was published in 1969 and has been reprinted more than twenty times. In addition to being a cookbook author, Fu was the first and most influential television chef and culinary educator in Taiwan and in overseas Chinese communities. With mainland China–style restaurants offering various Chinese foods, Fu's cookbooks and demonstrations of Chinese cuisine on television further accelerated the spread of Chinese cuisine in Taiwan. Her readers and audiences adapted her Chinese recipes for their family meals, introducing new ingredients into local dishes and mixing regional dishes, which had a profound effect on taste preferences and the culinary scene in Taiwan.[21]

All the practical and discursive transplantation of restaurants, dishes, food producers, consumers, and cookbooks composed a system of Chinese cuisine in Taiwan, consisting of major regional cuisines such as *Yue cai* (粵菜 Guangdong cuisine), *Lu cai* (魯菜 Shandong cuisine), *Chuan cai* (川菜 Sichuan cuisine), *Xiang cai* (湘菜 Hunan cuisine), and *Zhe cai* (浙菜 Zhejiang cuisine). Taiwanese cuisine exists within this linguistic context as *Tai cai*. Whereas Taiwanese cuisine was distinguished from Chinese cuisine during the colonial period, the boundary between them was redrawn in the postwar period. With the status of Taiwan shifting to that of a province, Taiwanese cuisine was included among Chinese cuisines as *Tai cai* and became a counterpart to other Chinese regional cuisines.

Similar to Japanese officials' use of the term *Taiwan ryôri* during the colonial period to distinguish colonial fare from Japanese cuisine, the naming of Taiwanese cuisine was clearly based on a need for differentiation. Without such a need, the type of dish ceased to be meaningful. After the end of the colonial period and the arrival of Chinese soldiers and civilians from the mainland, and in response to the changing political situation, the concept of regional Taiwanese cuisine was transformed into *Tai cai*, a subset of Chinese cuisine. In this context, new restaurants characterizing their dishes as authentic Taiwanese dishes emerged in the early 1960s and further proliferated during the 1970s, repositioning Taiwanese cuisine on a new culinary map.

Two types of Taiwanese restaurants emerged during the 1960s and 1970s. The first type provided sophisticated banquet foods similar to those served in exclusive restaurants during the colonial era. These restaurants also provided entertainment in the form of electronic piano performances, magic shows, and Taiwanese or Japanese song performances. The atmosphere in these restaurants was different from that of mainland China–style restaurants. However, this type of restaurant declined in popularity after the 1980s, becoming increasingly unable to compete with restaurants that served less sophisticated food by using modern bulk production methods.

The second type of Taiwanese restaurant emerged in the 1960s and offered light dishes such as *qingzhou xiaocai* (清粥小菜 porridge and small dishes). By contrast with banquet dishes, these light dishes consisted of food regularly eaten in homes, such as *digua xifan* (地瓜稀飯 sweet potato porridge), *cai pu dan* (菜脯蛋 fried eggs with dried radish), *danhuang rou* (蛋黃肉 steamed pork with salted egg yolk), and *jian shimuyu* (煎虱目魚 fried milkfish). The simple porridge and small dishes were categorized as Taiwanese dishes by various restaurants and were often served at night.

Through these processes, a condensed Chinese culinary map was transplanted to Taiwan through the substantial changes to restaurants and symbolic changes in culinary representations in cookbooks. This transplantation was influential on two levels. First, some regional Chinese cuisines, particularly Zhejiang cuisine, assumed the status of haute cuisine. Mainland officers had a preference for cuisines from their hometown regions, leading to their frequent gathering at mainland China–style restaurants. By contrast, Taiwanese cuisine was marginalized and reduced to being served at public canteens and family restaurants instead of at large restaurants.[22] When the Nationalist government proclaimed itself the legitimate government of China, Chinese culture, including traditional Chinese cuisine, was presented as the legitimate culture. Second, the situation of the forced migrants resulted in a diversity of Chinese restaurants, food stalls, and snacks in Taiwan, with the hybridization of regional cuisines observable in cookbooks, markets, and restaurants. In restaurants and family kitchens, cooks learned the cooking methods and foods of other regional cuisines and applied them to local dishes. During the hybridization process, regional Chinese cuisines became localized and distinct from their original tastes.

Thus, the first level pertained to haute cuisine and presentation, with the changing meanings of haute cuisine revealing a close relationship to changes in the ruling class and concept of nationhood. The state and ruling class influenced the discourse and presentation of cuisine through policy implementation and regulation at this level. The second level is the culinary adaptation and borrowing in daily life. As was the case in Mexico,[23] common foodways can be shaped and disseminated nationwide through communal cooking and media coverage.

The hybridization of Taiwanese cuisine also demonstrates the changing relationship between Taiwanese and Chinese cuisines. During the colonial era, Japanese colonial power differentiated Taiwanese cuisine from Chinese cuisine, with Taiwanese cuisine being presented as a selection of various Chinese cuisines. However, after the end of the colonial period, the new ruling class from the mainland did not seek to present itself as appreciating and enjoying Taiwanese cuisine; local dishes were therefore not promoted as haute cuisine. Because Taiwanese culture was considered a regional variation of Chinese culture, Taiwanese cuisine was regarded as a regional Chinese cuisine. The state power differentiated Taiwanese cuisine from other regional Chinese cuisines, including Zhejiang, Beijing, and Hunan cuisines. The notion of Taiwanese cuisine was thus contained within the concept of Chinese cuisine and was not independent of it.

TAIWANESE CUISINE AS A NATIONAL SYMBOL AFTER 2000

The authoritarian regime of the Nationalist government faced serious setbacks in relation to the international community and changes in domestic society during the 1970s. The CCP had consolidated its power in mainland China, and the international community recognized the necessity of making the People's Republic of China (PRC) a legitimate member of that community. In 1971, the Republic of China (ROC) delegation, which was led by the Nationalist Party, withdrew from the United Nations, advancing the growing diplomatic isolation of the ROC.[24] When the US government, the critical partner of the Nationalist government, forged a diplomatic relationship with the PRC in 1979, the Nationalist government could no longer rely on international support to maintain its authority and claim as the legitimate government of China.

Diplomatic failure prompted the Nationalist government to seek new bases of legitimacy in domestic society, with domestic groups advocating for more liberalization.[25] Consequently, the government under the new leader Chiang Ching-kuo (1910–1988), son of Chiang Kai-shek,[26] accelerated political reform in the late 1970s, leading to a shift "from hard to soft authoritarianism."[27] With increasing deregulation during the early 1980s, key steps toward liberalization were taken in 1986 and 1987. The first opposition party, the Democratic Progressive Party (DPP), was established on September 28, 1986, and the Nationalist government began adopting an attitude of toleration instead of suppression. On July 15, 1987, President Chiang Ching-kuo lifted martial law, which had been in effect since 1949 to assert the dominance of the Nationalist government over society.

During the process of change undertaken by the political regime, indigenization was a crucial political reform policy. The indigenization or Taiwanization policy had been implemented by Chiang Ching-kuo when he became the premier in 1972.[28] He widely recruited native Taiwanese into the political system and administrative offices, exemplified by his nomination of Lee Teng-hui (1923–2020), a native Taiwanese, as vice president in 1984. Indigenization became an increasingly crucial direction of Taiwan's political development under the presidency of Lee Teng-hui, who was inaugurated in 1988.[29]

These drastic political transitions during the 1980s and 1990s cultivated an institutional and discursive environment for proclaiming the nationhood of Taiwan. In this climate, DPP presidential candidate Chen Shui-bian achieved a surprise victory in the 2000 presidential campaign, ending the nearly half century–long rule of the Nationalist Party (1945–2000). After Chen's victory, state banquets became highly charged with symbolic references to indigenization and ethnic integration. Furthermore, Hakka and Indigenous cuisines emerged as "ethnic cuisines," representing a parallel development alongside political transformation.

Two prominent characteristics of the inauguration banquet of Chen Shui-bian held on May 20, 2000 were the use of local *xiaochi* and the symbolic cuisines of ethnic integration.[30] *Shimuyu wan tang* (虱目魚丸湯 milkfish ball soup) and *wan guo* (碗粿 bowl rice cakes), two local snacks from Chen Shui-bian's hometown of Tainan, were served at the banquet, marking the first time that local snacks were served at an inauguration banquet. Media reports highlighted that local snacks were receiving a national honor insofar as they were a main course at the state banquet, with the media praising the decision as an effective elevation of the status of local Taiwanese snacks.

In addition to the emphasis on local snacks, the idea of ethnic integration was embodied in the dessert of taro and sweet potato cake because taro had been regarded as the mark of Mainlanders in Taiwan, whereas the sweet potato represented native Taiwanese. Taro and sweet potato are popular ethnic symbols, and the dessert transformed their implicit meanings into an explicit and tangible meaning, further indicating that the state banquet was a field on which national rhetoric was expressed.

After the establishment of the DPP government in 2000, Taiwanese food became highly charged with symbolic meaning, an outcome that is crucial in two regards. First, food produced in Taiwan became prevalent at state banquets, and state banquets articulated political demands regarding indigenization. With the growing Taiwanese consciousness regarding Taiwan as a distinct nation, local delicacies and snacks became representative of Taiwanese culture and assisted in building connections between the Taiwanese inhabitants and their native land.

Second, Hakka and Indigenous cuisines were framed as ethnic cuisines, echoing the categorization of the "four major ethnic groups." The concept of the four major ethnic groups highlighted Taiwan's status as a consolidated community distinct from China. This concept was integrated into food consumption practices occurring in parallel with the development of Hakka and Indigenous cuisines. The government promoted ethnic cuisines as convenient symbols through which to present specific features of ethnic groups, such as frugality and endurance for the Hakka and purity for the Aboriginals. Through the mechanism of exhibitions, certification, and other forms of propaganda, belief in these ethnic characteristics has spread among the population.[31]

With Taiwanese local dishes being shaped into political symbols, they functioned as commodities in the marketplace, and the government actively played the role of a market agent in the process. This manifested in the cooperation between local governments and grand hotels in the promotion of state banquet menus and ethnic cuisines. The government was no longer merely establishing rules and enforcing regulations. Instead, the DPP government framed local food not only as an expression of native consciousness but also as a product representing local or ethnic characteristics.

In the 2000s, the government began to promote Taiwanese local delicacies, such as *xiaochi*, to international tourists. Evidence can be found in various official propaganda and cultural policies and in surveys of foreign tourists. The act of enjoying local snacks at night markets was framed as a vital activity for tourists in Taiwan. Local snacks are a core component of the contemporary discourse on Taiwanese cuisine. This is not a coincidence but a result of the social context in which this food discourse was established, whereby Taiwanization and indigenization/popularization were already articulated together through political and cultural discourse. Consequently, the fine dishes and extraordinary cooking that had been enjoyed by the upper class and urban restaurants were ignored.

CONCLUSION

This chapter describes the evolution of Taiwanese cuisine and its relationship with Chinese cuisine, exploring how shifts in the political zeitgeist have guided its development and how it is perceived.

During the Japanese colonial era, *Taiwan ryôri* was regarded as a selection of dishes from Chinese cuisine. These selected dishes were adapted in terms of local food resources and the tastes of the most privileged Taiwanese elites and the Japanese ruling class. Shortly after WWII, the Nationalist Party fled mainland China for Taiwan, where it established its authoritarian rule. Chinese cuisine was deemed the national cuisine, of which Taiwanese cuisine was a mere component. Because the new authoritarian regime and its dominant cultural assumptions presented Chinese cuisine as reflective of the national culture, *Tai cai* was marginalized on the culinary map generated through transplantation and was only vaguely defined during this period. However, the political liberalization of the 1990s prompted a newfound emphasis on the subjectivity of Taiwanese culture, in turn fueling challenges to the idea that Taiwan is a part of China. In this context, Taiwanese cuisine came to occupy a category notably different from that of Chinese cuisine. Today, Taiwanese cuisine is often regarded as a national symbol of Taiwan and comprises Hakka and Indigenous dishes, various regional Chinese dishes, some Japanese ingredients adopted in the colonial period, and some Southeast Asian ingredients adopted as a result of the increasing migration of Southeast Asian people to Taiwan. *Taiwan cai* thus became a more inclusive term to contain all these elements of Taiwanese cuisine.

In light of the changing definitions of Taiwanese cuisine reflecting its changing relationship with Chinese cuisine, Taiwanese cuisine is a relational concept. The definition of Taiwanese cuisine serves as a boundary demarcating the dietary culture of one group from that of others. The existence of a relational concept thus presupposes the existence of others from which it must be distinguished. The birth of all national cuisines involves the process of drawing boundaries. Defining national cuisine itself is also defining what a nation is.

The transformation and uncertainty of the nationhood of Taiwan also makes "Taiwanese cuisine" a highly relational concept. As Taiwan has shifted from being a colony of Japan to being a province of China and finally to existing as a sovereign nation, the boundary of Taiwanese cuisine has changed. In the context of such an unusual geopolitical status, the making of Taiwanese cuisine has been largely influenced by the complex relationships between the political entities and the self-consciousness of the Taiwanese people.

In addition to being a relational concept, Taiwanese cuisine is a performative concept. Cuisines imbued with national symbolism often perform the critical task of highlighting their distinctiveness. In the case of Taiwan, the Taiwanese cuisine during the Japanese colonial period acquired both a form and content that distinguished the dishes of the new colony and the position of the social elites. However, in the postwar period, specific distinctions relative to Taiwanese cuisine were unnecessary, and the definition of Taiwanese cuisine became vague. As such, the definitions of Taiwanese cuisine reflect the motives of the actors wielding the power to impose such definitions.

National cuisine is also a commercial product. A sufficient consumer base is crucial for Taiwanese cuisine's viability as a commodity. Although Taiwanese restaurants were replaced by mainland China–style restaurants in Taiwan after WWII, at which time various regional Chinese cuisines entered the market along with powerful producers and consumers of these cuisines, Taiwanese restaurants reemerged in the mid-1960s and proliferated during the 1980s; at this time, Taiwan was a place of growing tourism and impressive mercantile success. In addition, the economic growth in Taiwan expanded the local consumer base of people who could afford expensive food and who wished to emphasize their social status. High-priced foods, such as seafood, were established as a main feature of Taiwanese cuisine at this time.

For Taiwanese cuisine to be well established as a national cuisine, it requires recognition from the international community. Governments, scholars, cultural agents, and catering businesses have dedicated substantial effort to promoting and exporting Taiwanese cuisine to the world. This has led to increased anxiety in Taiwanese society to tell the story of Taiwanese cuisine to the rest of the world. The discourse surrounding Taiwanese cuisine reflects the desire of Taiwanese people to situate themselves within the global system. It is also a sphere of contested discourse on identity, connected to both nation-building and globalization.

NOTES

This chapter is based on the book "Taiwan cai" de wenhua shi: Shiwu xiaofe zhong de guojia tixian (「台灣菜」的文化史:食物消費中的國家體現 Cultural history of "Taiwanese cuisine": The embodiment of nationhood in food consumption) (Taipei: Linking, 2020) and partly on the author's PhD dissertation, "Embodying Nation in Food Consumption: Changing Boundaries of 'Taiwanese Cuisine' (1895–2008)" (Netherlands: Leiden University, 2010).

1. According to the *Annual Survey Report on Visitors Expenditure and Trends in Taiwan* published by the Tourism Bureau of Taiwan, cuisine/gourmet food in Taiwan has been a top tourist attraction since 2000, superseding "historical sites" and "scenery." See: https://admin.taiwan.net.tw/businessinfo/FilePage?a=14693 (retrieved 2/8/2023).

2. Taiwan was officially ruled by the Qing Dynasty from 1683 to 1895.

3. Pintsang Tseng, "Cong tianqi dao canzhuo: Qingdai Taiwan hanren de nongye shengchan yu shiwu xiaofei" [From farm to table: The agricultural production and food consumption of the Taiwanese Han People in the Qing Dynasty] (PhD diss., National Taiwan University, 2006), 176–177, 194–198. Tseng, "Cong huating dao jiulou qingmozhi rizhi chuqi Taiwan gonggong kongjian de xingcheng yu kuozhan (1895–1911)" [From private dining halls to drinking parlors: the formation and expansion of public spaces in modern Taiwan (1895–1911)], *Chinese Dietary Culture* 7, no. 1 (2011): 89–142.

4. *Taiwan nichinichi shinpô*, January 18, 1898, 3.

5. Yujen Chen, *"Taiwancai" de wenhua shi: shiwu xiaofei zhong de guojia tixian* [Cultural history of "Taiwanese cuisine": The embodiment of nationhood in food consumption] (Taipei: Linking, 2020), 43–47.

6. Katarzyna Joanna Cwiertka, *Modern Japanese Cuisine: Food, Power and National Identity* (London: Reaktion, 2006), 118–25, 139–44.

7. Crown Prince Hirohito (April 29, 1901–January 7, 1989) was the future *Shôwa tennô* (Emperor *Shôwa*) reigning Japan from December 25, 1926 until his death on January 7, 1989.

8. It is a deep-fried dish made with turkey, spring onion, and pork that is cut in the shape of coin.

9. "Eight treasure" means to cook with eight ingredients, which are all carefully processed and finely cut. Some frequently used ingredients are black mushrooms, bamboo shoots, ham, pork, chestnuts, and peanuts.

10. Frederick J. Simoons, *Food in China: A Cultural and Historical Inquiry* (Boca Raton: CRC Press, 1991), 427–32.

11. Chen, *"Taiwancai" de wenhua shi*, 70–77.

12. Tseng, "Cong huating dao jiulou," 89–142.

13. Wu Xi-shui, "Hagaki zuihitsu Taiwan ryôri [Notes on Taiwanese cuisine]," *Taiwan nichinichi shinpô*, March 29, 1938, 3; Chen Shui-tien, "Hagaki zuihitsu Taiwan ryôri [Notes on Taiwanese cuisine]," *Taiwan nichinichi shinpô*, July 6, 1939, 6.

14. Tseng's research on agriculture and daily food of the Han people in Qing Taiwan has identified three main food patterns by analyzing the close relationship between geographical factors and dietary practices, including dryland, paddy field and mountainous terrain. See: Tseng, "Cong tianqi dao canzhuo."

15. Pintsang Tseng, "Shengzhu maoyi de xingcheng: 19 shiji moqi Taiwan beibu shangpin jinji de fazhan (1881–1900)" [Emergence of pig trade: development of commercial economy in North Taiwan (1881–1900)], *Taiwan Historical Research* 21, no. 2 (2014): 33–68.

16. Hokkien is a language commonly used in Taiwan and the southern Fujian Province of China, which was the hometown of most Han immigrants to Taiwan during the Qing Dynasty. The language is also known as *Minnan* or Southern Fukienese.

17. Pintsang Tseng, "Banzhuo: Qingdai Taiwan de yanhui yu hanren shehui" [Banzhuo: banquets and Han society in Qing Taiwan], *New History* 21, no. 4 (2010): 1–55.

18. Jin-qing Liu, *Taiwan zhanhou jingji fenxi* [Analysis of the economy in post-war Taiwan], trans. H.-r. Wang, J.-W. Lin and M.-j. Li (Taipei: Renjian, 1992), 35–57.

19. Xing-tang Chen, ed., *Taiwan "228" shijian dang'an shiliao* [Historical archives of the 228 Incident in Taiwan] (Taipei: Renjian, 1992), 63.

20. Pintsang Tseng and Yujen Chen, "Making 'Chinese Cuisine': The Grand Hotel and Chuan-Yang Cuisine in Postwar Taiwan," *Global Food History* 6, no. 2 (March 2020): 110–127.

21. Research on Fu Pei-mei, see Michelle King, "The Julia Child of Chinese Cooking, or the Fu Pei-mei of French Food?: Comparative Contexts of Female Culinary Celebrity," *Gastronomica: The Journal of Critical Food Studies* 18, no. 1 (February 2018):

15–26 and "A Cookbook in Search of a Country: Fu Pei-mei and the Conundrum of Chinese Culinary Nationalism," in *Culinary Nationalism in Asia*, ed. Michelle T. King (London: Bloomsbury Academic, 2019), 56–72.

22. Chen, *"Taiwancai" de wenhua shi*, 159–197.

23. Jeffrey M. Pilcher, *Que Vivan Los Tamales! Food and the Making of Mexican Identity* (Albuquerque: University of New Mexico Press, 1998).

24. As one of the founding members of the United Nations, the ROC government represented the Chinese seat in the UN until 1971, when the passing of Resolution 2758 shifted the representation of China to the PRC authority.

25. Yun-han Chu, *Crafting Democracy in Taiwan* (Taipei: Institute for National Policy Research, 1992); Steven J. Hood, *The Kuomintang and the Democratization of Taiwan* (Boulder, CO: Westview Press, 1997); Wakabayashi Masahiro, *Taiwan: Fenlie guojia yu minzhuhua* [Taiwan: Split country and democratization], trans. J.-z. Hong and P.-x. Xu (Taipei: Yuedan, 1994).

26. Chiang Ching-kuo became the chair of the Nationalist Party after the death of his father in 1975 and assumed the presidency in 1978.

27. Edwin A. Winckler, "Institutionalization and Participation on Taiwan: From Hard to Soft Authoritarianism?," *The China Quarterly* 99 (1984): 481–499.

28. Shelley Rigger, *Politics in Taiwan: Voting for Democracy* (London: Routledge, 1999), 111–12.

29. Chiang Ching-kuo died in January 1988, with Vice President Lee Teng-hui succeeding Chiang in the office of the president and winning election to the presidency in 1990. On March 22, 1996, Lee won the first direct presidential election in Taiwan, with his term ending in 2000.

30. Other principles of this inauguration banquet include indigenization, popularization, and environmental protection, resulting in the disqualification of such dishes as shark fin and bird's nest.

31. Yujen Chen, "Ethnic Politics in the Framing of National Cuisine: State Banquets and the Proliferation of Ethnic Cuisine in Taiwan," *Food, Culture and Society* 14, no. 3 (2011): 315–333.

9 GETTING SMASHED: DRINKING AND ETHNIC SPACE CONSTRUCTION IN A HUBEI TOURIST SPOT

XU WU

In China, until the mid-1990s, foodways were not considered essential to defining a group's ethnicity, which was typically characterized in terms of dance, music, costume, language, and history. That food did not prima facie define ethnicity was partly due to the fact that an ethnic group's representative foodways are often shared by neighboring groups in the same region. But since the early 2000s, regional foods and foodways have been adopted into the promotion of ethnic tourism, to the extent that we now cannot imagine such tourism spaces without distinctive foods and food-related practices. Regional foodways are increasingly subsumed under the name of a particular minority group, while foodways from other regions are also appropriated and "repackaged." For example, certain *yiwei* (异味 unusual flavors/foods), such as *yangbie* (羊瘪), a famous stinky soup, are now being promoted for the development of tourism among many different ethnic groups in southwest China.

Creators of ethnic tourism spots in Chinese regions and cities have utilized a sensorial approach featuring foods, including various "unusual" foodways,[1] such as stinky foods, super spicy foods, and forced drinking. This sensorial approach has encountered criticism regarding the authenticity of these "ethnic" foodways. Yet, however "inauthentic" these foodways may be, they can also help create mutually supportive relationships between an ethnic space and tourism by stimulating people's interactions and enhancing their multisensorial experiences. An established ethnic space for tourism can function as a "repackaging space," turning foodways into "ethnic foodways." In turn, these repackaged foodways can reinforce the ethnic character of a newly built tourist spot.

This study of such a "repackaging space" is mainly based on my ethnographic fieldwork in the Enshi area in southwest Hubei, central China. On February 14, 2015, three days before the Lunar New Year's Eve, I participated in a reunion dinner party with friends on the third floor of a huge restaurant in Tujia Nü'ercheng (土家女儿城 Tujia Girls' City) in Enshi city. During the party, we heard the harsh sounds of bowls smashing on the floor. Curious, I left our dining room and followed the noise. I saw groups of customers sitting around at tables in the dining room on the first floor. Ceramic fragments were scattered on the floor near their tables. Two men stood up, and one of them drank with a toss of his head. When he finished a bowl of alcohol, he shouted excitedly and threw the bowl quickly, which smashed into pieces upon hitting the floor. At intervals between the loud, smashing sounds, I could hear the singing of performers, who wore colorful ethnic costumes as they sang and danced on a stage at the front of the dining hall.

Sound features in people's evaluations of a restaurant. Normally, the soundscape needs careful management in order to bring customers a sensorial experience of comfort and liveliness.[2] Too much noise is associated with customer annoyance, bad management, and poor hospitality.[3] In contrast, in the restaurants in Tujia Girls' City, noise is deliberately performed and enjoyed by customers. This noise performance of *he shuaiwan jiu* (喝摔碗酒 drinking smashing-bowl alcohol) has become popular in Enshi and has spread to cities across Hubei and China.[4] For example, according to a May 2017 report, a restaurant in Tianmen, a city in the middle of Hubei Province, 200 km away from Enshi, claimed to "have moved the whole Girls' City to its building."[5] What the restaurant had in fact "moved" (copied) were the foodways in Girls' City, specifically a dish made from Enshi-grown small potatoes and the drinking of smashing-bowl alcohol. The latter was also introduced into Xi'an, one of the most famous tourist cities in China, where this special way of drinking had become associated with a consumption hotspot on the city's Yongxingfang Street.[6] There, customers waited in line to experience this sensory stimulus of drinking and smashing. As reported in Xi'an, customers focused on bowl-smashing rather than on the quality of the alcohol. They bought bowls filled with alcohol, drank them in front of a wall, and then smashed the bowls. Smashing-bowl alcohol has also been introduced to leisure streets and tourist sites in Yunnan, Chongqing, Henan, and Hebei.

Tujia Girls' City is considered by some to be the eighth artificial ancient town in China.[7] It was built in the early 2010s and has been open since October 2013.[8] As indicated by the name, Tujia culture is its central theme. The Tujia are one of the fifty-five ethnic minorities that were identified and officially recognized in China in the 1950s. Most Tujia people live in the vast adjacent territory of southwest Hubei, west Hunan, northeast Guizhou, and Chongqing Municipality. The Tujia have experienced tremendous growth in their population nationwide, from 5.7 million in 1990 to 8.36 million in 2010. However, since the 1950s Tujia have been well known for their weak identifiability as a separate ethnic group, being relatively indistinguishable linguistically, culturally, and physically from the Han, the majority group in China.

Consequently, the symbolic construction of Tujia ethnicity is a complex task. Some of the few Tujia markers include the history of local chieftainships, the Tujia language (limited to a handful of villages in west Hunan, where about forty thousand elder people can speak it),[9] and Maogusi dancing (limited to west Hunan, too). Compared to many other ethnic groups, the Tujia have lacked representative symbols. Among Tujia areas, Enshi has stood out for its strategies for Tujia symbolism construction, characterized by hybridism through imitation, borrowing, and appropriation.[10] New trends in Enshi's Tujia symbolism construction are emerging as Tujia spaces are created in urban areas.

The Enshi Tujia-Miao Autonomous Prefecture has a population of nearly 3.5 million, of which 40 percent has been classified as Tujia.[11] The government has backed the construction of Tujia symbols and spaces there since 1983, when the autonomous prefecture was founded. The history of Tujia symbolism in Enshi can be divided into three stages: (1) searching for symbols in rural areas of Enshi prefecture (e.g., local chieftainship, local dancing, Girls' Festival, and traditional brocade) in the 1980s; (2) developing ethnic tourism in the rural areas in the 1990s; and (3) building Tujia spaces in cities (mainly in Enshi city) since the 2010s and creating urban enclaves with a densification of Tujia symbols.

In Enshi, the symbolic construction of the Tujia ethnicity in the 1980s was guided by the socialist theory of social evolutionism, which highlighted the local area's otherness and primitivity (marked by recent practices of slavery, matriliny/feminization, under-development, and exotic customs).[12] Due to the close connection with feminization, the Girls' Festival was a focus in Enshi.

During the second stage, the Tujia symbolism in Enshi was quickly transformed and became tourism-oriented because the Chinese government wanted to increase tourism and thereby the social development of rural ethnic villages.[13] Images of the Tujia have been heavily loaded with an imagined past of a mysterious land (in one of the tourist advertisements, Enshi is called "Hubei's Shangri-La").[14] The second stage was a period of in situ construction of scenic spots in the prefecture. For example, in 1996, the author revisited a village in Enshi prefecture, which had been turned into a scenic spot of traditional stone carving. In another village, I observed a meeting between a village head and three cadres from a county bureau, who were discussing the construction of a platform for staging tourism exhibitions and folk performances in the village. This kind of in situ construction also functioned to attach the officially recognized ethnonyms (mainly "Tujia") to local traditions in the villages.

Since the 2000s, the third stage has been one of "moving" (copying and transplanting) "Tujia" symbols from their original locations to a concentrated ethnic space in Enshi city. This kind of miniaturization or Shangrilazation has been widely applied in China's ethnic areas for tourism development.[15] Chieftain City and Girls' City are the representative examples in the Enshi area.

Girls' City is the most comprehensive demonstration of the ethnic symbols constructed for the Tujia since the foundation of a Tujia-Miao autonomous prefecture in 1983. As a Tujia ethnic space, Girls' City is open to almost everything that businesspeople want to bring in. The construction of Tujia space in this artificial ancient town has followed the principles of internal orientalism, emphasizing exoticization and eroticization, familiar from other ethnic tourism spots in China.[16] Yet the construction of Tujia ethnic symbols is also innovative. One of the innovations is the newly invented practice of noise performance through smashing alcohol bowls.

REPACKAGING A TUJIA SPACE

In the past, cities in ethnic minority areas represented modernization, serving as a contrast to the less-developed rural areas. Now, however, they have also been widely influenced by touristification, ethnicization, and villagization.[17] As the capital city of Enshi Tujia-Miao Autonomous Prefecture, Enshi city now has two famous Tujia symbolic spaces, namely Tusicheng (土司城 Chieftain City) and Tujia Girls' City. Both are archaized towns. The Chieftain City has a single focal point, the local traditional building style called *diaojiaolou* (吊脚楼 hanging foot building), assumed to have been used by chieftains in the past. Tujia Girl City, in contrast, has been a mixture, a repackaged space of "local cultures" for exotification and touristification.

The local *tusi* (土司 chieftain) has served as one of the most important symbols for the Tujia ethnicity since the mid-1950s, when anthropologist Pan Guangdan conducted research in central China in order to identify the Tujia as an ethnic minority group. In the 1990s, the Enshi government began to build an artificial town with the name of Tusi City. The inscription on the gate was written by the famous anthropologist Fei Xiaotong. When this town was built, some scholars criticized it for being artificial and located in an urban area. Since the early 2000s, however, the spot has become a must-see sight for tourists, marking the beginning of the spread of Tujia symbols into the urban space.

On October 19, 2013, there was a grand opening ceremony for Tujia Girls' City, the second "ancient town" in Enshi city. The first time I visited it was in 2015. I had assumed it was a new commercial area in the city until I heard someone say that this was a tourist spot. In contrast to Chieftain City, Girls' City was developed by nongovernmental capital. Heung Kong Group, a private enterprise from the Pearl River Delta with a strong background in building business centers throughout the country, invested 5 billion RMB in this project. Through integrating commerce and local culture exhibitions, Girls' City emerged as a repackaged ethnic space, consisting of multi-themed subspaces: stages for the performance and exhibition of the Girls' Festival, the streets of Tujia-Miao Customs, the Water Park, and areas of restaurants, stores, hotels, and residential buildings. To distinguish it from other theme parks and leisure streets, Girls' City highlights its Tujia affiliation. For example, the restaurants in the pseudo-old town sell local foods (daily meals, dishes from rural community banquets, smoked pork, snacks, and so on), in addition to popular foods from outside. In the district for folklore performance, consumers can watch and enjoy a traditional wedding ceremony, folk songs and dance, and other forms of heritage, such as traditional silverware, handcrafts, and textile machines.

Girls' City has been praised by several local scholars and officials as a perfect Tujia space, a successful example of a cultural product. It can be viewed as a displacement of the *nü'erhui* (女儿会 Girls' Festival), which in the local discourse has been tightly connected to Tujia identity. The Girls' Festival was the cultural background on which the whole artificial town was based. For example, many performances in the town are related to this festival: one piece of grassland is called *Xiangqin Caoping* (相亲草坪 dating lawn), and one hotel is directly named after the Girls' Festival—*Nü'erlou* (女儿楼 girls' tower).

The Girls' Festival, usually held on the twelfth day of the seventh lunar month, originated in Shiyao and other highland villages in the eastern part of Enshi prefecture. This area was not settled until the late Ming dynasty (1368–1644) and early Qing dynasty (1644–1911). Shiyao was originally called *Shigepeng* (十个棚 Ten Sheds), which referred to the ten families living in these highland areas.[18] A local chronicle called the *Jiayong zazhi* (家用杂志 Jottings for household demands) records the history of the ten families (Zhang, Xue, Li, Teng, Yang, Tian, Qin, Cao, Qian, Wang) in Shiyao, explaining that they were migrants from provinces across southern China. These pioneer settlers could occupy this "no-man's land" freely, by tying knots of grass as boundary markers and building sheds as shelters.[19]

The Girls' Festival may have emerged among the residents of this remote mountain area to provide an opportunity to exchange resources and meet with potential spouses. For the Ten Sheds people, it had been customary for parents to arrange marriages for their children. Later, according to legend, one local gentry man from the Xue family, returned from a long journey on the eleventh day of the seventh lunar month, a time when local families were busy celebrating the Ghost Festival.[20] Mr. Xue was warmly welcomed home by his wife with a wonderful dinner party. During the party, he became drunk and called for his nine daughters. He told his daughters that tomorrow would be a special day and that they could all go to the village fair. The girls were excited, because usually young girls were not allowed to go outside. On the following day, they dressed up and spent the day wandering through the streets of the village fair. Henceforth, the twelfth day of the seventh lunar month became a special festival for local girls, who would dress up in brand-new clothes. In the company of daughters from other families, they would bring medicinal herbs and other products to sell at the village fair. This festival provided a unique opportunity for young men and women to meet in a sanctioned way before marriage and became known as *nü'erhui* (女儿会

meeting of/with girls). This festival continued until the late 1960s when the Cultural Revolution (1966–1976) broke out. One local cadre said that it was the Girls' Festival in 1953 that helped him meet his wife.[21]

Today, the Girls' Festival has become the cultural basis for a newly repackaged commercial space—Tujia Girls' City. In this "ancient" town, there are several spaces dedicated to the Girls' Festival and its related themes. Close to the entrance, there is a stage for the performance of the Tujia wedding ceremony. Adjacent to the stage is a street called *Xiangqinjiao* (相亲角 matchmaking corner) with a wooden archway called *Ganchang Xiangqin* (赶场相亲 going to the fair and dating). Before 2013, tourists had to go to the remote mountain villages in Shiyao to observe or participate in the Girls' Festival in the seventh lunar month. Now, the pseudo-old town can stage theatrical pieces representing Tujia dating and wedding ceremonies almost daily. In this town, dating and wedding customs are also exhibited in the form of photos and written documents on *Feiyi Wenhuajie* (非遗文化街 Intangible Heritage Cultural Street) and in the *Tujia Minsu Bowuguan* (土家民俗博物馆 Tujia Folklore Museum). Local notable restaurants such as Barentang Restaurant also stage smaller-scale dating and wedding performances.

Choosing "Girls' City" as the name for this town provides one more example of how ethnic minorities in China are eroticized and feminized.[22] This feminization strategy can help exoticize the space of this town, but it hardly makes it unique in China's domestic tourist markets, since ethnic-tourist spots in China are widely equipped with similar themes and performances. Moreover, the dating and wedding spaces in Girls' City are generally outshone by the food spaces in the adjacent streets.

FOOD IN THE TUJIA SPACE

In China, food, marketing, and ethnicity have been entangled multidimensionally.[23] Walking in the streets of Girls' City, one can find that most of the stores are food related. There are several streets with stands or shops for selling local specialty foods, a feast for both the stomach and eyes. At a street near the stage for the wedding ceremony performances, there are various kinds of local cakes, including *mibaba* (米粑粑 cake made of fermented rice), *baogu baba* (包谷粑粑 cake made of tender maize wrapped with *tong* tree leaves), *ciba* (糍粑 cake made of sticky rice for the lunar New Year), and a type of nostalgia *bingzi* (饼子 cake) from the 1970s. Several stores sell *taopian gao*

(桃片糕 walnut cake) and demonstrate how they make it. The freshly made walnut cake is packaged and handed over quickly so that the customers can feel the warmth of the cake.

At a crossroad, there are performances of making *ciba* with sticky rice. This cake is made by rural people for the Chinese lunar New Year. Two young performers, wearing Tujia traditional clothing, put steamed sticky rice into a mortar and pound it with pestles until the rice becomes mash. One elderly performer removes the mash from the mortar to a large square table, on which he divides the mash into several small pieces. Then, other performers carry another similarly sized table and turn it upside down to press these small pieces of sticky rice into the form of a cake. When pressing the cake, the performers frequently invite the audience (tourists) to step on the table and tread on it heavily.

The snack street is a long street with several dozen small food stands and shops. This street was tightly packed with customers every time I conducted observations there. One can easily get full after going through just a few stores. On this street, tourists can taste famous local regional snacks, such as *youxiang* (油香 deep-fried rice-soybean-meat dumpling), *bagu baba, shaobing* (烧饼 cake baked over a charcoal fire), *kang yangyu* (炕洋芋 baked small potato), and so on. However, none of these local specialty foods can serve as straightforward markers of Tujia ethnicity, as all are typical of the region and common to all ethnic groups there. Moreover, many of the street stands sell nonlocal famous snacks, including *malatang* (麻辣烫 spicy hot pot), *mifen* (米粉 rice noodles), *shaokao* (烧烤 barbecue), *boboji* (钵钵鸡 Bobo chicken from Chengdu, Sichuan Province), *roujiabing* (肉夹饼 meat-filled bun from Shaanxi Province), and *choudoufu* (臭豆腐 stinky tofu from Changsha, Hunan Province), to name a few.

The eclecticism of the street food culture has rendered the specifically Tujia elements invisible in the foods consumed by tourists and other customers. One tourist I met thought that Girls' City lacked distinctive ethnic characteristics. Another of my interviewees considered the town a good night market. Another tourist thought there were some Tujia features hidden somewhere in the town, but one needed to be mindful in order to discover them. Even I didn't realize it was a tourist zone with Tujia characteristics when I participated in a dinner party in Girls' City for the first time, until I later observed the "small hill" made of broken alcohol bowls in the yard of a

FIGURE 9.1
A pile of broken bowls at Girls' City, Enshi. Photo by the author.

restaurant and heard the sound made by customers who drank and smashed bowls (figure 9.1).

ALCOHOL IN THE TUJIA SPACE

In the anthropology of food,[24] perspectives toward alcohol have been divided into two general schools of thinking: the conflicting and the integrative, as alcohol can cause social problems (individual pathology, alcoholism, or addiction) but can also be constructive (constructing collective identity). Alcohol and drinking set up boundary markers in terms of, for example, age, gender, class, lineage, profession, ethnicity, religion, regionalism, and nationalism.

In China, alcohol and drinking have been closely associated with ethnic cultures and tourism,[25] and have been deployed by both insiders and outsiders as ethnic boundary markers.[26] Along with *yiwei* (异味 unusual foods),[27] alcohol has been used by scholars and others to construct ethnic minorities as backward, primitive, and exotic. This has been done with reference to the raw ingredients used by minority groups for making alcohol, to tastes and smells, and/or to drinking practices (e.g., drinking with straws or through the nose, or the practice of proposing toasts in song). All these points of difference serve to further buttress the ethnic stereotype of backwardness and otherness. For example, the Wa people are closely associated with their *shuijiu* (水酒 rice beer[28]), Tibetans with *qingkejiu* (青稞酒 highland barley wine), and Mongols with *manaijiu* (马奶酒 kumis). Miao people in southeast Guizhou Province have the practice of drinking *lanmenjiu* (拦门酒 blocking-the-door alcohol) and *niujiaojiu* (牛角酒 drinking alcohol from bull horns). Shui people in Guizhou have a way of drinking that involves dropping pig bile into the alcohol. Qiang people in Sichuan Province have a unique alcohol consumed with long straws, which was locally called *zajiu* (咂酒 sucking alcohol). Dulong people in Yunnan Province have *zhutongjiu* (竹筒酒 a special alcohol contained in bamboo tubes), while Hani people in the same province have a custom of feasting and drinking in the middle of the village square together, which is called *jiexinjiu* (街心酒 middle-street alcohol).[29] Also in Yunnan, Lisu people drink a special alcohol known as *cujiu* (醋酒 vinegar alcohol).[30]

The provinces of Guizhou and Yunnan, which have a high concentration of minorities, have a reputation for having heavy drinkers and being good at persuading guests to drink—forced drinking with songs. Many tourists say that they have been warned not to drink at the invitation of ethnic minorities in southwest China, as they could become drunk easily or fail to understand the meanings associated with specific alcohols or manners of drinking.

Tourists often cannot distinguish between the alcoholic drinks of different ethnic areas. In Enshi prefecture, however, there is nothing especially distinctive about the local alcohol. According to some ancient records, there had once been a drinking practice called *zajiu* (咂酒 sucking alcohol, like the Qiang people have). It is said that this *zajiu* was made with fermented cereals, such as millet or sorghum, and stored in an earthen jar. When drinking *zajiu* alcohol, several people drank together with long drinking straws (small bamboo drains). But this practice disappeared many years ago and is believed to have been replaced by local spirits such as maize alcohol.[31] Villagers in the Enshi area are proud of this local maize alcohol. However, for many mountain areas and many different ethnic groups residing in different provinces (e.g., Hunan, Guizhou, Yunnan, and Guangxi), maize alcohol is also one of their daily necessities. For this reason, maize alcohol cannot serve as a unique ethnic marker differentiating the Tujia ethnicity.

GETTING BOWLS SMASHED

Though described by today's web writers and commentators as an important Tujia symbol, the practice of drinking smashing-bowl alcohol has never been recorded anywhere in local chronicles or other historical materials in Enshi. Nor is it mentioned in official publications since the 1980s concerning the Tujia ethnicity in the prefecture. None of my interviewees during fieldwork had heard about the practice before. Several thought that it sounded bizarre and wasteful as it involved destroying a bowl after drinking. Two interviewees simply called this way of drinking *weiminsu* (伪民俗 fakelore). One retired official from Enshi city, who had a rich experience of drinking, recalled that the first time he heard about the practice called "smashing bowls" was in 1998. At this time, the local government began to promote *nongjiale* (农家乐 farmhouse joy) tourism in rural villages, with restaurants serving farmers' foods being the key selling point. In the late 1990s, villagers in Wufengshan village (in Enshi city) got funding and opened several farmhouse joy restaurants. One of their unusual practices was smashing bowls alcohol. This retired cadre did not know what meanings were associated with this practice at that time. He just thought customers who could smash bowls must have been *erganzi* (二杆子 flaunting—but also stupid—persons, in local common saying). As to the relationship between this style of drinking and Tujia identity, he said he had never heard about it before.

Mr. Lin, who was born and raised in a mountain village in Enshi, had been a director of the general office of a government bureau in Enshi city for many years (this job involved a lot of liquor-soaked dinner parties). He claimed to be an expert on local alcohol culture since he liked drinking and had many opportunities to drink with his *jiuzhuo pengyou* (酒桌朋友 drinking friends). When I asked who invented this practice or from whom they had learned it, Mr. Lin thought that it was a myth. He said, half-jokingly, that this practice might be traced back to him and his friends in the 1990s, because he had never previously observed this practice anywhere in rural or urban areas of Enshi. At that time, he and his friends started to smash things after drinking a cup of alcohol at restaurant dinner parties. Initially, it had been small plates or dishes. He did not know exactly when the restaurant owners began to provide a kind of *tuwan* (土碗 locally made bowls) to replace cups for drinking and smashing.

One of Mr. Lin's friends related a story of a lively evening a couple of years earlier, when she had entertained her relatives from Chongqing at a restaurant in Girls' City. Although she could not recall what they ate or what kind of alcohol they drank, she still clearly remembered the harsh noise, laughter, and screams in the dining room. She said that the restaurant prepared a basket of bowls for them to smash. In the middle of dinner, she took a short video when one of the guests (her relative) proposed a toast. In the video, people all stood up and drank a full bowl of alcohol. Subsequently, all of them smashed their bowls. One of her relatives failed to break their bowl, and this guest quickly picked the bowl up and dashed it to the floor for a second time. Mr. Lin interrupted and explained that the proper way to deal with unbroken bowl was to ask the guest to drink another bowl of alcohol before smashing it again.

Mr. Lin said that the drinker who proposed a toast needed to drink and finish a bowl of alcohol first to demonstrate that there was no alcohol left in the bowl. Then they (as toaster) smashed the bowl behind their back with their right hand, while looking directly and attentively at the other drinkers. Under the toaster's gaze, the other drinkers had to finish their alcohol, an accomplishment finally marked by the harsh noise of smashing bowls. Mr. Lin also introduced what might be considered a beautiful way of smashing bowls. For a beautiful smashing, one needed to drink and empty the alcohol in the bowl first. Then he would hold the bowl upside down with three fingers to demonstrate that there were no liquids dripping from the bowl. After that, he just threw the bowl over the top of his head and let the bowl arc gracefully over his head. Finally, the bowl hit the floor with a great crash.

When I asked whether there was a connection between smashing bowls and Tujia ethnicity, Mr. Lin thought that such a relationship might exist. He told me that one of his table friends had once tried to verify it. This unverifiable connection hasn't prevented businesspeople from splicing the two "broken ends" together. According to local restaurant posters as well as stories spread widely online through mass media, there are three Tujia origin stories explaining this style of drinking. The first version involved Bamanzi, a legendary leader of the ancient Ba tribe living in today's Enshi area. The story took place in the ancient Zhou dynasty, when Bamanzi, a general of the Zhou dynasty, asked the Chu kingdom for military reinforcement and support. The Chu king stipulated his terms, which included getting three cities from the Ba territory. Later, after the Chu kingdom helped Bamanzi get through the crisis, the Chu kingdom demanded the three cities. Bamanzi would not cede any city to Chu, so he decided to give his life instead. Before committing suicide, Bamanzi drank a bowl of alcohol and smashed it. Then he cut off his own head as an expression of his appreciation for Chu's help. One person commented in his online short video about this origin: "Bamanzi was a person with *haoshuang* (豪爽 forthrightness)." The second version was related to the military ritual of ancient soldiers who fought for local chiefs. A chief was described as someone who could mobilize his soldiers by drinking alcohol. Once the soldiers had finished drinking, all of them smashed their vessels. The third origin story was a variation of the second one. It said that a local chief had two sons, and the two brothers grew up resenting each other. At the time of a war mobilization, the two brothers drank a bowl of alcohol together and smashed their bowls after finishing drinking. This activity symbolized that they had broken with the past and become reconciled.

These stories have given rise to several meanings added to the practice of smashing bowls, such as demonstrating a person's heroic spirit and friendship. Other meanings were provided by consumers I interviewed, including *suisui ping'an* (岁岁平安 wishing peace and good luck throughout the year). Mr. Lin confirmed that when you drank and smashed the bowl, you could demonstrate your character of *haoshuang*. He said that he and his friends smashed bowls because they want to demonstrate this trait; "We are people full of *haoshuang*." Lin believed that Tujia people shared this characteristic of forthrightness.

This new popular practice has received many comments from consumers and observers. First, even though so many people smash bowls in the restaurants, the meanings associated with bowl-smashing were not very clear to them. Some people just considered the invention of this fakelore as a thoughtless design, which one middle-aged man called a bit ridiculous. Quite a few commentators pointed out that before the 1990s, bowl-smashing after drinking had in fact already existed in China, in association with the special moments of death, such as before going to the battlefield or being executed. Smashing bowls symbolizes the saying of goodbye to life (as dead people no longer need bowls). Some suggested that smashing bowls was akin to "making a wish," with the wish being expressed as they smashed the bowls. One consumer in the Girls' City believed that there was a tiny difference between bowl-smashing and cup-smashing, as the former was toward oneself while the latter toward others. Some people even criticized the moral aspect of consumers as they believed that people who drank this way were those who lived an empty life.

Finally, some interviewees deemed drinking smashing-bowls alcohol to be vulgar, polluting, and dangerous. Quite a few commenters worried about the pollution resulting from this practice, as the dregs of broken bowls cannot decompose easily into soil. One lady said that she had to wear thick and heavy shoes when eating at the restaurants in Girls' City. One restaurant prepared detailed instructions for customers about how to smash bowls safely and warned customers that they would be responsible for any injury caused by their smashing.

Despite these comments, people agreed that during dinner parties they needed to follow certain rules concerning smashing bowls and performing noise in restaurants in Girls' City. Obviously, smashing bowls and performing noise have been integrated into the alcohol culture there, which have become ritualized practices for forging sociality.[32] Across the country, toasts and counter toasts are one of the features of Chinese banquets, which are guided by certain "prescribed conventions."[33] Moreover, performing noise is important as it is closely related to the successful production of *renao* (热闹 heat-noise) and people's sensory expression and sensory consumption of a social event.[34] In this sense, the newly invented custom of smashing bowls has been integrated into a fully-fledged "Tujia" tradition.

SMASHING BOWLS FOR THE PRODUCTION OF "HEAT-NOISE" (*RENAO*)

In China, alcohol drinking has long played a crucial role in producing "heat-noise" (热闹 *renao*, which might be translated as a lively hubbub bustling with noise and excitement).[35] Activities of making "heat-noise" in drinking such as *quanjiu* (劝酒 pressuring others to drink), *jingjiuge* (敬酒歌 toasting songs), and *caiquan* (猜拳 finger-guessing games) are well known nationwide, while the practice of drinking and smashing bowls has until recently remained unknown.

The introduction of the "hot and noisy" practice of smashing-bowl alcohol has effectively established Girls' City as a lively Tujia ethnic space for consumers. It integrates alcohol's destructive and constructive aspects and brings the urban night life into the space of this "ancient" Tujia town. Through making a soundscape full of harsh noise, smashing-bowl alcohol helps the Tujia ethnicity extend into the night-space, after the daytime staged performances have ceased.

Many restaurant staff and customers I met at Girls' City treated this invented custom as an authentic Tujia tradition. They drank and persuaded others to drink, and then smashed bowls following "traditional" rules. The sensory-rich experience of smashing-bowl alcohol strengthened the associations consumers made between Girls' City and "Tujia culture." It helped to integrate the diverse features at the site—architecture, textiles, handcrafts, foods, teahouses, bars, amusement activities, and performances—regardless of whether their origins were modern or traditional, ethnic, or mainstream. The connections of these features with Tujia culture have now been accepted by many consumers, despite some unfavorable comments.

CONCLUSION

Girls' City concentrates Tujia symbols identified or created in the Enshi area at different moments, together with symbols appropriated from other ethnic minority areas. With the help of foodways, it has been repackaged into a multisensorial space, where people use their bodies to gaze, touch, smell, move, hear, and exchange feelings. Therefore, it is not only businesspeople who have developed this tourist site. Through the heat-noise generated by smashing-bowls alcohol in the Girls' City restaurants, tourists and other consumers have been important contributors to the construction and maintenance of this multisensorial space. Consumers not only consume what they expect to experience in this town, they also provide noise or screams for other people to feel satisfied in their experiences. With continuous harsh noise, combining cracking sounds with people's laughter and screams, at the restaurants in this "ancient" town, people—locals and outsiders, customers and performers—have worked together to repackage and maintain an ethnic tourism space.

Obviously, Girls' City is also a contested space. There is a tension between the providers/consumers of Tujia foodways and critics of the site's claims to authenticity. Interestingly, several critics I met who joked about the foodways' authenticity in Girls' City were simultaneously eager consumers of this space. This is largely related to the repackaging function of this established ethnic space. The repackaged, noisy performance can bring people, both hosts and guests, an immersive, multi-sensorial experience that helps build friendships and *guanxi* (关系 social relations). *Guanxi* is a social resource highly valued by contemporary Chinese people, especially urbanites who have lived through dramatic social changes and uncertainties. In short, "smashing-bowl alcohol" has been appropriated and repackaged as both premodern (as a symbol associated with ethnic minority) and modern (as a very modern kind of Chinese foodway, a response to rapid modernization and urbanization).

NOTES

1. Xu Wu, "Ethnic Foods as Unprepared Materials and as Cuisines in a Culture-Based Development Project in Southwest China," *Asian Ethnology* 75, no. 2 (2016): 427.

2. Jochen Steffens, Tobias Wilczek, and Stefan Weinzierl, "Junk Food or Haute Cuisine to the Ear?—Investigating the Relationship Between Room Acoustics, Soundscape, Non-Acoustical Factors, and the Perceived Quality of Restaurants," *Frontier in Built Environment* 7 (2021): 1.

3. Flora Dennis, "Cooking Pots, Tableware, and the Changing Sounds of Sociability in Italy, 1300–1700," *Sound Studies* 6, no. 2 (2020): 174–195.

4. http://www.5h.com/yl/118216.html (accessed January 23, 2022).

5. https://www.sohu.com/a/143196317_196076 (accessed October 1, 2021).

6. https://www.sohu.com/a/240131399_99924162 (accessed October 1, 2021).

7. Many "ancient towns" have been created in recent years in China, such as the Song Town in Hangzhou, Zhejiang Province; Taierzhuang in Shandong Province; Qingming Shanghe Yuan in Kaifeng, Henan Province; Yuanmingyuan in Zhuhai, Guangdong Province; Luanzhou ancient town in Hebei Province; and Furong Garden of the Tang dynasty in Xi'an, Shaanxi Province.

8. http://news.focus.cn/enshi/2013-10-20/4164715.html (accessed October 10, 2021). In Enshi, there is also a private company named Enshi Girls' City Cultural Tourism Co., Ltd. The city government uses "Enshi Girls' City" as a brand to promote tourism.

9. https://www.zhihu.com/question/40014570 (accessed October 1, 2021)

10. Xu Wu, "The Farmhouse Joy (Nongjiale) Movement in China's Ethnic Minority Villages," *The Asia Pacific Journal of Anthropology* 15, no. 2 (2014): 158–177.

11. Melissa Brown, "Ethnic Classification and Culture: The Case of the Tujia in Hubei, China," *Asian Ethnicity* 2, no. 1 (2001): 55–72.

12. Stevan Harrell, *Ways of Being Ethnic in Southwest China* (Seattle: University of Washington Press, 2001); Stevan Harrell and Li Yongxiang, "The History of the History of the Yi, Part II," *Modern China* 29, no. 3 (2003): 362–396.

13. Jenny Chio, *A Landscape of Travel: The Work of Tourism in Rural Ethnic China* (Seattle: University of Washington Press, 2014).

14. https://www.meipian.cn/2aysajte (accessed March 3, 2022).

15. Chris Vasantkumar, "Dreamworld, Shambala, Gannan: The Shangrilazation of China's 'Little Tibet,'" in *Mapping Shangrila*, ed. Emily Yeh and Chris Coggins (Seattle: University of Washington Press, 2014), 51–73.

16. Louisa Schein, "Gender and Internal Orientalism in China," *Modern China* 23, no. 1 (1997): 69–98; Dru C. Gladney, "Representing Nationality in China: Refiguring Majority/Minority Identity," *The Journal of Asian Studies* 53, no. 1 (1994): 92–123; Eileen Walsh, "From Nü Guo to Nü'er Guo: Negotiating Desire in the Land of the Mosuo," *Modern China* 31, no. 4 (2005): 448–486.

17. Tim Oakes, "Villagizing the City: Turning Rural Ethnic Heritage into Urban Modernity in Southwest China," *International Journal of Heritage Studies* 22, no. 10 (2016): 751–765.

18. Shed people were migrant farmers in the region. Stephen Averill, "The Shed People and the Opening of the Yangzi Highlands," *Modern China* 9, no.1 (1983): 84–126.

19. Exi zhouminwei, *Exi shaoshu minzu shiliao jilu* [Collection of historical materials on minorities in western Hubei] (Enshi: Ethnic Affairs Committee of Exi (Enshi) Prefecture, 1986).

20. The whole seventh lunar month was considered the time period for ghosts to return to their homes.

21. Exi zhouminwei, *Collection of historical materials*.

22. Walsh, "From Nü Guo to Nü'er Guo."

23. Jakob Klein, "Heritagizing Local Cheese in China: Opportunities, Challenges, and Inequalities," *Food and Foodways* 26, no. 1 (2018): 63–83; Zhen Ma, "Sensorial Place-Making in Ethnic Minority Areas: The Consumption of Forest Puer Tea in Contemporary China," *The Asia Pacific Journal of Anthropology* 19, no. 4 (2018): 316–332; Megan Tracy, "Pasteurizing China's Grasslands and Sealing in 'Terroir,'" *American Anthropologist* 115, no. 3 (2013): 437–451.

24. Michael Dietler, "Alcohol: Anthropological/Archaeological Perspectives," *Annual Review of Anthropology* 35 (2006): 229–249.

25. Xu Zongyuan, "Zhongguo jiu wenhua yu lüyou" [Alcohol culture and tourism in China, *Jianghuai luntan* 4 (1993): 79–86; He Ming and Wu Mingze, *Zhongguo shaoshu minzu jiu wenhua* [The alcohol culture of ethnic minorities in China] (Kunming: Yunnan renmin chubanshe, 1999).

26. He and Wu, *Alcohol culture*.

27. Xu, "Ethnic Foods"; Susan D. Blum, *Portraits of "Primitives": Ordering Human Kinds in the Chinese Nation* (Lanham, MD: Rowman & Littlefield, 2001).

28. Magnus Fiskesjö, "Participant Intoxication and Self-Other Dynamics in the Wa Context," *The Asia Pacific Journal of Anthropology* 11, no. 2 (2010): 111–127.

29. Xu, "Alcohol Culture;" He and Wu, *Alcohol culture*.

30. https://www.sohu.com/a/211622612_535990 (accessed October 1, 2021).

31. Exi zhouminwei, *Exi zizhizhou minzu zhi* [Chronicle of ethnic groups in Exi Autonomous Prefecture] 91 (Chengdu: Sichuan minzu chubanshe, 1993).

32. Fiskesjö, "Participant Intoxication," 115.

33. James L. Watson, "From the Common Pot: Feasting with Equals in Chinese Society," *Anthropos* 82, no. 4 (1987): 390.

34. Adam Yuet Chau "The Sensorial Production of the Social," *Ethnos* 73, no. 4 (2008): 495.

35. Chau, "Sensorial Production," 493.

10 GASTROGRAPHISM IN CONTEMPORARY CHINA: FROM A LITERARY PASTIME TO A PROFESSIONAL ACTIVITY

FRANÇOISE SABBAN

KNOWING AND WRITING ABOUT TASTE

In Chinese history, many poets and writers have sung about the deliciousness of vegetable and meat dishes.[1] Some have even attached their name to a dish and immortalized it.[2] The archetypal model of this is, of course, Yuan Mei, whose treatise *Suiyuan shidan* (随園食单 *Recipes from the Garden of Contentment*) was published in 1796. Yuan Mei was a literary genius who "had fun" writing his recipe book, though he has been more praised for his literary and poetic work than for this peculiar text dedicated to cuisine and regulations on food preparation. In a different and more modern way, Lin Yutang (1895–1976) holds an important place as the author of a short text on the differences between Chinese and "Western" (mainly British) eating and cooking habits, in his book *My Country and My People* (1935). With humor tinged with irony toward the supposed habits of his Western readers, who are suspected of understanding nothing about the pleasures of life (unlike the Chinese), Lin describes the greedy habits of the Chinese, their eclecticism toward food, and their pleasure in talking about good food. Other early twentieth-century authors, such as Zhou Zuoren (1885–1967), the younger brother of the famous writer Lu Xun (1881–1936), agreed that taste is the very essence of culinary perfection and believed that the Chinese know how to appreciate this better than anyone else.

These books dealing with food love were generally anthologies composed of brief journalistic chronicles recounting an episode in the author's life in which a dish, a food, a taste, and so on, played a more or less important role. These authors often presented their views through nostalgia, humor, and comparison with Western habits. This literary vein, which could be considered a practice of *biji* (笔记 jottings) writing, was very much exploited by some journalists and writers in Taiwan in the 1970s, such as the renowned lexicographer, translator, writer Liang Shiqiu (1903–1987) and the emblematic Tang Lusun (1908–1985), author of at least a dozen volumes on his first gastronomic adventures in China before 1949.

Around the same time, after the economic reforms launched in the PRC at the end of 1978, China experienced a renaissance of the agri-food sector and an immense craving for good food. That resulted in many restaurants opening, from the simplest cafeterias to high-end gastronomic temples. This movement also resulted in the proliferation of a large number of publications dedicated to China's food, cuisine, culture, and history. Some of the works written in Taiwan, such as those by Liang and Tang, were republished in the PRC for a new readership. With these publications, food as a literary object regained a particular reputation. Other writers in the PRC added their voices to this conversation, as shown by the well-known writer Wang Zengqi (1920–1997) and his son Wang Lang, who together published a book on food and delicacies in 2003.[3] Wang Zengqi would become a major reference point in the reemergence of this form of culinary literary expression at the beginning of the twenty-first century.

How are we to understand this reemergence of culinary aestheticism and professionalized food writing in China's post-Mao reform era? On the one hand, it seems to be a continuation of earlier patterns of expression and on the other to be the symptom of a new style of life, resulting from radical social and economic transformations. In the premodern era, the public understood that gourmet authors came from good families. Their elite social backgrounds provided them with the means to enjoy material wealth, obtain the best foods, and savor excellent cuisines. For instance, Yuan Mei had a private cook. Thanks to their social advantages over two or three generations, their ability to appreciate and judge tastes naturally developed, making them innate tasters. Yet eating was only the first step: in the Chinese case, traditionally, it was the aesthetic act of fixing culinary experiences on paper through writing and sharing these writings with other literati that was the ultimate demonstration of taste.

For those self-anointed Chinese arbiters of taste emerging in the post-Mao reform era, what has been their claim to culinary legitimacy within a new context of marketization, branding, and increasing globalization? What can their debates about the meaning and self-definition of the neologism *meishijia* (美食家), first coined in 1983 and often translated in English as "gourmet," tell us about the broader impact of economic reforms on culinary discourse? In the remainder of this chapter, I argue that this reform-era rebirth of a professionalized gastronomic discourse not only aimed at reshaping Chinese consumers' tastes in a globalized era of dining but it also represented the creation of a new modern culinary identity, that of the professional culinary critic, which can be assumed by an individual of any background (no longer just elites), especially with the rise of social media.

FROM AESTHETICS TO EXPERTISE

The social transformations brought about by the post-1978 economic reforms enabled the Chinese to overcome the difficulties of daily provision: not only could they now eat their fill but they could also eat well, and sometimes even better than well. How did these economic changes influence the way food writers chose to write about tastes and food? One of the signature markers of this evolution is certainly Lu Wenfu's (1925–2005) novel, *Meishijia* (美食家 *The Gourmet*), published in 1983.[4] His novel features two opposing characters: the first, Zhu Ziye, whose only guide in life is the satisfaction of his gluttony, represents the old, feudal society. The second, Gao Xiaoting, is an employee of Zhu Ziye, yet he hates eating, is a convinced revolutionary, and one of his main objectives is to make his boss bite the dust. The plot is a subtle allegory of the troubles, disorders, and horrors of the political movements that affected China after 1949, in particular the Cultural Revolution (1966–1976). Neither of the two characters emerge unscathed over the course of the novel.

The title of this novel, *Meishijia*, attracts attention because although the word *meishi* (美食) in the sense of "choice food" is well documented in classical and modern texts, the term *meishijia* was heretofore unknown, as the well-known professor of aesthetics and literary criticism, Zhao Xianzhang, points out.[5] In today's Chinese dictionaries, etymologies refer to the novel of Lu Wenfu, indicating that the term did not exist before 1983. In modern Chinese, the suffix *-jia* (家) designates a specialist in a field of knowledge or given profession, such as *huajia* (画家 painter), *zuojia* (作家 writer), or *kexuejia*

(科学家 scientist). Thus, with this suffix, Lu Wenfu transforms the simple *meishizhe* (美食者 connoisseur of fine food) into a real "food scholar," or "food professional," literally deserving of the title "expert in gastronomy." The narrator of Lu's novel claims, for example, that *"pengren xue"* (烹饪学 the study of cooking) belongs to science. However, the English translator of the novel does not highlight this aspect of studied expertise, preferring to translate its title as *"The Gourmet."* (Perhaps the word "gourmet" brings a more literary flavor to the fiction than "expert in choice food"!)

Whether Lu Wenfu was influenced by his knowledge of French cuisine in creating the word *meishijia* is difficult to determine. He imagines his character, Zhu Ziye, establishing an International Culinary Association, whose chairman would be a Frenchman assisted by Zhu as vice-chairman. And yet, Lu Wenfu's inspiration to invent the neologism *meishijia* likely did not come directly from the French word "gastronome."[6] In French, "gastronomie" and "gastronome" are relatively recent words and date only from 1801 and 1803, respectively.[7] The French lexicon is rich in qualifiers designating greediness and gourmands. The appearance of these new terms testifies to a new social fact at the beginning of the nineteenth century: the steadfast commitment by greedy men to the pleasures of eating good food and the satisfaction of their appetites. Nevertheless, this specific expertise developed by a few privileged people did not stop at their plate—they wanted to share it, and they expressed it through writing intended for the pleasure and education of their readers. In the same way, according to Zhao Xianzhang, the Chinese neologism *"meishijia"* came at an apt time in modern Chinese culinary history. It diffused through society during a moment of renewed attention to cuisine, taste, and aesthetic judgment, after decades of political upheaval.

Against the backdrop of economic reform, Zhu Ziye, the anti-hero and as gourmand as ever, epitomizes the term *meishijia*. By the end of the novel, he becomes a real professor of gastronomy appreciated for his lectures and advice on culinary matters, from which he even manages to gain some financial profit! This new conception of the food lover and gourmand separates the hero of Lu Wenfu's novel from his predecessors, that is, talented writers belonging to the upper class of society. Zhu Ziye is certainly an heir but a low-class heir. He is the son of a Shanghai trader who made his fortune in real estate. In a way, Zhu was a parvenu who had the means to satisfy his appetite fully. After many ups and downs, including his financial setbacks due to the Cultural Revolution, from 1980 onward, he returned to his old life filled with laziness and food experiments that finally allowed him to earn a living. The inveterate, greedy Zhu Ziye claims another skill than the mere expertise provided by food science: he can show knowledge of tastes formed by accumulating eating experiences.

However, this application of the term *meishijia* to Zhu Ziye is somewhat ambiguous. It should be a positive attribution given its suggestion of a foreign gastronomic tradition. And yet, it contrasts with the traditional Chinese practice of aesthetic appreciation of food, which results from the inspired work of aesthetes capable of capturing the elusive essence of tastes in writing, which has nothing to do with a professional activity or expertise. Without expressing it directly, Lu Wenfu suggests this ambiguity through the slightly despising attitude Gao Xiaoting holds toward the gastronome Zhu Ziye, whose activities are always described under a veil of light mockery.

As early as 1983, then, Lu Wenfu announced that the appreciation of taste would no longer be restricted to a wealthy and literate elite. Indeed, almost anyone would be able to demonstrate this skill, because it is natural. Gao Xiaoting's young grandson, for example, knows that chocolate is the most delicious of confectioneries without being told.

Judging by the numerous publications on *chi* (吃 eating) that appeared in China after 1980, especially from the 1990s onward, the elite aesthetes of yesteryear who praised refined cuisine without any ulterior motive or career plan did not disappear. However, they were increasingly overwhelmed by gourmet entrepreneurs who blended taste and financial ambition. This opening up of food writing, an activity that, pre-1949, was essentially considered a refined, elite relaxation activity for a minority of recognized writers, is due to several factors. The new economic environment that opened the country to entrepreneurship after 1978 played an important role in everyday life. During the Maoist period, food loving and even speaking of food were considered sins, because eating was deemed as purely technical means for maintaining one's ability to work in the service of communism. After Mao Zedong's death in 1976, this strict social and political orthodoxy that forced lifestyles of extreme frugality and translated poverty and scarcity as morally valorous qualities was largely abandoned.

Given that context, the creation of the China Commercial Publishing House in July 1980 helped usher in new perspectives about food and eating, enabling the future renaissance of the food sector. Among the main objectives of this new publishing house, which prioritized disseminating basic knowledge in economics, commerce, and finance deemed valuable to ordinary life, the art of cooking and its techniques received special attention. *Zhongguo pengren* (中國烹飪 Chinese cuisine), officially defined at the time as "art, culture, and science" according to a new recurring motto,[8] was the subject of a government policy aimed at rehabilitating a sector that had suffered greatly.[9]

The period between 1983 and 1993 witnessed a revival of culinary activities and the publication of specialty guidebooks and the like as new professional cooks were trained, educated, and given confidence about the value of their work. Major personalities from the world of arts, letters, and the sciences supported the idea that cooking and cuisine constituted forms of culture and encouraged the public to see themselves as gourmets. For example, the China Commercial Publishing House published Zhou Zuoren's *Zhitang tan chi* (知堂谈吃 *Talk on food*) and *Xueren tan chi* (学人谈吃 *Personalities' talk on food*) in 1990 and 1991—anthologies of short texts on food, cooking, and taste, some of which date back to the 1920s.[10]

These publications testify to the spread of a new sensibility, one replacing communist thrift and frugality. The Chinese now could not only eat their fill but, more importantly, eat well and conceive of new ideas and ways of life as they wished. A new era opened for the appreciation of cooking, tastes, and flavors, but in contrast to past food writing, the context for how and where to appreciate food had shifted. No longer did an aesthete depend on his private cook; today's gourmands appreciate and judge the cooking of professional chefs at restaurants. And with the growth of the middle class, contemporary food lovers often possess relatively deep pockets, can count on friends' support, and are willing to share the best tables. Thus, with the twenty-first century, a new movement of publications has been born, whose authors could proclaim themselves food experts or connoisseurs of good eating.

TO BE OR NOT TO BE A *MEISHIJIA*?

The publication of Lu Wenfu's novel created new conceptions of food writing and culinary cultures, and at the same time, opened space for a new type of author to occupy this field. Indeed, with the spread of Lu Wenfu's neologism, the result has been that, among the fifteen or so texts explored here, almost every author has felt compelled to define their position in relation to the term *meishijia*.

Let us look at a few portraits of today's food writers to better understand why it matters so much for them to claim or reject the title of *meishijia*. The tone was set by Zhong Shuhe, the editor of the anthology of Zhou Zuoren's essays on food. Between 1980 and 2016, Zhong Shuhe launched a mammoth undertaking to publish the writings, testimonies, and diaries of Chinese personalities who traveled to different foreign countries between 1840 and 1919. He is a great scholar whose erudition and cultural insight is evident in his brief preface to Zhou Zuoren's anthology in which he lists and quotes poets and writers known for their food references. Nevertheless, Zhong concluded his preface by saying that he himself was *fei-meishijia* (非美食家 not a *meishijia*), knew nothing about actual cookery, and had never opened a cookbook. By aligning *meishijia* with the act of cooking, Zhong Shuhe contrasts his position against those new food writers who consider themselves as gourmands attached to the materiality of food. For Zhong, genuine writers know how to bring their audience aesthetic pleasure and cultural satisfaction through their discourse on food. Zhong Shuhe's claim that he has never read a single recipe book has been adopted by other authors who also seek to cast themselves as knowledgeable tasters, unburdened by the technicalities of cooking itself.

Shen Hongfei, a well-known journalist and author, adds to this distinction by pointing out that appreciating tastes requires sharing in discussion and exchange *with friends.* Born in Shanghai in 1962, Shen attended Jinan University in Canton and spent many years in that city. He is known for his intelligent, refreshing, humorous, and inspired chronicles on food, which were published in various newspapers, especially in the weekend supplement of *Nanfang zhoumo* (南方周末 *Southern weekly*). His 2000 book titled *Xieshi zhuyi* (写食主义), a neologism that can be translated as "Gastrographism,"[11] can also be considered as representative of this choice of not being a *meishijia*. Like Zhong Shuhe, Shen emphasizes his lack of cooking skills. For Shen, the genuine connoisseur and lover of good food of the past is a gourmet born and bred. These wealthy people possess the right genes passed down through generations. Therefore, Shen Hongfei believes that he does not fit the portrait of an authentic *meishijia* but is instead simply a *hen chan de ren* (很馋的人 very greedy man). But while past culinary connoisseurship may have been predicated on birth and wealth, the contemporary food lover needs only a love of eating delicacies and sharing this pleasure with friends. And yet, as a food writer, successfully transcribing the exquisite tastes he has enjoyed into words and sentences remains paramount. In *Gastrographism*, Shen does not seek to play the guide; he wants to arouse the readers' enthusiasm and emotion because *"xieshi"* (写食 writing on food) matters to him more than other abilities. Perhaps this is why the question of accepting or rejecting the label of *meishijia* was not so crucial for Shen. Even if this designation represents a higher position in a putative hierarchy of food lovers, as he claims, he is happy to be placed at a lower rank of the simple gourmand.

Was Shen Hongfei's view unique, or was it shared by other food lovers of his time? In a small book titled *Haochi* (好吃), a title that means either "It is good!" or "I like to eat," the poet, painter, and writer Che Qianzi, born in 1963, recounts how he fell into a kind of melancholy after reading Shen's *Gastrographism*. Having been provoked by the book to ponder the meaning of *meishijia*, Che Qianzi concludes that the activity of a *meishijia* has nothing to do with day-to-day eating or sustenance; to be a *meishijia* is to let pure sensuality arise in oneself. To be a *meishijia* is to embody a kind of desire that only comes after having eaten one's fill.

For Che Qianzi, Shen Hongfei evaded the taint of professionalization because Shen presented himself as an authentic gourmet devoted to *writing* about food-related topics. "Gastrographism" is a solitary literary activity, essentially an expression of individualism, which Che Qianzi liked. Insofar as Shen Hongfei and Che Qianzi agree with each other, both found it challenging to reconcile a true, yet disinterested love for food with the professional exercise of tasting, even if in both cases, these activities lead to writing papers, articles, and novels for pay.

Thus, they do not accept the title of *meishijia* and instead prefer to claim the status of sensitive gourmands able to express their feelings on food matters through texts.

In contrast, the journalist Dai Aiqun, born in 1967 in Beijing, boldly asserts his status as a *meishijia* in his 2013 book *Chun jiu qiu song* (春韭秋菘 *Spring chive, autumn cabbage*), whose title was drawn from a classic quotation referring to the freshness of vegetables.[12] The subtitle of his book leaves little doubt of his identification: *Yige meishijia de xunwei biji* (一个美食家的寻味笔记 *Jottings on food by a meishijia*). For Dai Aiqun, *meishijia* could be construed in two ways, both of which applied to his situation but nonetheless existed in tension with each other. On the one hand, he is a *"gongzuo de meishijia"* (工作的美食家 professional gastronome), a role he does not particularly like and that consists of arguing about nutrition, famine, or cooking techniques, interacting with cooks and restaurant owners, and speaking authoritatively in the media about the whims of ever-changing food fashions. However much he may dislike these activities, he cannot do otherwise because he has to make a living. Drawing and applying his knowledge about cooking techniques, food culture, and food practices of the past is a job that pays the bills. On the other hand, he also sees himself as a *"shenghuo de meishijia"* (生活的美食家 gastronome of life), a role he loves from the bottom of his heart because it was not imposed on him. In this role, he can get along with great chefs and converse freely about sublime dishes or enjoy drinking tea or wine with friends under a floral arbor—such activities forming the heart of a "gastronome of life." Dai Aiqun insists that he always enjoys good food without preconceived ideas, masks, and orders from above or any other affiliation. And yet, he also feels indebted to his masters in gastronomic writing, including scholars such as Liang Shiqiu, Wang Zengqi, and others whom he thanks for the inspiration they have given him. Having cultivated his expertise of food and food culture through encyclopedias, digital sites, conferences, and repeated visits to restaurants and bistros whose bosses generously shared materials with him, Dai Aiqun finds himself wavering between his practical situation as a professional gastronome and his emotional commitment to being a "gastronome of life."

These three portraits of *meishijia* born in the 1960s are limited, but they give an overview of how various authors dealt with the shifting social and cultural landscape of meanings associated with writing on *meishi* (美食 food and cuisine). The term *meishijia* did not denote a stable set of values, although we can nonetheless discern certain persistent concerns generated by the term. For both Zhong Shuhe and Shen Hongfei, *meishijia* bore the traces of culinary work and cooking, which they distinguished from the realm of literary and aesthetic appreciation. They were eaters who had cultivated their tastes and sensibilities but who used language to convey and share such experiences. Lacking both skills and interest in cooking per se, their contributions to gastronomy, by their own estimations, were literary. Che Qianzi complicated this negative assessment of *meishijia* by suggesting that a *meishijia* preoccupies himself with desires and sensibilities beyond those of immediate need and hunger, and yet he too emphasized the literary dimension of true food appreciation. Only Dai Aiqun of this group of writers sought to directly address the tensions contained by the word *meishijia*. But even his treatment preserved this opposition between *meishijia* as a professional occupation deserving a salary and a purely cultural and artistic activity, free and disinterested. The merit of Dai Aiqun is to have tried to reconcile these two forms of activity by claiming that he is two kinds of *meishijia* at the same time.

One can observe a further wrinkle in the semantic and affective terrain of *meishijia*. Wang Lang, born in 1951 in Beijing, is a food writer and son of the respected novelist Wang Zengqi, who wrote essays on food and had known Dai for about ten years. Wang Lang, the son, wrote the preface to Dai Aiqun's *Chun jiu qiu song*. To introduce Dai's book, Wang sketches Dai's biography: Dai was a deserving man who, after much toil and various ups and downs, finally obtained the means to earn a living by writing about food. He became a freelance food writer, thanks to the connections he cultivated within the restaurant industry and media circles, and succeeded in earning a living through his publications and appearances on television shows. By emphasizing Dai's commercial and professional experiences, Wang implicitly separates Dai's food writing from those authors and poets who wrote about their personal experiences with food, sprinkling their texts with philosophical reflections, without depending on monetary compensation. Even if Dai Aiqun's writing was less affected by such stylistic concerns, he was still bound by professional ethics, that is, to be honest and to provide measured judgments while avoiding sycophancy toward certain known specialists. In this regard, Wang Lang approves of Dai Aiqun's restraint. He did not hesitate to denounce fashions that had no future, such as molecular cuisine, or the eccentric findings of a so-called creative cuisine.

Interestingly, when Wang Lang wants to identify those who call themselves specialists in appreciating food taste and quality, he does not use the term *meishijia*. Instead, he prefers the expression "*zhiye meishi pinglunjia*" (职业美食评论家 professional food critic), which supposedly applies to Dai Aiqun. The threshold to enter this profession is very low, because anyone can tell if a dish is good or not. But according to Wang Lang, what distinguishes actual food critics such as Dai Aiqun from the average opinionated food eater is their ability to express themselves frankly, but always in a civil or polite manner, backed by precise knowledge of the nature of ingredients and foods and mastery of specific culinary techniques. A professional food critic needs to know the history of Chinese and foreign cuisines. An ability to express one's feelings and the possession of literary skills will further elevate the professional food critic from the crowd, but for Wang Lang, the skills of the professional food critic are more a matter of factual knowledge than of literary sensitivity.

Without denigrating Dai Aiqun's achievements, Wang Lang nonetheless demarcates Dai's work as a professional food critic from those engaged in gastronomy as a literary pursuit. At the end of his preface, Wang, somewhat hypocritically, adds that he hopes that by reading this book, readers will better understand the unceasing work of its author to make people recognize the importance of the current development of China's restaurant and catering industry. In what can only be described as a backhanded compliment, Wang's preface succeeds in recognizing Dai's work without affording him the status of a *meishijia*, because for Wang, a real *meishijia* is a food writer whose work can be recognized for its literary beauty. Why Wang Lang chose this path in his preface is difficult to explain, although it may be that Wang considers himself a *meishijia* in its positive sense. Having written his own work, *Diao zui* (刁嘴 *To be picky about food*), published in 2014, he may have implicitly sought to cast himself as a genuine writer like his father and would not concede this title to Dai Aiqun, whom he considers a philistine who does not belong to this literary world.[13] In any case, all these writers demonstrate the semantic slipperiness of *meishijia*. Perhaps it will take some time for the word *meishijia*, a recent lexical invention, to acquire weight before we know whether it will be pejorative, laudatory, or otherwise.

A DIFFERENCE OF GENERATIONS, TIMES, AND SOCIAL CLASSES

The debate on the meaning of *meishijia*, and its attendant complexities of identification, is built on a difference of generations, times, and social classes. In addition to the socioeconomic transformations that have taken place since the 1980s, a social hierarchy has emerged behind the various positions of these writers. This was already evident in Lu Wenfu's novel, insofar as the *meishijia* Zhu Ziye was depicted as a greedy glutton who could satisfy his vices as much as possible, thanks to his financial means. His character's reversal at the end of the novel into a professor of gastronomy somehow saved him from the writer's contempt.

Indeed, Lu Wenfu offers himself as a poignant example of such transformations. In "Chihe zhi dao" (吃喝之道 "The way of eating and drinking"), which was likely a speech and dated December 6, 2001, Lu Wenfu expressed what he meant by *meishijia*:

In the past, I wrote a book called *Meishijia*—perhaps it was not a good idea. Now I have become a *meishijia*, and when people introduce me, they say 'here is Mr. Lu, the *meishijia*.' But in reality, this *jia* is not mine; mine is the one of the *zuojia* (作家 writer). That is who I am. But, when I hear this kind of introduction, I let it go, and I think that I have climbed a rung [on the social ladder]. Because to be a writer is simple; you just need a pencil and a sheet of paper. While to be a *meishijia*, it is not the same.[14]

With this humorous explanation, Lu Wenfu makes fun of those who have attached so much importance to status and qualifications by reversing the cultural values attributed to the two occupations: writing and tasting. Society has changed, and one can no longer afford to be simply a writer, with his head in the clouds and detached from the material world of economic and financial interests. This little verbal manipulation may have been a subtle way to pay tribute to the memory of the writer Zhou Shoujuan (1895–1968), also known as Eric Chow. Prior to the Cultural Revolution, Zhou regularly organized fine meals for his small group of friends that followed a well-established ritual with the chef of Suzhou's most famous Songhe Lou restaurant. These gourmet gatherings were moments of gastronomic theater wherein participants could taste the flavors of dishes conceived and designed by an exceptional chef at the top of his game. This cuisine was so elaborate that it required several days of preparation and a careful selection of quality and fresh ingredients.

For Lu Wenfu, the designation of *meishijia* applies primarily to a connoisseur of good food who knows the art of appreciating tastes. Not all writers possess this additional quality. That is how Lu Wenfu justifies accepting the attribution of a *meishijia*, because, in the end, it is more complicated than being just a writer! If this anecdote serves as a roundabout way of paying homage to Zhou Shoujuan, who committed suicide during the Cultural Revolution, it also serves as a literal account of Lu Wenfu's cultivation of tastes. These kinds of free aesthetic activities for which the instruments are indeed the mouth, tongue, and palate have nothing to do with the daily need to eat, but they are also no longer possible today. Who could afford to spend several days preparing such a fine meal that would be so expensive? "If we were to organize this kind of meeting again today, I would have to write an essay of 10,000 characters each time to cover the costs of my participation, at least 400 yuan," laments Lu Wenfu.

A NEW WAVE IN THE SPIRIT OF OUR TIME

The *meishijia* identity still exists, but it no longer seems to arouse the same anxiety or debate. Contemporary food writers still refer to it, but without attaching the same importance as their predecessors in the 1990s and 2000s did. A good example is the writer, founder, and editor of the online magazine *Yueshi* (悦食 *Epicure*), Shu Qiao, who was born in 1980 and can be seen as representative of her generation. In her book, *Chi, chidexiao* (吃, 吃的笑 Let's eat and laugh about eating), published in 2009, she condemns *meishijia* for their eccentricity and gluttony, explaining that they no longer fit in with the times.[15] People today do not want to eat with such pomp and are increasingly concerned about animal welfare. She declares that she is not a *meishijia* but a *"meishi gongzuozhe"* (美食工作者 worker for good food).[16] Her rejection of the title *meishijia*—both as a personal identifier and as a profession—does not prevent her from appreciating oysters and champagne, nor does it prevent her critics from calling out her consumption of the food luxuries enjoyed by China's golden youth.[17] Still, her assessment flags a broader shift in cultural understandings about food, good food, and eating.

The huge success of the CCTV-produced documentary series, *Shejian shang de Zhongguo* (舌尖上的中国 *A Bite of China*), which began in 2012, underscores Shu's points. If *meishijia* connoted an excessive preoccupation with luxury and high class, *A Bite of China* devotes similar energy and attention to rustic, local food cultures and practices of a rapidly changing China. Close visual attention was paid to food preparations, culinary techniques, and products from the countryside, some of which are on the verge of extinction or have already disappeared. The series was followed by a flood of books dedicated to grandmothers' and mothers' cooking and memories of rustic childhood food, sometimes written by very young authors.

This trend has not yet disappeared. Consider, for example, Zhou Huacheng's book, *Yi fan yi shijie* (一饭一世界 A world in a bowl of rice), first published in 2012 and then revised in 2018, in which he evokes the foods and plants grown without fertilizers or pesticides in his home village that he had left to work in the city. He too rejects the label of *meishijia*, since his book is not about fine food but about the simple foods of the countryside. And yet his writing style, use of literary references, and even themes follow a long tradition of writers touched by the memory of their native rustic food or the succulent blandness of *doufu* (豆腐 tofu). The difference between Zhou and his predecessors is that his writing manifests equal concern for local ecology and the preservation of nature and animals as well as ordinary life and its simple pleasures. In this, he is like so many of today's young people born in a rising middle class, for whom the label of *meishijia* is not an issue of utmost concern.

THE GASTRONOMIC BUSINESS: A WORLD OF IMAGES FOR EVERYONE

The field of food publishing and promoting gourmet foods has changed significantly since the 2010s, especially with the rise of the digital world. Gastronomic reviews and comments of yesteryear are competing with digital sites, such as Dazhong dianping wang (大众点评网), offering tourist recommendations, restaurant reviews, and cultural activities. Gastronomic guides such as the *Michelin Guide*, which was first published in 2017 for Shanghai, have also entered the Chinese marketplace, although it was not initially accepted by the Chinese gastronomic cognoscenti.[18] Food appreciation has become integral to Chinese tourism with, for example, the creation of the network "Cities of Gastronomy" by UNESCO in which four Chinese cities (Chengdu, Yangzhou, Macao, and Shunde) are included in the list of sixty-nine cities around the world.

Within this digital landscape, the growth and prominence of food bloggers is of special note. One in particular, Li Ziqi, the young and beautiful blogger born in 1990, has the most subscribers for a Chinese-language channel on YouTube: 14.1 million according to the last count of January 2021, for which she earned a Guinness World Record.[19] Her large number of followers demonstrate her allure and appeal to international audiences. Known as "a Chinese food and country-life blogger, entrepreneur and internet celebrity," since 2017 her content has focused almost exclusively on creating dishes from essential ingredients and traditional tools using Chinese cooking techniques.[20] According to what can be found on her biographical background, Li Ziqi swims against the tide of larger social trends that have emptied the countryside of young people who have moved to cities in search of work and better futures. Having failed to find a job in the city, she returned to the countryside to take care of her sick grandmother, where she successfully developed her digital cooking business. It is often said that she is the best ambassador of Chinese traditional culture and soft power. To her foreign audience she has become a symbol of Chineseness, and for her domestic audience her videos fit both into a traditional vision of *quan nong* (劝农 encouraging agriculture) and into the current policy of agricultural development with the aim of revitalizing rural China. (There has been the return to farming by emigrants who came from the countryside and settled in the cities since 1980.[21])

What makes Li Ziqi interesting, in light of our consideration of the vicissitudes of meanings associated with *meishijia*, is how her popularity and social media success have altered the semantic terrain. The term *meishijia* has been overtaken and replaced by *meishi bozhu* (美食博主), the "good food bloggers." Li Ziqi can be considered one of the key representatives of this new identity, which promotes a new field of endeavor, "rustic gastronomy," whose distinction and accessibility begins with the homemade.

With Li Ziqi, for the first time, it is no longer the high skill of the great chefs, but the remarkable know-how of the housewives in the Chinese countryside that is praised. Li Ziqi plays all the roles—an excellent cook, a discerning taster (in fact, a *meishijia!*)—when she sits down with her grandmother to eat the dishes she has prepared. For the viewers, it is no longer a question of reading, but of watching and enjoying vicariously the many occupations—cook, gardener, cabinetmaker, and so on—of this young and beautiful woman who lives in a wonderful Edenic Sichuanese countryside, all set to gentle music. (At no time does Li Ziqi actually speak, so the audience has no idea of the sound of her voice.) Sometimes the ingredients of a recipe are briefly printed on the screen as an indication, but this is the only reference to the written word in these videos. Everyone can understand her videos through images, even foreigners without any knowledge of the Chinese language! It is a new nontextual way of presenting Chinese culinary skills that has made her immensely popular in and outside China. With her wordless films, Li Ziqi signals the end of a conception of the *meishijia* that belonged to a literary world that itself has largely disappeared.

Of course, Li Ziqi has imitators. One of them, the young Dianxi Xiaoge, was a police officer who returned to her village in Yunnan Province when her father became ill. Note that the illness of a family member is presented as a determining factor in the decision to return to live in their village of origin for the two bloggers. Dianxi Xiaoge too has become famous thanks to her cooking videos, which are also very popular and are followed by more than eight million subscribers. She seems to be as good a cook as Li Ziqi, but her style is less sophisticated; it is casual and talkative. In particular, she makes a point of explaining aloud what she is doing while cooking.[22] She shows a country family's life, whose members sometimes help her in her culinary activities. If Li Ziqi's elegant dresses make her look more like a fairy, according to her admirers, Dianxi Xiaoge looks like a genuine, charming country girl full of life and talent. They constitute a new phenomenon in Chinese food culture that deserves to be studied and analyzed more. Indeed, these videos raise many questions about their overall production and the actual cooking involved, and they would require an objective investigation to understand the perspectives and motivations of their creators.

CONCLUSION

These bloggers have changed the expression of traditional Chinese food in two ways. They have taken their subject seriously and made it a profession and a source of income. Some have even gone so far as to obtain professional qualifications, defining themselves simply as "gastronomy workers." Moreover, this transition from elite hobby to profession has occurred through a change in the means of expression. These new gastronomy workers have ushered in a new genre of food "writing," expressed through film and video, thus making the so-called "food vlogging"[23] a rich medium that touches the heart and taste of gourmets. Are writings on paper gradually disappearing in favor of filmed images? The example of Li Ziqi's videos is emblematic because they are not only well-informed "cooking films" but they touch the viewer with the beauty and poetry of their images, as a poet can with his verses. This is indeed the reflection of a new fashion, more or less ecological, aiming to showcase some traditions of ordinary Chinese life and addressed to the youngest members of the newly emerging Chinese middle class, convinced that Gaia can be saved.

Moreover, Li Ziqi and her competitors have constructed a more "feminine" and domestic vision of the pleasure of cooking, especially in rural areas. It's no coincidence that young, attractive women have feminized the food vlogging field. But they don't just rely on their charms to make their business a success, as they also demonstrate their know-how, determination, and willpower—all qualities that are not spontaneously attributed to the female gender. It is also interesting to note that when they use a male assistant, he always appears as a subordinate. Is this a new positive model of a strong Chinese woman to be compared with the image of the well-to-do city woman whose main activity is writing, like Shu Qiao? Only time will tell.

NOTES

The author would like to express her gratitude to the editors whose comments and insightful suggestions have helped to improve the text.

1. This text is part of a long-term research project, some of which has already appeared in print. Cf. Françoise Sabban, "Les nouvelles figures du gastronome chinois," in *L'imaginaire de la gastronomie*, ed. Julia Csergo and Olivier Etcheverria (Chartes: Menu Fretin, 2020).

2. Isaac Yue and Siufu Tang, eds., *Scribes of Gastronomy: Representations of Food and Drink in Imperial Chinese Literature* (Hong Kong: Hong Kong University Press, 2013) and Cheng Wing Fun and Hervé Collet, *Dans la cuisine du poète* ([Millemont]: Moundarren, 1998).

3. Wang Zengqi and Wang Lang, *Sifang shishi. Hu jiao wenren* [A literary gourmand speaking gibberish] (Nanning: Guangxi Renmin chubanshe, 2003).

4. Lu Wenfu, *Meishijia* [The Gourmet] (Chengdu: Sichuan Renmin chubanshe, 1983).

5. Zhao Xianzhang, "Xingshi meixue zhi wenben diaocha—yi 'meishijia' wei li" [A textual analysis of formal aesthetics, using *Meishujia* as an example], *Guangxi shifan daxue xuebao* (*Guilin*) 3 (2004): 54–59.

6. Zhao Xianzhang mentions the foreign origin of the word *meishijia*, but he says nothing about its potential source; Zhao, "Xingshi meixue zhi wenben diaocha." In France, Lu Wenfu's novel achieved great success after the publication of its French translation in 1988.

7. Joseph Berchoux, *La gastronomie, ou L'homme des champs à table* (Paris: Imprimerie de Giguet et Cie, 1801); Simon-Célestin Croze-Magnan, *Le gastronome à Paris: épître à l'auteur de La Gastronomie ou L'homme des champs à table* (Paris: Suret, 1803).

8. Cf. Françoise Sabban, "Art et culture contre science et technique. Les enjeux culturels et identitaires de la gastronomie chinoise face à l'Occident," *L'Homme* 137 (January–March 1996): 163–194.

9. The China Commercial Publishing House was commissioned by the State Council to publish thirty-three small volumes for the *Zhongguo pengren guji congkan* [Collection of Ancient Culinary Texts], which lasted from 1982 to 1994.

10. Zhong Shuhe, ed., *Zhitang tanchi* (Beijing: Zhongguo shangye chubanshe, 1990); Yu Jun, ed., *Xueren tanchi* (Beijing Zhongguo shangye chubanshe, 1991).

11. Shen Hongfei, *Xieshi zhuyi* [Gastrographism] (Chengdu: Sichuan wenyi chubanshe, 2003).

12. Dai Aiqun, *Chunjiu qiusong. Yige meishijia de xunwei biji* [Spring chive, autumn cabbage: jottings on food by a meishijia] (Beijing: Shenghuo Dushu Xinzhi Sanlian shudian, 2013).

13. Wang Lang, *Diaozui* [To be picky about food] (Beijing: Shenghuo Dushu Xinzhi Sanlian shudian, 2014).

14. Lu Wenfu, "Chihe zhi dao" ["The way of eating and drinking"] *Rain8* (September 26, 2021): https://www.rain8.com/wenzhang/47568.html.

15. Shu Qiao, *Chi, chidexiao* [Let's eat and laugh about eating] (Hong Kong: Sanlian shudian, 2009).

16. https://www.baike.com/wikiid/7455519998489210664?view_id=32zfv1g6gou000.

17. See, for example, her talk, "Ren yu shiwu de meihao guanxi" [The interconnectedness of food and life] (uploaded September 23, 2017: https://www.youtube.com/watch?v=ctK97JNdiFQ) and the critical comments made by viewers.

18. Cf., Hu Yuanjun, "Yiwei qian chushi dui Miqilin zhinan de zhuguan jiedu" [A former chef's thoughts on the Michelin Guide], *Zhongguo pengren* 459, no. 11 (2019): 136.

19. All videos of Li Ziqi are available on Liziqi Channel on YouTube, https://www.youtube.com/@cnliziqi. To know more about Li Ziqi's life and her incredible success, see Liang Limin, "Consuming the Pastoral Desire: Li Ziqi, Food Vlogging and the Structure of Feeling in the Era of Microcelebrity," *Global Storytelling: Journal of Digital and Moving Images* 1, no. 2 (2022): 11, and William Thomas Whyke et al., "An Analysis of Cultural Dissemination and National Image Construction in Chinese Influencer Li Ziqi's vlogs and Its Impact on International Viewer Perceptions on YouTube," *Journal of Chinese Sociology* 9, no. 14 (2022), https://doi.org/10.1186/s40711-022-00173-2.

20. On Li's life, see "Li Ziqi. Year in Review: Most Inspirational Figures of 2019," *China Daily*, December 29, 2019.

21. Liang, "Consuming the Pastoral Desire."

22. See, for example, "Who Is Dianxi Xiaoge, the Chinese Youtuber with Millions of Fans?," https://www.youtube.com/watch?v=rC8duU_tbC0.

23. Liang, "Consuming the Pastoral Desire."

IV "CHINESENESS" IN MOTION: MIGRATION AND MOBILITIES

11 JAPANESE CUISINE IN CHINESE FOODWAYS

JAMES FARRER AND CHUANFEI WANG

WHAT IS THE MOST POPULAR CUISINE IN CHINA?

In recent years, if you ask a young Chinese urbanite what type of restaurant cuisine they associate with an image of success, with a casual date with a friend, or a quick healthy meal, the answers to all three questions might be *riliao* (日料 Japanese cuisine). With the explosive growth of an international dining scene in the China's coastal cities in 2000s, Japanese restaurants not only dethroned French and Italian as the most popular foreign restaurant cuisine but they also passed most regional Chinese cuisines in sheer numbers, including venerable favorites such as Cantonese (see Table 11.1). By 2020 there were more than eighty thousand Japanese restaurants in China, up from ten thousand in 2013.[1] In major cities, Japanese restaurants range from pricey *omakase* sushi shops in the city center to cheap *takoyaki* stands where middle school students stop for afterschool snacks.

According to 2020 China Restaurant Industry Survey Report, the average price per person spent in the Western and Japanese restaurants at the mass, middle, and high level was respectively 75 yuan, 207.6 yuan, and 537.5 yuan in the fiscal year of 2019.[2] This paper attempts to explain this rise of Japanese cuisine to popularity in urban China and its place in "Chinese foodways." Ultimately, we challenge the idea that "Chinese foodways" have to be "Chinese," and the related idea that culinary globalization and localization are opposing processes. Rather, we show how contemporary Chinese foodways are inclusive of diverse "foreign cuisine" and how localization processes drive culinary globalization.

Table 11.1

The Numbers of Restaurants for Major National and Regional Cuisines in Shanghai and Tianjin (Top Three Are in Bold)

Cuisine	Shanghai	Tianjin
Shanghai-Jiang-Zhe	**7808**	865
Sichuan	7750	2975
Japanese	4954	1538
Cantonese	3866	579
Hunan	2351	476
Korean	1707	1355
Dongbei	1214	**1812**
Xinjiang	746	409
Yunnan	644	470
Italian	591	564
Jiangxi	414	45
Taiwan	313	92
Shandong	238	339
French	186	34
Guizhou	158	22
Xibei	133	18

Searched in Dianping.com.cn on January 31, 2022, in Chinese *riben cai* (日本菜 Japanese cuisine), *yidali cai* (意大利菜 Italian cuisine), *faguo cai* (法国菜 French cuisine), *hanguo cai* (韩国菜 Korean cuisine), *yue cai* (粤菜 Cantonese cuisine), *lu cai* (鲁菜 Shandong cuisine), *sichuan cai* (四川菜 Sichuan cuisine), *xiang cai* (湘菜 Hunan cuisine), *benbang jiangzhe cai* (本帮江浙菜 Shanghai, Jiangsu, and Zhejiang cuisine), *Xinjiang cai* (新疆菜 Xinjiang cuisine), *Yunnan cai* (云南菜 Yunnan cuisine), *dongbei cai* (东北菜 Northeast cuisine), *guizhou cai* (贵州菜 Guizhou cuisine), *xibei cai* (西北菜 northwestern cuisine), *Taiwan cai* (台湾菜 Taiwanese cuisine), *jiangxi cai* (江西菜 Jiangxi cuisine). Japanese cuisine is the third most popular cuisine in both cities.

Stepping back, what, we might ask, are "Chinese foodways" to a young Chinese person raised on KFC, pizza, ramen, and sushi, alongside more traditional Chinese fare? Such patterns of omnivorous consumption call into question the ethnonational boundaries of modern Chinese foodways assumed in most writing about food in China. Looking broadly at urban Chinese foodscapes, we suggest that foreign foods—with Japanese cuisine as the most prominent example—have become an intrinsic part of modern Chinese foodways. This is not unusual. Many scholars have noted how Chinese foods have become part of everyday foodways around the world, including historic chop suey restaurants in the United States, Indian-Chinese restaurants throughout urban India, and "neighborhood Chinese" eateries in urban Japan.[3] Lok Siu's chapter on *chifas* in Peru (chapter 12) exemplifies how Chinese restaurants are indigenized as local culinary heritage. If Chinese food is part of Indian, Japanese, American, and Peruvian foodways, often positioned as familiar comfort food, what is the position of such "foreign" cuisines in urban Chinese foodways? Is it analogous to the role of Chinese food abroad or something rather different?

Foodways and tastes for foods are not just the agglomeration of individual consumer choices, but also the products of history and policy. In China we see the mid-twentieth-century policies to rid China of foreign influences. The varied gastronomic scenes of semicolonial Tianjin and Shanghai were denuded of their cosmopolitan flavor in the Maoist decades of the 1950s and 1960s. After the reform and opening policies beginning in 1978, these policies were reversed. By the 1990s, the state began actively reinventing a cosmopolitan foodscape in Chinese cities. This history is briefly sketched out in the first section of this chapter, from the flourishing Japanese foodscapes of prewar Shanghai and Tianjin, to the reemergence of high-end restaurants since the 1990s, and the explosion of Japanese restaurant cuisine in both cities since the 2010s.

Chinese consumption of foreign foods can also be examined in a comparative context that includes notions of ethnicity and identity. "Ethnic" or "foreign" foodways are part of the foodways of cities globally. When asked in a survey, Americans typically name Chinese as their favorite ethnic cuisine, making it a central part of American foodways. At the same time, Chinese food in the United States is still regarded as an immigrant cuisine, and sometimes regarded suspiciously as unsafe or unclean.[4] In both India and Japan, in contrast, Chinese cuisine restaurants are often owned by members of the dominant ethnic group and regarded as localized comfort foods, with a faded identification with Chinese migrants.[5] The position of foreign foods in China is, at first glance, different from either of these cases. They are neither regarded as immigrant cuisines nor as variations of a distinctly "Chinese" culinary culture. We will describe how a foreign cuisine forms a major part of modern Chinese urban foodways, while not being regarded as "Chinese."

Culinary consumption is associated with social stratification and strategies of social distinction shaped by social inequalities. In many societies, consuming foreign cuisines may be a means of expressing social distinctions and self-identities, with some cuisines (such as French haute cuisine) associated with conspicuous luxury consumption and the consumption of more "exotic" cuisines a way of displaying alternative or non-mainstream identities.[6] In both cases, the stereotypes associated with various "ethnic" cuisines play a role. In the United States, the status of a cuisine has much to do with the status of its migrant producers, with cuisines such as Mexican stigmatized by the low-wage image of Mexican migrants.[7] In China, with fewer international migrants, this association of foreign-themed restaurants with migrant populations is less evident. Rather, beginning with Western fast food in the 1980s, foreign cuisines are seen as reflecting cosmopolitan tastes and aspirations and thus as distinct from the domestic sphere of Chinese cuisine.[8] Next, we show how Japanese cuisine has come to represent both high-status fine dining for urban elites and a type of individualized casual dining culture for young people seeking escape from the pressures of social competition in China's coastal cities. In times of political tensions between China and Japan, the Japanese restaurant also becomes a target of political expression, an issue we explore in more detail elsewhere and turn to briefly at the end of this chapter.[9]

This chapter is based on interviews, participant observation fieldwork, archival data, and internet research conducted by the two authors with Farrer focusing on Shanghai and Wang on Tianjin. Interviews were conducted between 2010 and 2020. Both authors have much experience in these cities. Farrer has conducted research in Shanghai since 1993, and Wang grew up in Tianjin in the 1980s and 1990s. We have thus experienced the full arc of the reemergence of Japanese cuisine restaurants in these cities. This research is part of a larger study of the globalization of Japanese cuisine across six continents, based at Sophia University in Tokyo.[10]

THE LOST LEGACY OF JAPANESE COLONIAL FOODSCAPES

By the early twentieth century, there were flourishing Western and Japanese food scenes associated with foreign settlements, starting in Shanghai in 1842 after the first Opium War and in Tianjin in 1860 after the second Opium War. In Shanghai, early Western restaurants, known as *fancaiguan* (饭菜管), were often run by Cantonese entrepreneurs. Sima Lu (now Fuzhou Road) became Shanghai's first culinary contact zone in which Western food was associated with the companionship of trend-setting courtesans.[11] Less known, many of the earliest eateries in Sima Lu in the 1880s were run by Japanese, who served Western, Chinese, and Japanese specialties. Known as *dongyang chaguan* (东洋茶馆 Japanese tea houses), they operated as brothels as well as eateries. In a period in which very few Japanese yet lived in the city, these establishments primarily served Chinese and Western men.[12]

In the years following World War I, the Japanese population of Shanghai and Tianjin grew rapidly, tourism flourished, and eateries became more sophisticated and oriented toward Japanese tastes. In Shanghai, Japanese businesses were concentrated in the Hongkou District, which became known as Shanghai's Japantown, though officially part of the International Settlement. By the 1910s, the area had over fifty restaurants run by Japanese, with about half serving alcohol and featuring the companionship of Japanese geisha and prostitutes along with food. A similar growth can be seen in Tianjin. By 1928, Tianjin's Japanese Concession featured thirty-two restaurants and twenty-five food ingredient suppliers. In the 1930s, Japanese restaurants included prominent urban landmarks such as the Shanghai's Rokusan Gardens (六三花園), a Japanese-style garden and teahouse, and its older sibling, the Rokusan-tei Restaurant (六三亭), which attracted the patronage of both Japanese and Chinese elites in the city. In Tianjin, the Shikishima (敷島) kaiseki restaurant was regularly patronized by Pu Yi, the last emperor of the Qing Dynasty, who dined with Japanese officials from the Tianjin consulate.[13]

This brief sketch of prewar Japanese restaurants in these two cities offers a point of comparison with the cosmopolitan foodscapes that emerged in the "reform and opening era" after 1979. They were colonial-style foodscapes primarily organized to serve Japanese settlers. They thus were migrant community institutions run by Japanese who established roots in these cities. The fancier *ryōtei* (料亭) and club restaurants, such as Shanghai's Rokusan Gardens, were also an expression of Japanese colonial "hard power" and the business activities of Japanese elites. Over time, they attracted Chinese urbanites and resident Western expatriates intrigued by culinary novelties such as sukiyaki, an interest that even survived into the late 1940s.[14]

However, the legacy of Japanese foodways was eliminated (and Western foodways severely diminished) by the nationalist cultural politics of the new Communist government. A handful of Western restaurants, such as Tianjin's

Kiessling or Shanghai's Red House and De Da survived as state-owned enterprises and were revived as "famous local brands" in the 1980s, but none of the famous Japanese restaurants of the prewar period were revived or memorialized.[15] This has less to do with their erstwhile popularity than with policies to rid Chinese cities of the legacies of Japanese colonialism and militarism. Therefore, unlike Japan where localized *chūka* (中华 indigenized Chinese) and *yōshoku* (洋食 indigenized Western) restaurants are celebrated as part of modern Japanese cuisine, in urban China, the reintroduction of Western and (especially) Japanese cuisine restaurants in the 1990s was received primarily as foreign culinary influences that remained separate from "Chinese" cuisine.[16] Foreign foods only became familiar and affordable for most people with the opening of McDonald's and KFC in the early 1990s.[17] Beyond fast food, a general sense of foreign cuisine restaurants as novel, unapproachable, and even unpalatable prevailed in China through to the end of the twentieth century.

JAPANESE AS THE BLING CUISINE IN URBAN CHINA

In the early 1960s, a state-owned Japanese restaurant Wafu (和風) opened in Beijing, with fixtures imported from Japan. Though it would not survive the Cultural Revolution, this unique project was supported by Japan-educated Liao Chengzhi, a high-level official who later played a key role in re-establishing relations between China and Japan. With the "opening and reform" policies beginning in 1978, a few small independent restaurants opened to serve Japanese expatriates and visitors in major cities, including the Beijing restaurant Hyakuyun (白雲),which opened in 1983.[18] However, beginning with the Beijing Hotel in 1977, hotel restaurants dominated the early spread of Japanese cuisine in China, catering to Japanese business travelers and their clients.[19] In Tianjin, newly built Western hotels such as the Crystal Palace Hotel and Sheraton opened Japanese restaurants. In Shanghai, two restaurants opened at the Okura Garden Hotel in 1990 on the premises of the former French Club of Shanghai, fronted by the club's elegant gardens. These hotels pioneered the import of Japanese ingredients. Head chefs from Japan trained the first generations of Chinese chefs making Japanese food, and these went on to staff independent Japanese restaurants that would appear later in the decade.[20] From these beginnings, Japanese restaurants were associated with an aura of luxury and modernity, and the image of Japan as a developed and "civilized" Asian country.[21]

China's proximity to Japan, the rapid growth of Japanese inbound investment, and the fashion for Japanese food in other developed countries helped Japanese restaurants rise to the top of the culinary prestige hierarchy in Chinese cities. As Krishnendu Ray argues in his research on the United States, price can be taken as one approximation of the status of a cuisine. Table 11.2 shows two recently published lists of the "most expensive restaurants" for each city. Japanese cuisine is now the most prominent fine dining cuisine in Shanghai and Tianjin, surpassing French and Cantonese cuisines. While an experimental French restaurant remains the most expensive one in Shanghai, Japanese restaurants were, as a whole, the priciest in both cities (see table 11.2).[22]

Our fieldwork on high-end restaurants focused on Shanghai, where there are far more such restaurants than in Tianjin. Shanghai is a top-tier global city attracting high-earning professionals and entrepreneurs. By 2020, ultra-high-end Japanese restaurants had opened throughout the city, and no longer only in hotels. The majority served omakase-course sushi, but others served modern kaiseki, teppanyaki, or other specialized dishes. In the early 2000s, this sector still catered to the Japanese business community, but by the late 2010s, this market was oriented to high-flying Chinese consumers. Before the COVID-19 pandemic, these consumers often visited Japan, and they valued chefs who could reproduce the atmosphere and tastes they enjoyed in Tokyo or Kyoto. Japanese chefs and restaurateurs from Japan thus found a niche in this high-end sector. (As discussed next, Japanese were displaced by Chinese cooks and restaurateurs at the low and midrange in the city.)

The most expensive Japanese restaurant in Shanghai in 2020 was Kurogi (黒木), which opened in 2018 in a new luxury hotel complex near Shanghai's waterfront. Operated by the Japanese retail giant Laox, Kurogi Shanghai was the first overseas branch of a small luxury Tokyo restaurant famed as the most difficult to book restaurant in Tokyo. Kurogi Tokyo specializes in *kappo* cuisine, a style of fine dining in which chefs serve customers at the counter (50,000 yen a person exclusive of taxes and drinks in 2020). Aiming at Shanghai's biggest spenders, Kurogi Shanghai, was conceived as an "upgrade" of the Tokyo branch with a more formal kaiseki-style of table service and more obvious touches of luxury on the menu than in Tokyo. In the first Spring seasonal menu for 2018, some of the added luxury items included sea cucumber (rarely served in Japan) in the opening *hassun* (八寸 the second course) and foie gras on top of grilled sesame tofu, a signature dish of Chef Kuroki Jun.

Table 11.2

The Most Expensive Restaurants in Shanghai and Tianjin

	Restaurant name [type of cuisine claimed by the restaurant] (average price per person in RMB yuan)
Shanghai 2020	1. Ultraviolet by Paul Pairet [French] (6000)
	2. Liang She Ye Yan 良设夜宴 [Chinese] (4000)
	3. Kurogi [**Japanese**] (3683)
	4. Huang Gongzi 黄公子 [Creative Western] (3500)
	5. Utsusemi Kaiseki Cuisine 空蝉怀石料理 [**Japanese**] (3000)
	6. Wang Jiang Ge 望江阁 [French] (2946)
	7. Sassa 佐々 [**Japanese**] (2587)
	8. 8 1/2 Otto e Mezzo [Italian] (2340)
	9. Sushi Naoki 鮨直辉 [**Japanese** omakase] (2210)
	10. Sushi Kokorowa 鮨心和 [**Japanese**] (2136)
	11. Maison Lameloise [French] (2084)
	12. Ginza Onodera 鉄板焼き [**Japanese**] (2080)
	13. 泰安门 TaianTable [Creative Western] (2000)
Tianjin 2021	1. Kokorowa [**Japanese**] (1950)
	2. Nobori Sushi 登·寿司 [**Japanese**] (1520)
	3. ZhaiJi Guoyan Sifang Cai 佗寂国宴私房菜 [Chinese] (1345)
	4. Kimura Sushi 木村·鮨 Omakase 料理 [**Japanese**] (1192)
	5. Wei Lan Hai Restaurant 蔚蓝海餐厅 [French] (972)
	6. Er Shi Si Qiao 二十四桥 [Chinese] (905)
	7. Ganen Meat Cuisine 岩塩·肉料理 [**Japanese**] (745)
	8. MAXIM'S [French] (720)
	9. Crab House 蟹府 [Chinese] (659)
	10. Xi'er steak house 囍er牛排馆 [steak] (555)

Shanghai zui gui canting paihang bang (上海最贵餐厅排行榜 List of the most expensive restaurants in Shanghai), accessed on February 10, 2022, https://www.360doc.com/content/20/0331/08/741756_902799139.shtml; *Tianjin chaogui canting TOP10! Kankan ni yi ge yue gonzi neng chi ji jia?!* (天津超贵餐厅 TOP10!看看你一个月工资能吃几家?!TOP10 super expensive restaurant in Tianjin! How many of them can you afford with your salary?!), accessed on February 10, 2022, https://www.163.com/dy/article/FVM2FKQ10545BQ6G.html.

Abalone would be used in later versions of this dish, the manager said. Other upgrades include the addition of sea urchin to the raw eggs served with the sukiyaki. In general, Shanghai fine-dining customers expect such luxurious ingredients (such as foie gras, sea cucumber, and abalone), the Japanese staff explained, and the new restaurant did not want to disappoint their guests. The attentiveness to detail and luxury at Kurogi seems to have been appreciated by wealthy Shanghainese, and in 2020 the restaurant received a rare diamond award from the Chinese Black Pearl guide, designed by Dianping.com as a rival to the Michelin guides. Dinner cost 3,000 RMB per person including a drink pairing with sake, Japanese cocktails, and wines.[23]

The only Japanese restaurant in Shanghai to receive a star in the initial Michelin guide in Shanghai in 2017 was Kanpai Classic, a restaurant specializing in yakiniku located in the prestigious shopping complex Three on the Bund.[24] Owned by a Taiwanese restaurant company, Kanpai Classic also emphasized imported luxury ingredients and the atmosphere of an authentic Japanese restaurant, though a livelier version befitting a yakiniku establishment. On our visit in 2017, wait staff welcomed customers with hearty shouts of *"irashaimase!"* A Chinese waiter circled the restaurant with a bucket of hot coals filling grills inset into the tables and repeatedly warned *"atsui des"* ("this is hot") in Japanese (despite the restaurant being full of Chinese customers). Even requests for the check were relayed to the kitchen in Japanese, though the head chef was Taiwanese. Servers attentively introduced the Australian wagyu-style beef, with marbled slabs lined in a row in glass-fronted refrigerators at the entrance. (The Michelin award plaque is also prominently displayed there [see figure 11.1].) Our beef was graded M-9, the highest grade in Australia, we were told. The other specialties included rice steamed in chicken broth and premium sake imported from Japan. Most tables ordered all three specialties. The emphasis on expensive and imported ingredients appealed to Shanghai customers who were most acquainted with these concepts of culinary luxury. With the sake marked up significantly from Japan prices, dinner prices were about 600 RMB per person.[25]

Sennou Atsushi, a Japanese restaurateur in Shanghai, opined that the local emphasis on luxury ingredients did not always produce the best results on the plate. First, options are limited, he said, with only two major suppliers providing fish to all the high-end restaurants (in 2017). "Wagyu" was also imported from Australia, and not Japan. (Japanese beef imports were not allowed in China because of reports of bovine spongiform encephalopathy

FIGURE 11.1
At the entrance of Kanpai Classic in Shanghai, the plaque of the Michelin star is displayed alongside an array of refrigerated cases of expensive imported beef. Photo by James Farrer, March 2017.

in Japan.) Other luxury items, such as foie gras, were overrepresented on menus, he said, and not at all common in Japan, but reflect the influence of Western fine dining culture. Also, by 2017, the market was becoming saturated. The ubiquitous all-you-can-eat buffets were influencing practices even at the high-end restaurants. For example, customers expected thick cuts of fish, even when this was not the prescribed thickness for good sushi. Already, he said, profit margins had been hit by rising rents and labor costs, a complaint echoed in other interviews with restaurateurs.[26] So, despite the high profile of Japanese cuisine, it was not easy to succeed in a saturated market.

Luxury dining as conspicuous consumption is nothing new in China. But the preference for Japanese fine dining may represent a broader trend toward simpler and more photogenic forms of luxury. One influence was the government anti-waste and anti-corruption campaigns. Japanese restaurants are notable for their small portions and elegant presentation. Worldwide, Japanese cuisine, particularly sushi, seems favored by a new highly educated

JAPANESE CUISINE IN CHINESE FOODWAYS 239

urban elite interested in lighter meals that fit into a busier schedule and that display cultural capital (a sense of cosmopolitan taste) rather than crude economic capital.[27] Finally, the rise of women as consumers also drives trends (see King, chapter 7, and Watson, chapter 13, in this volume). Japanese restaurants appeal to women through healthier and visually stylish foods and a style of intimate sociability among diners seated along a dining counter, in contrast with the male-dominated sociability and ritualized drinking of Chinese-style banquets.

JAPANESE RESTAURANTS AS URBAN THIRD PLACE FOR THE INVOLUTED GENERATION

While expensive sushi and kaiseki restaurants represent luxury consumption for the rich, a more casual and informal register of Japanese restaurants appeals to younger people who increasingly see such an opulent lifestyle as unattainable. One type of Japanese restaurant that has become a part of the everyday foodscapes of young people in both Shanghai and Tianjin is the izakaya, which is a casual gastropub in Japan but is treated more like a restaurant in China. Similar casual restaurants specialized in yakiniku or yakitori and are often indistinguishable from izakaya. Izakayas proved especially flexible in adjusting to local markets. They can also serve as a type of everyday urban third place—a space between home and work that can be a refuge from both, a place to escape from the natal family, or especially for women with children, a place to outsource cooking. They are, however, not a typical space for expensive business entertainment and, thus, can serve as an escape from instrumental social relations.[28]

Urban Chinese youth became familiar with these establishments through consumption of Japanese "two-dimensional culture"—manga and anime that often depict Japanese restaurants as a space for individualized consumption free from burdensome social obligations and economic competition. This includes the manga and television program *Shen ye shitang* ("Late Night Diner;" Jp: *Shin ya shokudō*) about an izakaya in Tokyo's Shinjuku district where the laconic master serves home cooking to a motley retinue of urban outcasts who gather nightly, sharing their troubles in a loosely structured sociability that appeals to many urban Chinese young people. The manga and television program *Gudu de meishijia* ("Solitary gourmet;" Jp: *Kodoku no gurume*) depicts a traveling salesman who enjoys delicious meals alone in modestly priced eateries, silently communing with other customers. Beyond such popular media representations, even middle-class Shanghainese and Tianjin customers can travel to Japan and experience affordable Japanese eateries. Before COVID, two low-cost carrier flights flew four times a week between Tokyo and Tianjin, offering baggage-free ticketing. Their major customers were budget travelers.

Izakayas thus became a kind of affordable and familiar third place in urban China. In Shanghai, we frequented the izakaya Kamon (花门) on the Dagu Road, a cosmopolitan food street that includes a wide variety of Western restaurants, including some serving Indian, Bulgarian, New Zealand, and Israeli fair as well as pubs with bar food. By 2020, these other types of restaurants were outnumbered by twelve Japanese restaurants, specializing in teppanyaki, yakiniku (two), kushi-yaki izakaya (three), washoku cuisine, izakaya (generic), poke bowls, sushi, a yakitori sake bar, and one that advertises itself as yakiniku wedding location. Kamon's manager and part-owner was a migrant from Zhejiang Province who found a job in 2002 in one of the first Japanese-owned izakayas in Shanghai. After five years, the Japanese manager there suggested they pair up to open their own restaurant, which became Kamon. In 2019, Kamon still attracted Japanese customers from the nearby offices of the Japanese broadcasting company NHK, but most customers were families from the high-priced apartments fronting Dagu Road or young people working in the central city. The shift to Chinese customers entailed adjusting the menu. Despite Kamon's specialization in yakitori, Chinese customers asked for sashimi, so they added it to the menu. One Shanghainese regular described eating there several times a week with her two daughters, since her husband seldom ate at home during the week and the daughters preferred Kamon to nearby Western or Chinese restaurants. From her perspective, Japanese cuisine was relatively light and healthy and a break from cooking.[29] According to Dianping, the average expense per person at Kamon was 107 yuan, an affordable price in central Shanghai.

In Tianjin, we visited similarly mid-priced izakayas on central Shanxi Road, a less upscale culinary area than Dagu Road but equally cosmopolitan. As in the Dagu Road area, ten Western restaurants were outnumbered by thirty Japanese restaurants. In 2021, seventeen were listed on the Dianping.com website as serving generic "Japanese cuisine" while six were listed as specializing in yakiniku, two as izakaya, two serving shabu-shabu (*rishi huoguo* 日式火锅), two specializing in sushi, and one specializing in something called foie gras rice (*egan fan* 鹅肝饭). Tianjin's Shanxi Road is one of the major side roads in the largest shopping area of the urban center. Between 1900 and 1945, it was the elite residential area of the Japanese concession. Many Japanese-style buildings remain. However, by the 2010s, the area was a rundown urban district known for its accessibility to domestic rural migrant workers, who could easily rent sections of old Japanese houses, living there while running small businesses in the area. Over the 2010s, Shanxi Road was built by young restauranters into an area known for its diverse izakaya culture. The first Japanese restaurant Okachimachi (御徒町) opened on this road in 2012. Because of the architectural scale, Japanese restaurants on this road were small, and all were owned by Chinese.

In this area, Yi Lu Yi Xian (一爐一鲜) Seafood Izakaya was one of the most popular izakayas in Tianjin, as ranked on Dianping.com, offering both lunch and dinner. It was owned by a young couple in their late thirties. The wife was from Xianjiang, and the restaurant marketed itself as *qingzhen* (清真), referring to Islamic or halal food in the Chinese context. No pork was served, though alcohol was on the menu. Facing a middle school and four large-scale residential buildings, it was a popular lunch place for students and female workers. Its moderately priced menu was heavy on chicken and seafood. Like izakayas in Japan, it offered a *teishoku* (a lunch plate with a main dish, soup, rice, and salad) from 11 a.m. to 2 p.m., featuring teriyaki chicken and grilled fish, priced between 10 and 20 yuan per plate, reflecting considerably lower prices in Tianjin than in Shanghai. It also provided delivery of *teishoku* lunch boxes to workers in the office buildings on the main roads nearby.[30]

The uses of these mid-priced izakayas on Shanxi Road change throughout the day. In the daytime, young parents with children are major customers. The owner of Izakaya He Feng (和風) told us that his izakaya was especially crowded with young families for dinner between 5 and 7 p.m. on weekdays. On the weekends, families reserve tatami rooms for celebrations of the parents' birthday or marriage anniversary. At nighttime, they serve as an urban third place for local young people to gather. At the izakaya Er Fan Zhu, we observed only young women in small groups of two to four people on a Friday night at 10 p.m. in March 2017.[31] Both owners told us that women were the major nighttime customers. They said Japanese popular culture was gaining popularity and their restaurants were places where young people could experience this culture in person without going to Japan. Local restaurateurs strove to replicate a Japanese izakaya atmosphere and interior. For women,

FIGURE 11.2
Young women enjoy late night dining in the neighborhood izakaya Er Fan Zhu in the Shanxi Road area of Tianjin. Photo by Chuanfei Wang, March 2017.

besides the Japanese cultural experience, the izakaya was also a place where they could hang out late at night with their friends without worrying about sexual harassment (see figure 11.2).

For such young people, the cheap izakayas in Tianjin became places where they could feel at ease financially and socially competent. Moreover, they were comparatively cheaper than restaurants in more prosperous southern cities. For example, a young man from Ningbo, who had graduated from a four-year university in Tianjin, posted on Zhihu, a popular blogging site: "For most people, . . . Tianjin is a city where one can live a comfortable life even if he does not want to work hard but instead wants to *tangping* (摊平 lie flat)." Turning to food, he compared the Japanese restaurants in Tianjin to those in Ningbo:

> I went on a date at a Ningbo Japanese restaurant, and I had to pay 400–500 yuan for half-finished fried shrimp sukiyaki. The amount served was for the birds . . . I had to go downstairs to fill my stomach. At this price, I could have eaten until

my stomach bulged at the Japanese restaurants on Binjiang Road in Tianjin. The menu is full of family-style dishes like what we can see in the "Solitary Gourmet," and the portions are solid in Tianjin. After the meal, there is a good-looking waitress who always says goodbye in Japanese. Isn't that sweet?[32]

In this discussion we can thus see two themes: one is the use of the izakaya as a place of escape and social distinction, and the other is a new-found regard for the slower lifestyles in second-tier cities, including Tianjin, that were not on the cutting edge of China's economic boom.

For young customers, mid-priced Japanese restaurants provide a third place in which they can gain respite from the fierce social competition of urban life, sometimes described as *"neijuan"* (內卷 involution). *Neijuan* was one of China's top ten internet buzzwords of 2020. Borrowed from anthropologist Clifford Geertz, it describes a situation in which urban young people were locked in forms of academic, economic, and social competition that they regard as increasingly futile.[33] Faced with a hopeless rat race, young people sought ways of opting out, described colloquially as *"tanping"* (躺平 lying flat), often seeking solace or escapism in consumer culture.[34] Mid-priced Japanese restaurants with their fantastical and playful designs, informality, and individualized eating became one such respite from the serious adult world. They were spaces where young people could consume authentic (as seen by them) foreign food in an inexpensive, familiar, but still exotic atmosphere. Their ability to discern authentic foods and spaces was a way that the involuted generation distinguished themselves from other members of society without needing to spend much money. As seen in other countries, consumption of foreign foods is a way in which young cultural omnivores can create a social distinction vis-à-vis less cosmopolitan consumers such as older people or rural migrants.[35]

THE DOMESTICATION AND CORPORATIZATION OF JAPANESE RESTAURANTS

From the 2010s onward, two trends stand out in the Japanese food sector in China: domestication and corporatization. The domestication of the Japanese restaurant scene in China began in the mid-2000s.[36] As described in the previous two sections, high-end restaurants opened by Japanese established the reputation of Japanese restaurant cuisine as fashionable and luxurious. By 2017, however, only 3 percent of Japanese restaurants were owned by Japanese, with the rest domestically owned. The domestication of Japanese restaurants extended to tastes, including an emphasis on the salmon-based sushi rolls with elaborate sauces, all-you-can eat buffets, and Chinese-style hot dishes. For example, in 2020, the Beijing-based Murakami House (村上一屋) izakaya chain launched mango prawn rolls, using fried shrimp and fresh mangoes, flavors that appealed to young diners. Such restaurants maintained a Japanese image less through "authentic" foods than through décor. Murakami House employed a busy and colorful mélange of lucky cats, daruma figures, paper lanterns, and Japanese-style porcelain, in line with young people's playful aesthetic sense.[37]

The second trend was corporatization. The explosion of Japanese restaurants in the past five years was driven increasingly by restaurant groups. Already in the 2000s, Japanese-owned chains hastened the popularization of Japanese tastes throughout China.[38] However, as the market for Japanese restaurants became saturated in top-tier cities after 2015, Japanese-managed companies stumbled in localizing their products and lowering prices to meet growing competition. By 2020, of the top ten Japanese food chains in China, only the market leader Genso Sushi was a Japan-based brand. The others were domestic Chinese brands.[39] Corporatization and brand franchising were trends throughout the Chinese restaurant industry.[40] Moreover, with 19 percent of Japanese restaurants in China owned by chains, the rate of corporatization in this sector surpassed most other culinary genres.[41]

Chains aimed squarely at the youth market by emphasizing elements of Japanese popular culture and encouraging social interactions among customers. Chatty staff tried to recreate the atmosphere of izakayas in Japanese media representations.[42] One example in Shanghai was the Heiseiya (平成屋), a locally owned chain with seven branches throughout Shanghai. Priced to be competitive with fast-service restaurants, the interior was saturated with colorful images from Japanese postwar pop culture, including a big screen running Japanese programs. Like Kanpai Classic, the staff imitated the noisy greetings of a Japanese izakaya. Customers might be asked to share tables, fostering interaction. A group of young Chinese customers we spoke with explained that the regulars were familiar with this culture through their consumption of Japanese pop culture, though the dishes were adapted to Chinese tastes. "People come here not to consume food really, but to consume culture," one said. He also said Japanese food is popular because of its similarity to Chinese food, and is easy to accept, especially for young people. The average cost was about 100 yuan per person, a price within the reach of many Shanghai consumers.[43]

TENSIONS, PANICS, AND CRISES

Since the first Sino-Japanese war of 1895, Japanese-owned businesses in China have been a focus of political expression, ranging from boycotts of Japanese products to attacks on Japanese businesses. For example, Japanese restaurants became a target in anti-Japanese demonstrations in Shanghai in 2005 and 2010, both times related to maritime territorial disputes and controversies over Japanese teaching of wartime history.[44] Even beyond overtly political incidents, when food-safety-related panics occur, Japanese restaurants seem especially vulnerable.

The COVID-19 pandemic, which erupted in January 2020, posed particular challenges for Japanese cuisine restaurants. A report from the Beijing Xinfadi wholesale fish market on June 12, 2020 found that the new coronavirus was detected on the chopping board of imported salmon.[45] Immediately, customers began avoiding Japanese restaurants, for which salmon was by far the most popular and well-known ingredient. Although their business was drastically hit, domestic chains adapted more quickly than Japanese-owned ones, introducing *aburi* sushi (seared with a gas torch) and other hot foods to reassure consumers of the safety of products. They also pioneered the use of "private traffic" social media, such as WeChat and Douyin, to reach young consumers. A number of Japanese-owned food brands exited the market in 2021, including the former market leader Watami, which sixteen years earlier had pioneered the idea of the izakaya in China as a place for everyday eating.[46]

In the wake of the pandemic, the falloff in international travelers and resident expatriates impacted high-end dining in Shanghai in particular. The city saw the closure of several famous Western restaurants frequented by expatriates, including the pioneering M on the Bund.[47] More focused on Chinese consumers, Japanese restaurants seemed more resilient. New openings continued. In December 2022, the prestigious Tokyo restaurant Narisawa (No. 45 on the World's 50 Best Restaurant list in 2022) announced its first overseas branch in Shanghai.[48] And as China ended its zero-COVID policy in January 2023, "eating Japanese food" topped many wish lists. Some bloggers could not wait to show off their Japanese all-you-can-eat dinner after "reopening."[49]

A larger crisis emerged in the summer of 2023 when Japan began releasing wastewater from the Fukushima nuclear reactor disaster into the Pacific Ocean on August 24, 2023. China initiated a ban on all seafood imports from Japan, a move some attributed to ongoing political tensions between Japan and China.[50] Regardless of the motives, the decision had a profound impact on Japanese restaurants operating in China. According to a survey conducted by Sina Weibo, when asked, "Do you still want to dine at Japanese restaurants?" 281,000 of the 522,000 respondents answered that they would no longer patronize Japanese restaurants. It was estimated that Japanese restaurants would experience a 30 percent reduction in overall profits due to the loss of customer confidence in the safety of Japanese food. High-end Japanese restaurants seemed to be hit hardest. They relied heavily on imported food ingredients from Japan compared to other Japanese restaurants, which had already been using local ingredients and seafood from Norway and Chile.[51] No one can yet predict losses across the wide range of Japanese restaurants, but, clearly, the association of Japanese foodstuffs with health and safety was damaged. However, Japanese restaurants have weathered many such storms for over a century and will likely survive this one also.

CONCLUSION: GLOBAL URBAN FOODSCAPES AND CHINESE LIFESTYLE PROJECTS

To return to the questions at the beginning of this chapter, how do we understand the growing consumption of "foreign"—especially Japanese—cuisines in contemporary China? Needless to say, Japanese cuisine's popularity in China can no longer be understood as an expression of Japanese economic or political "hard power" as it was in the prewar period. Nor is it a very convincing case of contemporary "soft power," since Japan seems unlikely to reap broad political benefits from the spread of sushi and izakayas throughout China. Moreover, from a comparative perspective, China's love affair with Japanese food also seems different from the consumption of Chinese cuisine in many other contexts. First, with the exceptions of restaurants in a few migrant enclaves in Chinese cities, such spaces are rarely understood as serving "immigrant" cuisines (as is the case with most ethnic cuisines in North America, including Chinese). Nor can Japanese and other "foreign" cuisines be regarded as fully domesticated cuisines, such as the "neighborhood Chinese" restaurants in Japan, whose dishes, such as ramen, gyoza, *hiyashi chūka* (cold Chinese noodles), and *Tenshin don* (Tianjin rice), are even exported back to China as "Japanese dishes."[52] Localization of dishes in Japanese restaurants is common in China, and ultimately this may produce a novel genre of Chinese-Japanese cuisine. However, for now, Japanese restaurants in China generally market themselves as explicitly "Japanese"—including in their interior design—to attract Chinese customers interested in a distinctively Japanese cultural space.

Much like Yan Yunxiang's analysis of McDonald's in Beijing a generation ago, we emphasize the novel social practices that "foreign" restaurant spaces facilitate.[53] Foreign restaurants represent an alternative to the social relations embedded in traditional Chinese foodways, especially the formal banquet space with its hierarchical and patriarchal relations. Consumers use the spaces of the Japanese restaurant to stage novel projects of status signaling, individualization, escapism, and urban placemaking. These appropriations of "foreign" culinary culture are by no means opposed to the simultaneous processes of localization, because it is precisely through modifying "foreign" spaces to suit Chinese consumers that they become useful for their personalized and contextualized meaning-making projects. Cities also develop different cosmopolitan foodscapes depending partly on their position in the larger national economy and the types of consumers who dine there.

Two major uses—or meaning-making projects—of Japanese restaurants have been emphasized in this chapter, and they relate somewhat differently to our two cities. First of all, in Shanghai we see how Japanese fine dining represents a novel form of highly exclusive consumption. Visitors to the restaurant Kurogi know that this is a restaurant few people even in Tokyo could dream of entering. Such restaurants are characterized by their locations in top-tier global cities, and this alone means that Tianjin, now a second-tier Chinese city, generally lacks such highly exclusive spaces. At the same time, this form of status consumption also reflects broader social trends in urban China. As discussed earlier, the preferences of women for lighter and less alcoholic meals drives an interest in Japanese fine dining among the urban upper classes. Well-off youth who grew up on fast food also seem to have developed a taste for quick-service luxury—such as high-end sushi—that is also visually stylish and shareable on social media.

The other meaning-making project we have observed is the use of low- or mid-priced izakayas to stage a type of individualized, egalitarian, and backstage sociability that Chinese consumers imagine as genuine and carefree.

In this imagined space, consumers can "lie flat" (not struggling to succeed) and enjoy casual, horizontal (or "flat") relationships with friends and strangers rather than vertical relationships typically cultivated in formal banquets. Tianjin's Shanxi Road has become a rich territory for developing this spatial imaginary, while Shanghai lacks a similar concentration of cheap izakayas in one district.

Viewed historically, the special position of Japanese cuisine in China may be puzzling, given the strained relations between the countries. At the same time, the explosion of Japanese restaurants also shows that Chinese foodways are flexible and dynamic, and not fully encapsulated by a tendency toward culinary nationalism. The foreign but now familiar flavors of Japanese foods have proven useful for many place-making and self-making projects of Chinese urbanites.

NOTES

1. Chen Lan, "Bentu pinpai kuaisu jueqi, qianyi riliao shichang ye yao xia chen?" [With the rapid rise of local brands, the 100 billion Japanese restaurant market will sink?], *Hongcanwang*, November 4, 2021, https://mp.weixin.qq.com/s/qhXLsojWFy5FIatI0Kke_g.

2. China Restaurant Association, "2020 Zhongguo canyin ye niandu baogao" [2020 China Restaurant Industry Survey Report], 2020.

3. Yong Chen, *Chop Suey, USA: The Story of Chinese Food in America* (New York: Columbia University Press, 2014); Amal Sankar, "Creation of Indian–Chinese Cuisine: Chinese Food in an Indian City," *Journal of Ethnic Foods* 4, no. 4 (2017): 268–273; James Farrer, "The Decline of the Neighborhood Chinese Restaurant in Urban Japan," *Jahrbuch für Kulinaristik—The German Journal of Food Studies and Hospitality* 2 (2018): 197–222.

4. Chen, *Chop Suey, USA*; Emelyn Rude, "A Very Brief History of Chinese Food in America," *Time Magazine*, February 8, 2016, https://time.com/4211871/chinese-food-history/.

5. Farrer, "The Decline," 197–222; Sankar, "Creation of Indian–Chinese cuisine."

6. Alan Warde, Lydia Martens, and Wendy Olsen, "Consumption and the Problem of Variety: Cultural Omnivorousness, Social Distinction and Dining Out," *Sociology* 33, no. 1 (1999): 105–127.

7. Krishnendu Ray, *The Ethnic Restaurateur* (London: Bloomsbury Publishing, 2016).

8. Yunxiang Yan, "Of Hamburgers and Social Space: Consuming McDonald's in Beijing," in *The Consumer Revolution in Urban China*, ed. Deborah S. Davis (Berkeley: University of California Press, 2000), 201–225.

9. James Farrer, "Domesticating the Japanese Culinary Field in Shanghai," in *Feeding Japan: The Cultural and Political Issues of Dependency and Risk*, ed. Andreas Niehaus and Tine Walravens (Cham: Palgrave Macmillan, 2017), 287–312; Chuanfei Wang, James Farrer, and Christian Hess, "Colonialism and Its Culinary Legacies: Japanese Restaurants in East Asia," in *The Global Japanese Restaurant: Mobilities, Politics and Imaginaries*, ed. James Farrer and David Wank (Honolulu: University of Hawai'i Press, 2023), 21–66.

10. See Farrer and Wank, eds., *The Global Japanese Restaurant*.

11. Mark Swislocki, *Culinary Nostalgia: Regional Food Culture and the Urban Experience in Shanghai* (Stanford, CA: Stanford University Press, 2008), 133.

12. Wang, Farrer, and Hess, "Colonialism and Its Culinary Legacies," 32.

13. Wang, Farrer and Hess, "Colonialism and Its Culinary Legacies," 22–40.

14. Wang, Farrer, and Hess, "Colonialism and Its Culinary Legacies," 39–40.

15. James Farrer, "Imported Culinary Heritage: The Case of Localized Western Cuisine in Shanghai," in *Rethinking Asian Food Heritage*, ed. Sidney Cheung (Taipei: The Foundation of Chinese Dietary Culture, 2014), 75–104.

16. Katarzyna Joanna Cwiertka, *Modern Japanese Cuisine: Food, Power and National Identity* (London: Reaktion Books, 2006).

17. Yan, "Of Hamburgers," 201–225. As explored elsewhere, there was a legacy of indigenized Western foods in Chinese coastal cities, but restaurants celebrating this heritage were rare by the 2000s; see Farrer, "Imported Culinary Heritage."

18. "Nitchū kōryū ni hitoyaku: Pekin no nihonryōriten tojiru mise fukkatsu shita mise . . . sono eikoseisui" [A Japanese restaurant in Beijing that played a role in Sino-Japanese exchanges: Some restaurants closing and some re-opening . . . their rise and fall], *Television Asahi*, Feb 24, 2024, https://news.tv-asahi.co.jp/news_international/articles/000338301.html.

19. Atsushi Hamamoto and Shigeto Sonoda. Gendaichūgoku ni okeru nipponshoku denpa no rekishi to rikigaku ichi Pekin no nihonryōriten keiei-sha o taishō ni shita intabyū kara [Dynamism of localization process of Japanese food in contemporary China: preliminary analysis of interviews with Japanese restaurant wwners in Beijing] *Journal of Asia Pacific Studies* 9 (2007), 5–6; "Washoku denjyu de moutokkun" [Special training in teaching Japanese food], *Yomiuri Shinbun*, August 10, 1985: 13.

20. Farrer, "Domesticating the Japanese Culinary Field."

21. Jamie Coates, "Between Product and Cuisine: The Moral Economies of Food among Young Chinese People in Japan," *Journal of Current Chinese Affairs* 48, no. 3 (2019): 381–399.

22. The results in table 11.1 were consistent with other lists we found online. In February 2021, we used the keywords "Shanghai zui gui canting" [Shanghai most expensive restaurants] and "Tianjin zui gui canting" [Tianjin the most expensive restaurant]

to search through the search engine Baidu. Combining four lists for each city from 2014 to 2020, we found Japanese restaurants made up 24 percent of the entries for Shanghai and 48 percent of the entries for Tianjin. The smaller proportion in Shanghai could be explained by more entries in the "French" and "other" categories.

23. Interview with Kurogi staff by James Farrer, March 10, 2018.

24. The reasons for this snub by Michelin are beyond the scope of this paper, but several factors are the emphasis on French and Cantonese cuisine by the judges and a perceived lower quality of Japanese cuisine in the city in comparison to its regional rivals, Hong Kong and Singapore.

25. Fieldwork at Kanpai Classic by James Farrer, March 26, 2017.

26. Fieldwork and interview with Atsushi Sennou at Uo Kura Restaurant by James Farrer, March 26, 2017.

27. Farrer and Wank, eds., *The Global Japanese Restaurant*.

28. James Farrer, David L. Wank, Chuanfei Wang, and Mônica R. de Carvalho, "The Izakaya as Global Imaginary," in *The Global Japanese Restaurant*, 255–288.

29. Interview and fieldwork by James Farrer, March 28, 2016.

30. Interview with the owners, Mr. and Mrs. Zhang, by Chuanfei Wang in March and September 2017.

31. Interview with the staff and owner by Chuanfei Wang, March 17, 2017.

32. "Zenme pingjia Tianjin zhege chengshi?" [How would you evaluate Tianjin?], Zhihu, January 21, 2022, https://www.zhihu.com/question/371319318.

33. Wang Qianni and Ge Shifan, "How One Obscure Word Captures Urban China's Unhappiness: Anthropologist Xiang Biao Explains Why the Academic Concept of "Involution" Became a Social Media Buzzword," *Sixth Tone*, November 4, 2020, https://www.sixthtone.com/news/1006391/how-one-obscure-word-captures-urban-chinas-unhappiness.

34. *China Daily*, "'Neijuan' 'tangping': Zhongguo nianqing yidai de xin shenghuo guan" ['Involution' and 'lie flat': The new outlook on life of China's younger generations], September 1, 2021, https://cn.chinadaily.com.cn/a/202109/01/WS612f23e4a3101e7ce97616bd.html.

35. Warde, Martens, and Olsen, "Consumption and the Problem of Variety;" N. Simay Yalvaç and Irmak Karademir Hazır, "Do Omnivores Perform Class Distinction? A Qualitative Inspection of Culinary Tastes, Boundaries and Cultural Tolerance," *Sociology* 55, no 3 (2020): 469–486, https://doi.org/10.1177/0038038520967257.

36. Farrer, "Domesticating the Japanese Culinary Field," 287–312.

37. Chen, "100 billion Japanese restaurant market."

38. Iwama Kazuhiro, "Shanghai no nihonshoku bunka: menyū no genchika ni kansuru hiaringu chōsa hōkoku" [Shanghai's Japanese food culture: a hearing survey of the localization of the menu], *Chiba shōdai kiyō* 51, no. 1 (2013): 1.

39. Chen, "100 billion Japanese restaurant market."

40. Thomas DuBois and Xiao Kunbing, "China's Food Culture Is Changing: Why It Matters," *Asia Global Online*, December 15, 2021, https://www.asiaglobalonline.hku.hk/chinas-food-culture-changing-why-it-matters.

41. "Zhongguo canyin jiameng hangye baipishu" [White Paper on China's Catering Franchise Industry], CCFA and Meituan, July 28, 2021, https://blog.csdn.net/u011948420/article/details/119178249, 10.

42. Chen, "100 billion Japanese restaurant market."

43. Fieldwork by James Farrer, March 25, 2017.

44. James Farrer, "Domesticating the Japanese Culinary Field;" Wang, Farrer, and Hess, "Colonialism and Its Culinary Legacies."

45. "Wei faxian shiyong haixian ganran xinguan zhengju, riliao hangye xia yi bu neng fou chengfeng polang?" [No evidence of COVID-19 infection from edible seafood has been found. Can the Japanese food industry ride the wind and waves next?], Hualalaguanfang, June 19, 2020, https://baijiahao.baidu.com/s?id=16699160163688868848&wfr=spider&for=pc.

46. Chen, "100 billion Japanese restaurant market."

47. Farrer, "Imported Culinary Heritage," 75–104.

48. Robbie Swinnerton, "With Borders Reopened, Tokyo Dining Closed 2022 on a Welcome High," *Japan Times*, December 31, 2022, https://www.japantimes.co.jp/life/2022/12/31/food/tokyo-food-file-2022-in-review/.

49. Sanfentangqubing, "Jiefeng hou de di yi dun riliao zidong" [The first Japanese all you can eat after reopening], Xiaohongshu, December 17, 2022, https://www.xiaohongshu.com/discovery/item/639dcc10000000001f026e33.

50. BBC, "Fukushima: The fishy business of China's outrage over Japan's release," August 25, 2023, https://www.bbc.com/news/world-asia-66613158.

51. Yu Le, "Chen wuran hai mei guolai, zhe ji ge yue ganjin chi?" [Eat as quickly as possible in the next few months before the pollution comes?], Sina, August 25, 2023, https://k.sina.cn/article_6724296984_190cca1180190151la.html?vt=4&wm=5005?type&from=food.

52. James Farrer, "The Decline of the Neighborhood Chinese Restaurant in Urban Japan," *Jahrbuch für Kulinaristik—The German Journal of Food Studies and Hospitality* 2 (2018): 197–222; Yagexiaoxi, "2021 you yi zhong 'yansheng,' jiao riben liaoli dianli de 'zhonghua liaoli'" [There is an "unfamiliar item" in 2021, called "Chinese cuisine" in Japanese restaurants], March 31, 2021, https://baijiahao.baidu.com/s?id=1695733360362254910&wfr=spider&for=pc.

53. Yan, "Of Hamburgers," 201–225.

12 CHIFAS: HOW CHINESE FOOD BECAME A PERUVIAN NATIONAL TREASURE

LOK SIU

Chifas have become ubiquitous throughout Peru, the South American nation that has the largest population of ethnic Chinese in Latin America.[1] The term *chifa*, a Spanish transliteration of Chinese words, was first documented in newspapers in the late 1920s, though its usage in the Peruvian vernacular to reference the food cooked by ethnic Chinese most probably emerged much earlier. Over time, *chifa* became synonymous with not just the food itself but also the various kinds of eating institutions that served it. Today, it also broadly refers to the enduring culinary style and logic, developed by Chinese cooks since their arrival to Peru in the mid-nineteenth century, that innovates the fusion of Peruvian local ingredients with Chinese culinary techniques. What began initially as subsistence food for former Chinese coolies in Lima quickly became adopted by the Peruvian working-class, and *chifa(s)* over time diffused into Peruvian quotidian life, transforming itself and leaving an indelible mark that has forever shaped not only Peruvians' palates but also the foodscape of Peru.

Today, *chifa* is celebrated as a national treasure of Peru. Even the internationally acclaimed chef-entrepreneur and ambassador of Peruvian cuisine Gastón Acurio Jaramillo has developed his own rendering of *chifa* with the opening of Madam Tusán, an upscale franchise of restaurants dispersed throughout Lima, Peru. While Chinese food and Chinese restaurants have flourished everywhere that diasporic Chinese have settled, what is most fascinating about the story of the Chinese Peruvian *chifa* is not just its spectacular rise from its most humble roots to become a nationally recognized feature of Peruvian cuisine—certainly a kind of culinary "rags-to-riches" story like no other in the Chinese diaspora. But what intrigues me more is its uniquely Peruvian formation in culinary and linguistic terms—made recognizable by specific dishes like the *lomo saltado*, among others, and the distinct coinage of *"chifa."* Both in name and substance, this distinctly Chinese Peruvian formation of *chifa* sets itself apart from other diasporic Chinese foodways in the Americas (and perhaps globally), even as it remains legible as part of a general diasporic Chinese culinary-social phenomenon. To a large extent, the similarities and differences reflected in diasporic Chinese food cultures globally evidence what Paul Gilroy calls the "changing same,"[2] a concept that acknowledges the persistence of shared diasporic cultural forms that take on localized specificities through time, interaction, and transformation. Hence, while *chifa* remains endurably recognizable as a diasporic Chinese culinary form and social-economic institution, it also undeniably reflects its specific Peruvian roots/routes.

So, what does *chifa*—its emergence, characteristics, and evolution over time—tell us about the ways in which diasporic Chinese have navigated and negotiated Peru's cultural, social, and culinary worlds? As a contact zone[3] where culinary, social, and geopolitical forces converge, how does *chifa* serve as a site of contest and conflict, a reservoir for building resilience, and a source for ethnic uplift and collective flourishing? How has *chifa*—both as food and as social institution—shape Chinese Peruvian life, their sense of being and belonging, and their transformation as a community in Peru? While these are the broader questions that motivate and guide my research, this essay focuses on tracing the emergence of *chifa* and the assemblage of social entanglements that make possible its layered rendering.

CHINESE DIASPORA FOODWAYS: MIGRATION AND FOOD

So, what do we know about diasporic Chinese foodways, and what is distinctive about the Peruvian case? With few exceptions, the literature on Chinese diasporic foodways has focused primarily on the historiography of Chinese American food. They explore the relationship between migration and the making of Chinese American foods,[4] the popularization and diversification of Chinese American food since the 1850s,[5] restaurants as ethnic enterprises,[6] and the ongoing challenges faced by Chinese restaurant workers.[7] In the United States, we can trace the rise of Chinese eateries and restaurants in the San Francisco Bay area to as early as the mid-nineteenth century gold rush period. With the influx of migrants—Chinese, European, and East-Coast Americans—pouring into California, Chinese merchants created an infrastructure to facilitate food production, import, distribution, and service. The restaurants that opened in the mid-nineteenth century catered to a range of socioeconomic groups. Some were tailored to the Chinese commercial and political elite and served elaborate banquet-style Chinese food. Others served simple inexpensive meals that appealed to the working class. The existence of this range of Chinese restaurants points to the varied socioeconomic positions of workers, merchants, and transnational commercial elites that populated the area during this period.

Despite the shared regional origins of Chinese migrants in both the United States and Peru, the different circumstances that facilitated migration to these sites determined the kinds of migrants that arrived there and their socioeconomic backgrounds. In California, the first wave of Chinese migrants came in search of gold: some went into gold mining, while others set up industries to serve the growing population. By 1860, the census counted thirty-five thousand Chinese miners, workers, and merchants in California. It was not until 1865 that twenty thousand Chinese laborers were recruited to work on the railroad. None came as indentured laborers, as the United States had prohibited the importation of Chinese "coolies" in 1862. In contrast, Peru, along with Cuba, received the largest number of Chinese indentured laborers in the Americas. Between 1849 and 1874, about one hundred thousand coolies were brought to Peru to mine guano, work on sugar and cotton plantations, and perform various kinds of domestic work. Certainly, the scale of Chinese migration was much larger in Peru than in the United States, and Chinese merchants—unlike their California counterparts—represent only a small percentage of the entire Chinese migrant population. Chinese coolies, inserted into Peru's plantation economy and domestic service sphere, were placed in close proximity to indigenous and Afro-Peruvians. These early migratory and social differences between Chinese migrants in Californian and Peru have shaped their differential access to and relationships with other racial-ethnic communities, which, in turn, have also influenced their cultural and culinary intersections. These different circumstances would cast a long shadow on the cultural, social, and culinary formation of these communities, which we will discuss later in the chapter.

Hence, while Cantonese-style cooking has dominated diasporic Chinese foodways throughout the Americas (given that almost all early Chinese migrants came from the province of Guangdong), the particular ways in which certain dishes are made, flavors are blended, and local ingredients are used all reflect particularized versions of Cantonese food produced through the ongoing exchanges between Chinese cooks and their broad range of food consumers. In *chifa*, we recognize the deep social entanglements among the Chinese, indigenous, Afro-, and mestizo Peruvians. Where official historical narratives may not include or recognize their social intimacies, *chifa* prompts us to imagine otherwise. As a style of food, a culinary practice, and a social institution, *chifa* provides important insight into their possible encounters, overlaps, and sustained coexistence. In this sense, *chifa*, as a form of diasporic Chinese food, reflects not only the shared culinary techniques and aesthetics used by Cantonese migrants throughout the Americas but also the distinctive cultural, culinary, and agricultural elements of Peru that have come to shape what is unique about "Chinese Peruvian" food.

CHIFA AS RESTAURANT: ITS APPEARANCE IN PERUVIAN PRINT CULTURE

The earliest documented usage of the word *chifa* can be traced to a 1920 article that appeared in the daily *El Comercio* announcing the construction of a luxurious Chinese theater in Lima, which would serve Chinese and non-Chinese alike and would have a gambling parlor, a billiards room, and "an elegant restaurant for *chifa* [my italics]."[8] Without any explanation of what *chifa* is, the article quickly moves on to discuss other aspects of this enterprise. The fact that *chifa* needed no explanation reflects both the broad vernacular usage and understanding of the term and the widespread familiarity of Chinese food among Peruvians. What is noteworthy in the article is not the existence of *chifa* itself but that *chifa* would be served in this new "luxurious" and "elegant" setting. According to the journalist Mariella Balbi, it was not until 1935 that it appeared again in written form, this time as *"chifan"* in the Peruvian weekly *Cascabel* that published an ad for the Gran Chifan, a popular Chinese restaurant that also announced its plans to expand its accommodations for its "distinguished" patrons. The addition of the "n" in *chifan* can represent both the restaurateurs' attempt to bring the transliteration closer to its proper Cantonese pronunciation and their emphasis on the restaurant's grandness and roominess to distinguish its status from "lesser" and smaller *chifa* establishments that had long served the working class. Despite these being *chifa*'s earliest documented appearances in print media, what they indicate is less about the newness of *chifa* and much more about both its widespread familiarity among Peruvians at large and its anticipated entry into the social world of Peruvian middle class and elite dining.

The newspaper ads of that period show that upscale Chinese food establishments predominantly referred to themselves as "restaurants," which at the time specifically denoted an institution of "fine dining." They described their restaurants as *"sitio aristocrático"* (aristocratic establishment) serving *"comida exquisita"* (exquisite food). In the 1920s and 1930s, we witness the rapid emergence of Chinese restaurants, beginning with the inauguration of Kuong Tong in 1921, followed shortly by San Joy Lao, Ton Po, Men Yut, Tonquin Sen, among others. All were located on *Calle Capón* (otherwise known as Chinatown). What we can infer from this are the following. First, these establishments reflect a sudden spurt in economic growth among the Chinese Peruvian population, probably a result of increased migration of Chinese merchants and transnational elites. The Chinese names of the restaurants suggest a strong sense of cultural pride, as reflected in the restaurant name of Kuong Tong (Guangdong), the place of origin for most diasporic Chinese of the nineteenth and early twentieth centuries. The self-designation as a "restaurant," their descriptions suggesting elite status and cultural refinement, along with their grand décor with formal dining rooms all indicate their explicit attempt to represent themselves as comparable to existing upscale Italian and French restaurants. As such, they aimed to attract the bourgeois and elite classes of all Peruvians, Chinese and non-Chinese alike. With the exception of the two occasions mentioned earlier, newspaper ads publicizing these upscale restaurants rarely used *chifa* self-referentially.

These restaurants made a decisive move to diverge from their nineteenth and early twentieth century antecedents in style, form, and affect. Yet, what is particularly interesting is their *eventual* adoption of the word, *chifa*, to describe the food they serve. Since its emergence, the word *chifa* has been associated with the legacy of Chinese indentured servitude. As such, it is a racialized and classed term that conjures Chinese inferiority. The restaurateurs' initial reluctance to use *chifa* as a form of self-representation might have proven futile. By the 1920s, it is very possible that *chifa*, through its embeddedness in the senses, memory, language, and social imaginary, had reached all sectors of Peruvian society to become a generalized aspect of Peruvian cultural life. It may be that the newly arrived Chinese merchants who had established these new, upscale restaurants had underestimated *chifa*'s profound infusion into Peruvian palates. It may also be the case that they wanted to distinguish themselves from the working-class roots of *chifa*. Whatever the case, *chifa* continued to reference the kind of food served in these restaurants, regardless of its targeted class of patrons. To a large extent, these restaurants had to adapt to the cultural and gastronomic context in which they found themselves. *Chifa*, as a racialized-classed culinary tradition closely associated with Chinese workers, was so deeply embedded in the Peruvian imaginary that these new Chinese restaurants, despite their aspirations, could only be made legible through the category and concept of *chifa*. And to some degree, these restaurants had to incorporate dishes that were emblematically *chifa* dishes, thus also incorporating the logic and ethos of culinary innovation based on mixing ingredients, cooking styles, and flavors to satisfy the growing desires of an ever-broadening eating public.

Exactly what the term *chifa* means in Chinese and how it originated and gained popular usage among Peruvians are impossible to trace. Two interpretations persist. Depending on whether we use Mandarin or Cantonese to interpret its Spanish transliteration, it can have two different meanings. If using Mandarin, *chifa* resembles *chi fan* (吃饭), which is "to eat rice/a meal," a phrase that is used much like *"buen provecho"* or *"bon appétit."* In Cantonese, it would be *zyu fan* (煮饭) or "to cook rice/a meal."[9] The translation in Mandarin would suggest that the Chinese servers were using *chifa* as a greeting to wish their patrons a pleasant meal. If spoken in Cantonese, it most likely was used by the Chinese server or waiter to tell the cook to start cooking. The former emerges from the interaction between the server and its patrons, while the latter is communicated between the server and the cook. At any given time, then, depending on how it is used and to whom it is spoken, and who is hearing it, *chifa* can mean either "to eat" or "to cook." The polyvalence of the term gives it a quality of capaciousness that aptly captures the multiple dimensions of food production, consumption, and mediation. Also, it is entirely possible that the term *chifa* shifted over time or was repurposed and resignified as it circulated. Given the triangulation of languages involved (Cantonese/Mandarin, Spanish, and English) and the potential of error in transliteration, one can see the complicated nature of hearing, recording, and translating sounds into meaning. Admittedly, transliteration and translation are never perfect, but understanding the social context of that time may help provide some clues.

We should remind ourselves that almost all Chinese migrants in Peru, coolies and merchants alike, came from the Guangdong Province of China, and they would be communicating in Cantonese (or some dialect thereof) with one another. Also, the eating institutions that preceded the formation of these restaurants mainly saw themselves as fulfilling a social-economic function—feeding the expanding working class in Lima—rather than creating a "dining experience." With regard to the sociality of these eateries, a number of court records and newspaper archives document tense and often hostile interactions between the Chinese workers and non-Chinese patrons. Based on these factors and the cooking style used in *chifa*, we can be quite certain that *chifa* emerged from a diasporic Cantonese context where Cantonese migrants settled, cooked, ate, interacted with others, and created community. It is possible, perhaps even likely, that *chifa* was resignified in the twentieth century to convey a more elegant and refined origin story. Let us turn now to a brief overview of Chinese migration to Peru, followed by a discussion of nineteenth-century working-class Chinese food enterprises that provided fertile ground for food experimentation.

CHINESE MIGRATION TO PERU: COOLIES, MERCHANTS, AND COOKS

With the 1849 passage of *Ley China* or the "Chinese Law," the first wave of Chinese indentured laborers began arriving to Peru. Triggered by the abolition of the slave trade in 1821, this Chinese Law allowed for the mass importation of Chinese (male) labor that began in 1849 and ended in 1874, when China finally put an end to the Chinese coolie trade by signing the Peace, Navigation, and Friendship Treaty with Peru. According to migration historians, over one hundred thousand Chinese contract laborers, or coolies, came to Peru between 1849 to 1874, with the largest number arriving after the abolition of slavery in 1854. To a large extent, the Chinese coolies came to replace slave labor and went to work on sugar and cotton plantations, railroad construction, and the islands where guano, with its poisonous fumes, was mined. In fact, by 1879, it was documented that 90 percent of laborers on sugar and cotton plantations were Chinese. As part of their work contract, Chinese laborers received specific rations of rice, dried meats, and vegetables. Moreover, Chinese cooks were allocated per a set number of Chinese workers. It was clear that the kind of food they consumed and the way it was to be prepared were of paramount importance to the Chinese indentured workers. Cooks, even at this early time, had a higher status than the other laborers, and they were paid more.

The infungibility of the Chinese cook increased their value as a category of workers. The Chinese cook took on a status of its own, and it was further elevated with the growing demand that came with continued Chinese migration to Peru and as surviving coolies found their way out of the plantation system and into Lima. Many, in fact, transformed themselves into cooks and domestic servants. According to Lima's 1860 census, Chinese represented 35.4 percent of the domestic service labor force and 27 percent of cooks. In the late 1880s, having a Chinese cook became fashionable among elite Peruvian society, and households aspired to include a Chinese cook among their trio of racialized domestic workers: butlers and maids were often indigenous or Afro-Peruvian, and cooks were almost always Chinese.

As early as 1854, Chinese-owned *fondas* (taverns) were established on Lima's *Calle Capón*, which quickly became the hub where both former Chinese coolies and new Chinese migrants congregated.[10] By 1858, according to the Shop Registry of the Conception Market, Hong Kong-based transnational companies had set up satellite branches there.[11] Although Chinese labor migration was prohibited from 1874 onward, a steady flow of Chinese merchants and skilled laborers continued to arrive, opening work opportunities and diversifying the demographics of Chinese Peruvians. As a variety of Chinese-owned shops grew along *Calle Capón*, former coolies found work as peddlers or performed menial labor for shopkeepers. As more Chinese people and shops moved into the area, the demand for Chinese cooks also increased. Both the shopkeepers and workers needed to eat, and they wanted to eat Chinese food. Many of the companies, in fact, hired their own cooks to serve all their workers, and kitchens were built at the back of the stores. Smaller establishments also began serving meals to non-Chinese workers.

CALLE CAPÓN: THE BUSINESS OF FOOD PROVISION AND THE EMERGENCE OF "*CHIFA*" FOOD

As Chinese migrants found their way to *Calle Capón* in Lima and opened businesses of modest scale, several notable food establishments played a crucial role in preparing the groundwork for the emergence of *chifa*, the kind of food associated with Chinese Peruvian (fusion) cooking. *Fondas, cocinerías* (eateries), and *mantequerías* (butcher shops) were some of the earliest institutions that sold food served by Chinese Peruvians. As Chinese Peruvians moved into other neighborhoods, they established bodegas (small general stores), which also began serving food. Then, beginning in the 1910s, *cenas* (dinner eateries) emerged to serve food only in the evenings through the late nights. All these institutions preceded the upscale Chinese restaurants I discussed in the previous section. Moreover, as modest businesses themselves, all were tailored to the working class, though they also attracted curiosity from a wide range of Peruvians. The *fondas* were among the earliest and longest-lasting institutions. Some were tailored specifically to working-class Chinese and were marked with ubiquitous Chinese characters and served typical Cantonese dishes. Others invited a broader public and seemed to include other more Peruvianized foods, though it is unclear the variety of dishes they served. One popular dish or snack that keeps appearing in historical records is fried fish served with a sauce and bread. A number of newspaper articles and police records reported conflicts and sometimes physical altercations that emerged between Chinese shopkeepers and Peruvian patrons who refused to pay.

Fondas gained wider popularity in the early twentieth century when recurring economic crises in Peru made meat consumption particularly scarce. These Chinese-managed *fondas*, using Cantonese-style integration of sliced or diced meat into soups, stews, and dishes, provided the poor and the working class at least some access to meat consumption. These *fondas* had always served as crucial contact zones where Chinese and non-Chinese Peruvians came in contact with one another. As they gained popularity and appeal in the early 1900s, the *fondas* became less ridden with conflict and more accepted as communal sites for food consumption. With the shortage of meat during the depression era, Cantonese-style food preparation and cooking became more appealing and accepted. The ubiquitous style of integrating small amounts of meat in dishes, such as *chaufa* (fried rice), not only satisfied people's cravings for meat but also kept them satiated and the dishes affordable. Moreover, it was largely through these public-serving eating institutions that Peruvians—Chinese and non-Chinese alike—began to recognize their culinary overlaps, such as their shared consumption of broths and stews, and to experiment with the integration of local Peruvian ingredients with Chinese culinary styles and techniques.

Chinese Peruvian *mantequerías* (butter shops) also began selling prepared pork dishes. Given the importance of pork in Chinese cooking, Chinese Peruvians incorporated their own products to meet the demands of their community. The Cantonese-style *cha-siu* (*cha shao* 叉烧) or *cerdo asado* and the roasted pig or *chancho a la caja china* quickly seeped into Peruvian foodways. There, they sold not only the meat for cooking but also prepared pork dishes such as *chicharrones de prensa* (deep-fried pressed pork served with bread). While the Chinese principle of using all parts of the pig may have been shunned before, it became a means of consuming meat during times of food scarcity. What was discovered in the process was that pork belly, pork rind, and pig ears also tasted quite good.

While the *fondas* and *mantequerías* were establishments that served food or sold food products during the day, the Chinese also innovated *cenas*, which began to appear in the Hygiene Bureau archive starting in 1901 and disappeared around the 1930s. They were situated in downtown Lima, not *Calle Capón*, and they opened only at night and served primarily the working-class population desiring an inexpensive but hearty late meal. The food that was served included "meat, unseasoned rice, accompanied with crackers, butter, and pickled turnips ... Broth or a stew made with potatoes, manioc, and meat [were also available]."[12] *Cenas* provided yet another eating institution where the Chinese might have experimented with mixing Cantonese and Peruvian ingredients and styles of cooking.

As the Chinese moved outside of *Calle Capón* and into various neighborhoods throughout Lima and beyond, they opened *bodegas*, small-scale general food stores that sold everyday food items such as bread, cheese, olives, beans, tea, and liquor, among other things. Food distribution was an important aspect of Chinese entrepreneurship in Peru and, arguably, throughout the Americas. Many were and are still involved in retail and wholesale food distribution. Some scholars have characterized *bodegas* as extensions of what once served as warehouses on haciendas. No longer confined to *Calle Capón*, these geographically dispersed networks of bodegas spread throughout Lima and became important food provisioning institutions for neighborhoods near and far. Their ubiquity and close association with Chinese Peruvians gave rise to the vernacular usage of *"el chino de la esquina"* ("the Chinese at the corner") to refer to these *bodegas*. In these *bodegas*, as they did in *fondas*, Chinese Peruvians served well-tested Peruvian favorites.

Because these small-scale but widespread enterprises were tailored to working-class everyday Peruvians, they provided the foundational contact zones (drawing on Mary Louise Pratt's term) for interethnic and intercultural exchange. They became intimate public spaces that facilitated social interactions, cultural exchanges, and culinary experimentations. It is not always possible to trace the origins of particular *chifa* dishes, but these social institutions certainly provided the necessary conditions for exchange, experimentation, and innovation.

While historical documents, such as police and court records, provide ample evidence of conflict between Chinese retailers and Peruvian patrons in these various sites, newspaper articles and cartoons reflect the endless rehearsal of racist portrayals of Chinese food and their establishments as unsanitary and serving questionable types of meat.[13] Despite these popular discourses and images, Chinese food institutions continued to proliferate and thrive. Their popularity reflect not simply their crucial role in fulfilling a social-economic need within Peruvian society, but it is also worth noting that Chinese cooks were willing and interested in incorporating local ingredients and taste preferences in order to appeal to the different publics they served. In this manner, *chifa* emerged as a distinctly Chinese Peruvian culinary form, and it is emblematic of a kind of fusion cooking that is considered neither completely Chinese nor "native" Peruvian. What it recognizes is the enduring Chinese migrant infusion and its continued process of cultural-social mixing within the Peruvian context. Employing Gilroy's concept of the "changing same"[14] to diasporic culinary culture, I suggest that while Chinese food remains recognizable as "culinarily Chinese" no matter where it is found in diaspora, it nonetheless is continually changing and transforming itself through its engagement with local contexts. In this way, while *chifa* remains within the framework of diasporic Chinese culinary culture, its specific manifestations reflect its particular flourishing within Peruvian cultural, social, and economic life. The previous discussion of the different kinds of public eating institutions that facilitated the possibility of social interactions and culinary experiments provides insight into the emergence of *chifa*. We turn now to the particular food dishes where culinary mixing is immediately apparent in order to speculate on the possible social and cultural intimacies that might have given *chifa* its form.

TRACING SOCIAL AND CULINARY INTIMACIES IN *CHIFA*

Three particular *chifa* dishes open up questions about the social intimacies among Chinese, indigenous, black, and mestizo Peruvians. *Lomo saltado* is perhaps considered the most quintessential Peruvian dish. In Peru and abroad, no self-proclaimed Peruvian restaurant can do without featuring *lomo saltado* on its menu. *Lomo saltado* is a stir-fried dish made with slices of beef and onions, mixed with deep-fried potatoes. It is flavored with soy sauce, vinegar, and finely chopped garlic. The dish is often served with white rice. The origin of the dish remains unsettled, although most people, including internationally acclaimed Peruvian chef Gastón Acurio, agree that the stir-fry technique used in cooking the dish is unmistakably Cantonese, along with the use of soy sauce and the style of cutting the beef, onions, and potatoes into strips. The incorporation of potatoes into a stir-fried dish is distinctly Peruvian. Indeed, the history of the domestication of potatoes can be traced to the Andes region between six thousand and ten thousand years ago.[15] The Inca empire built upon the agricultural advances of previous indigenous populations in cultivating and diversifying the potato plant, and potatoes served as the major staple food that fed its entire population.[16] Potato cultivation and consumption, hence, are deeply embedded and endure in Peru's agricultural production, diet, and gastronomy. The varieties of potatoes and the wide range of possibilities in preparing them reflect the food's continued significance in Peruvian cooking. Its incorporation into a Cantonese-style stir-fried dish is *the* defining characteristic of *lomo saltado*, and it is what makes the dish a quintessential *chifa* dish that is distinctly different from any other stir-fried beef dish found anywhere else in the Chinese diaspora. While there is no written documentation illustrating how the dish came about, the combination of ingredients, flavors, and technique of cooking reflects a well-accepted experimentation that draws on Chinese and indigenous culinary traditions.

Another dish, the *tacu-tacu*, is a fried rolled ball or thick pancake made of mashed rice and beans or lentils. *Tacu-tacu*, now a typical Peruvian dish, is generally attributed to Afro-Peruvian foodways. However, the name comes from the Quechua word, "taku," meaning mixed. Its main ingredients are rice and beans or rice and lentils, seasoned with onions, garlic, tomatoes, cumin, oregano, chili peppers, and cilantro. It is known to be a dish made with leftovers. Today, we see it served by itself or topped with breaded steak, fried eggs, or other proteins. *Tacu-tacu* appeared in historical records as one of the food items served in *fondas*. Given the growing population of the working class in Lima in the latter half of the twentieth century, we could imagine that the *fondas* were serving a multiracial population, who might have requested the inclusion of *tacu-tacu* as part of their offerings. Its appearance in the *fondu* "menus" opens up a set of questions around the kinds of interracial encounters and everyday interactions that existed among the Chinese, indigenous, and Afro-Peruvian populations and prompts us to speculate on the cultural intimacies that made possible this culinary inclusion.

Certainly, the haciendas and plantations offered one important location for Chinese, indigenous, and Afro-Peruvian interactions. While indentured Chinese labor was used primarily to replace Afro-Peruvian slave labor, there was a period of overlap. Not much is known about their interactions, except that they were housed in separate quarters and placed in different parts of the plantation. In cases where Chinese coolies escaped the plantation labor system, many were known to have integrated into indigenous communities where they were accepted. I suspect that most (if not all) instances where social and intimate relations emerged between Chinese men and indigenous, Afro-, or mestiza Peruvian women took place outside the plantation system. Most likely, they involved Chinese domestic workers based in Lima or small-scale Chinese merchants, and in some cases, former or fugitive coolies.

Chinese migration to Peru was overwhelmingly male, and it is well documented and accepted that many Chinese men formed unions with indigenous, Afro-, and mestiza Peruvians. It is very possible that some of the Chinese men learned to cook from their partners. It is also likely that these sustained intimate relations facilitated the mixing and blending of ethnic food items and tastes, as well as culinary practices and techniques. Another possibility of interracial interaction is in the domestic spaces of elite Peruvian households, where the Chinese were often hired as cooks and indigenous and Afro-Peruvians as domestic servants. Their daily interaction with one another can present opportunities for learning about local ingredients and food preparation, especially when it is likely that the Chinese cooks had to prepare food not only for household members but also for the domestic staff. In this context, varying degrees of social intimacy developed. Indeed, the high rates of mixed-race families formed between Chinese and indigenous, Afro-, or mestiza Peruvians may very well be the cultural source and wellspring for the innovation of *chifa* dishes. The intimate spaces of family and domestic households, along with the various Chinese food enterprises discussed previously, provide ample opportunity for culinary exchange and elaboration.

The third dish, *chancho con tamarindo* or *cerdo en salsa tamarindo* (pork with tamarind sauce), is the Peruvianized version of sweet and sour pork. Substituting vinegar with the abundantly locally grown tamarind is another example of culinary adaptation made by Chinese migrants. Rather than using imported vinegar, tamarind not only was a more flavorful and durable substitution, it also was less expensive. Using tamarind, a common ingredient in Peruvian foodways, also provided a form of cultural recognition by making the unfamiliar Chinese food a bit more familiar. It made the dish more accessible to non-Chinese Peruvians. One story behind how tamarind came to substitute vinegar in this dish recalls the many incidents where food inspectors mistook vinegar as bad wine and insisted on pouring it out. In its place, tamarind was used, and a new dish was born. Chinese specialty items were often misunderstood by food inspectors, and some were probably influenced by racist images that depicted Chinese use of questionable ingredients. Items like dried preserved sausages and thousand-year-old eggs were often mistaken as spoiled food. Nonetheless, it is also these kinds of social interactions between Chinese and non-Chinese Peruvians that gave way to unexpected *chifa* dishes, like the tamarind pork.

CHIFA AS CULINARY CONTACT ZONE: EXPERIMENTATION AND THE ECONOMIES OF TASTE

While *chifa* is now associated with all forms of Chinese food and the culinary establishments that offer it, its popularization among Peruvians actually stems from the early interactions and experimentations that occurred in the small-scale, modest food institutions like the *fondas, cenas, mantequerías,* and *bodegas* that emanated from *Calle Capón* and that served primarily the working class. Through everyday encounters with indigenous, Afro-, and mestizo Peruvians, Chinese Peruvian cooks infused the culinary influences of these groups into their own to innovate certain dishes that have become known as distinctively *chifa*. With time, the food and the eating establishments that formed along *Calle Capón* also attracted the attention of travelers and curious Peruvians of middle and upper classes. These establishments became critical sites of interracial contact, conflict, and sociability. However, because of its historical roots, *chifa's* characterization as working-class food has endured over the decades and generations, even as it has broken down class borders and expanded its consuming public.

With the critical mass of Chinese merchants coming to Peru and the founding of more upscale Chinese restaurants in the early 1920s, the status of *chifa*, the food, gained wider appeal and social approval. Almost overnight, the new Chinese *restaurantes* transformed popular reception and perception of Chinese Peruvian food by realigning it with the sensibilities of bourgeois culture, imbuing it with a sense of prestige and notoriety, and cultivating new palates among the Peruvian elite. If the early *chifa* establishments served as a site of food mediation among the multiethnic Peruvian working class, these new Chinese restaurants undertook the process of translating Chinese food across economic class lines. The broadened social acceptance of *chifa*, then, can be traced to shifts in migration, culinary innovations that sprang from different forms of cultural and social intimacies, and the economic diversification of Chinese Peruvians. While these restaurants initially distanced themselves from the existing Chinese *fondas* and bodegas, they nonetheless eventually adopted the term *chifa* to make their food legible to the larger Peruvian public. *Chifa*—as a style of food produced by early Chinese labor migrants—had become so deeply embedded in Peruvian vernacular and culinary culture that it was impossible to unhinge from its ethnic working-class roots. The popular perception of Chinese as former coolies and working-class people would persist through their alimental contribution to Peruvian cultural life. Of course, the Chinese Peruvian community, now with more than seven generations, has diversified over the last 170 years. At the same time, Peru's relationship to China has improved significantly, and the two countries since 2013 have embarked on a "comprehensive strategic partnership." Given these changes, *chifa*, a racialized and class-conscious reference, may also be changing.

The recent shift in Peruvian culinary culture and its elevation of indigenous, African, Japanese, and Chinese elements is transforming the perception and reception of these "ethnic" foods. Some food scholars may refer to this process as cultural appropriation (a topic for another essay). If there is any indication that *chifa* is entering yet another phase of transformation, we may only point to Chef Gastón Acurio's latest Chinese Peruvian franchise, Madam Tusán. *Chifa* as both a specific reference to a particular food phenomenon and a general concept of Chinese Peruvian mixing remains malleable and ever-changing. It offers an apt allegory to explore the shifting conditions, the enduring struggles and contests, and the layered interactions and intersections that come to shape the possibilities of Chinese Peruvian cultural life.

NOTES

1. Estimates of Peru's population of Chinese descent vary dramatically. According to the Chinese embassy, about 5 percent (1.2 million) of Peru's population have Chinese roots and ancestry. Scholars, such as Peruvian anthropologist Humberto Rodriguez Pastor and Latin Americanist Eugenio Chang-Rodriguez, posit a much higher estimate of 2.5 million Peruvians of Chinese descent or approximately 10 percent of the population. The different estimates likely reflect how narrowly or broadly they define "people with Chinese roots and ancestry" and "people of Chinese descent."

2. Paul Gilroy, *The Black Atlantic: Modernity and Double Consciousness* (Cambridge, MA: Harvard University Press, 1993).

3. Mary Louise Pratt, "Arts of the Contact Zone" *Profession*, MLA, 1991, 33–40; James Farrer, ed., *The Globalization of Asian Cuisines: Transnational Networks and Culinary Contact Zones* (London: Palgrave McMillan, 2015).

4. Haiming Liu, *From Canton Restaurant to Panda Express: The History of Chinese Food in the United States* (New Brunswick, NJ: Rutgers University Press, 2015).

5. Andrew Coe, *Chop Suey* (Oxford: Oxford University Press: 2009); Yong Chen, *Chop Suey, U.S.A.: The Story of Chinese Food in America* (New York: Columbia University Press, 2014).

6. During the period of Chinese exclusion in the United States (1882–1952), Chinese cooks were considered "skilled" workers shielded from blanket exclusion. The restaurant, then, served the dual purpose of bringing family members to the United States as well as providing a means to earn a livelihood. Now, as before, Chinese restaurants, like many ethnic businesses, still function as economic landing pads where newly arrived immigrants with no English language or translatable professional skills can get work.

7. Jennifer Lee, *The Fortune Cookie Chronicles: Adventures in the World of Chinese Food* (New York: Twelve Press, 2008); Jenny Banh and Haiming Liu, *American Chinese Restaurants: Society, Culture, Consumption* (Abingdon: Routledge Press, 2019).

8. Mariella Balbi, *Los Chifas en El Peru: Historia y Recetas* (Lima: Universidad de San Martín Porras 1999), 124.

9. Katie Birtles, "Chifa: 9 Facts You Never Knew about Peru's Chinese Fusion Food." The Real World. September 19, 2020. https://www.trafalgar.com/real-word/chifa/#:~:text=Chinese%20food%20is%20only%20called,translates%20to%20%E2%80%9Ccooked%20rice%E2%80%9D.

10. Patricia Palma and José Ragas, "Feeding Prejudices: Chinese Fondas and the Culinary Making of National Identity in Peru" in *American Chinese Restaurants*, 44–61.

11. Balbi, *Los Chifas en El Peru*, 52.

12. Balbi, *Los Chifas en El Peru*, 88.

13. As economic competition between Chinese and local Peruvians increased, general anti-Chinese sentiments also grew. Political cartoons of Chinese food preparation raised questions about their use of questionable meat. Yet, according to the *Municipal Journal* of 1874, the Municipality of Lima's Hygiene Inspection Bureau found Chinese taverns and *fondas* in good standing. In other words, these representations had no evidence to support it. Balbi, *Los Chifas en El Peru*, 57.

14. Gilroy, *The Black Atlantic*.

15. David M. Spooner and Wilbert L. A. Hetterscheid, "Origins, Evolution, and Group Classification of Cultivated Potatoes," in *Darwin's Harvest: New Approaches to the Origins, Evolution, and Conservation of Crops*, eds. Timothy J. Motley, Nyree Zerega, and Hugh Cross (New York: Columbia University Press, 2006), 285–307.

16. Hielke De Jong, "Domestication of the Potato," *Spudsmart.* February 3, 2015. https://spudsmart.com/domestication-of-the-potato/.

13 PIGS FROM THE ANCESTORS: CANTONESE ANCESTRAL RITES, LONG-TERM CHANGE, AND THE FAMILY REVOLUTION

JAMES L. WATSON

PRELUDE: THE PERILS OF LONG-TERM ETHNOGRAPHIC RESEARCH

Standing behind the tomb and looking out toward the sea, I was completely disoriented—and not a little unsettled: Was this the same place? How was it possible that the *fengshui* (風水 geomancy or "wind and water") had changed so much?[1] The original view featured a distant blue sea, glimpsed through coastal mountains. The scene was now dominated by several high-rise apartment blocks and a six-lane highway. One could hear the traffic in the distance, and the hill surrounding the tomb had been landscaped to create a public park. A sparkling new polished-granite wall (bearing an elaborate inscription in classical Chinese) stood nearby. Was I the only one in the assemblage who noticed these changes?

It was the sixteenth day of the ninth lunar month (November 2, 2009), exactly thirty-nine years to the day since I first stood on this site with old friends—elders of the Man lineage from San Tin, a village in the northern extremities of Hong Kong's New Territories. On that day in 1970 my friends had come to commemorate their founding ancestor, Man Sai-go, born in 1367 during the collapse of the Mongol empire. Like so many other refugees of that era, he had settled along the south China coast and proceeded to raise six sons—four of whom became successful farmers and constituted the foundation of a proud patrilineage. Today, Sai-go's lineage is forty-three generations deep (by most reckonings) and incorporates at least six thousand people worldwide.[2]

On that brilliant November day in 2009, approximately four hundred of those descendants boarded buses in San Tin, twenty miles away, and made the trek to this tomb. Most were diasporics, people born and reared in Britain, Germany, Holland, and Canada. San Tin had been transformed from a farming village into a full-scale emigrant community in the 1960s and 1970s. For some in the crowd, this was their first-ever visit to Hong Kong. They were intrigued by everything they saw that day. Several were keen to perform what they referred to (in English) as "respect and greetings" at the tomb—under the guidance of their grandfathers.

The highlight of the day was the presentation of offerings by the oldest surviving male of the most senior generation of descendants, known (in Cantonese) as *jok-jeung* (*zuzhang* 族長 lineage master/head).[3] In 1970, the male descendants who gathered at the tomb performed two rites, which they referred to as the *kau-tau* (*kou tou* 叩頭 bowing one's head, a traditional act of homage) and the *bai-sahn* (*bai shen* 拜神 worshipping the spirit, with burning incense). In this regard little had changed.

However, it was not just the tomb's surroundings and its view to the sea that were different in 2009: I was startled to see large numbers of women—wives, mothers, daughters, daughters-in-law—performing these acts of worship and supplication, alongside their fathers and grandfathers. Even a few sons-in-law[4] (including a Caucasian who spoke with a British working-class accent) offered incense at the tomb under the guidance of an obviously amused San Tin elder.

When I first attended Man Sai-go's rites thirty-nine years earlier, there were no adult women present at the tomb, even as observers.[5] In that era, ancestor worship was a male-only activity dominated by *fulao* (父老 elders, alternatively *qilao* 耆老)—men aged sixty and older. But the surprises (for me) did not end with the worship activities: the centerpiece of the centuries old event—*fen zhurou* (分豬肉 pork division)—had also changed into something completely new and "modern" (to quote a participant who lived in London). Everyone, irrespective of gender or age, joined a long queue and waited patiently to pick up a small picnic-sized portion of roast pork that had been prepared by hired contractors. The assemblage then dispersed into small family gatherings and consumed the pork along with accompanying bags of fruit and sweets.

Many of the younger people present spoke only a few words of Cantonese (greetings and salutations) and were obviously curious about the site and the event. I overheard one father, an emigrant I had first met in 1969, tell them (in English): "Go ask that man standing over there. He writes books about this."

I was immediately besieged with questions: "Why is the tomb up here on this hill?" "Do people think that the spirit is still alive inside that tomb?" "Do you lecture about this in your classes?" "Why did they have to carry those heavy pigs up the hill and eat them here?" "What does that tomb inscription say?" They were unfailingly polite and obviously intrigued by the ritual performances of the day. After an hour of good-natured inquisition, I was rescued by an old friend (a journalist and distant relative of Man Sai-go) who drove me back to Kowloon—exhausted, perplexed, and recharged for yet another attack on the old question of Cantonese ancestor worship and the division of pork.

Such are the perils of long-term, multi-sited, never-ending ethnography. If you wait long enough, the surprises of real life will upend and confound every effort to generalize and all claims of authority. Ritual systems—like their performers—are embedded in history and, as such, they are constantly changing.

SETTING: TWO VILLAGES IN RURAL HONG KONG

The western sector of the New Territories, like other areas along China's southern coast, was (until the late twentieth century) dominated by large, single-surname/single-lineage villages that held the best land and controlled local commerce. Two of these villages, San Tin and Ha Tsuen, are the focus of this essay. In 1911 San Tin had a population of just under 1,100; Ha Tsuen had approximately 1,200 people.[6] All resident males in these villages (save for a handful of hereditary servants[7]) shared the same surname: Man in San Tin, Teng in Ha Tsuen. All daughters married out, and all wives married in from other communities—reinforcing an androcentric culture that rivaled anything found in southern Italy or northern India.

Ha Tsuen and San Tin constituted branches of larger, multi-community surname alliances known in the anthropological literature as higher-order lineages.[8] The Teng higher-order-lineage (H-O-L) included four major village-complexes, all of which fell under British control in 1898 and continued to cooperate (and, at times, feud among themselves) throughout the twentieth century and beyond. The Man H-O-L has a different and more complex history. In 1898, the new Hong Kong-China border was demarcated along the

Shenzhen River, half a mile north of San Tin. Regular social interaction with the six other Man villages immediately north of the border was severed in 1941, following the Japanese invasion of Hong Kong. After the war, communist troops closed the border and British authorities responded by building barbed-wire fences along the river. Social relations across the border did not resume until the early 1990s—during the leadup to Hong Kong's *huigui* (回歸 repatriation) to China in 1997. San Tin elders (many of whom are returned emigrants from Europe) were soon recruited as ritual instructors, or guides, by their H-O-L relatives in Chinese territory who began reinventing ancestral rites during the 1990s and early 2000s.[9]

This chapter draws primarily on first-hand information (based on interviews and observations) collected during several decades of ethnographic research.[10] Rubie Watson and I lived in San Tin (1969–1970) and Ha Tsuen (1976–1977); we then returned to the New Territories for follow-up research many times during the following four decades. In both communities we were privileged to attend, observe, and photograph dozens of ancestral rites—including many long treks into the hills of the New Territories. We are indebted to the many elders (Man and Teng) who not only tolerated our hundreds of questions but were unfailingly helpful in making sure that we did not miss any rites. It soon became apparent that our elder friends and confidants were aware that these events were fast disappearing in the heat of economic development and social change. When we returned for follow-up studies in the mid-1980s and early 1990s, many of the rituals described in the next section were no longer performed. By the early 2000s, the nature of the Man lineage ancestral rites had changed so fundamentally that our 1960s confidants would have been amazed, puzzled, and perhaps even pleased by what had transpired: the 2009 picnic at the tomb demonstrated that Man Sai-go was still "alive" in the minds and memories of his far-flung descendants who now constitute a worldwide diaspora.

PORK DIVISION: THE FOUNDATION OF ANCESTOR WORSHIP (ANNUAL RITES IN THE ANCESTRAL HALL)

On the morning of February 3, 1970 (the twenty-seventh day of the twelfth lunar month), a large truck filled with caged pigs arrived in front of Dun Yu Tong, San Tin's main ancestral hall—an austere three-chambered building dedicated to Man Sai-go. This was the beginning of the annual *tuan nian*

FIGURE 13.1
Pigs butchered on the steps of the ancestral hall. Photo by James L. Watson.

(團年 full year) celebrations that local elders had been planning for months. The commencement of the rites was broadcast by the high-pitched, piercing squeals of the pigs as they were systematically slaughtered behind the hall. Once heard, this is a sound that can *never* be forgotten and is forever associated (in villagers' memories and my own) with the annual rites. My neighbor, a highly respected elder, stood with me near the hall and said in Cantonese *"wai tou wa"* (*wei tou hua* 圍頭話 the local Cantonese dialect): "The cries of the pigs tell the ancestor that his annual rites are about to begin."

San Tin has three other ancestral halls dedicated to Sai-go's most important descendants whose wooden tablets[11] are displayed at the top of elaborate altars. Each of these halls was the center of major "full year" pork division rites in 1969 and 1970, with dozens of pigs sacrificed on the same day.[12] "In the past," I was told, the pigs were raised locally, but by the late 1960s they were purchased in the nearby market town of Yuen Long. The slaughtered carcasses were allowed to bleed out before being moved to the veranda of the ancestral halls, where they were dismembered (see figure 13.1).

FIGURE 13.2
Man lineage elders (in background) evaluate pork shares. Photo by James L. Watson.

Then, as illustrated in the accompanying Figure 13.2, slabs of raw pork were carried into the hall and arrayed in front of the ancestral altar. The pork was cut into pieces—a carefully regulated procedure known in Cantonese as *fen ju-yuk* (*fen zhurou* 分豬肉 to divide pork) that involves the weighing of *fen* (分 shares) to ensure equal distribution.[13] For example, a share might constitute one *jin* (斤 catty, 1.3 pounds) of lean meat, one quarter of a trotter (lower leg with hoof attached, sliced lengthwise), one small section of tail, one piece of snout, one slice of liver, and a portion of intestine. Each share had to be judged as equal by a panel of three or four elders, who sat nearby (see the background of figure 13.2). This was important work and only the most respected men of the lineage were trusted to regulate the procedure.

When the shares were deemed to be acceptable, they were laid out on reed mats, and the distribution began. The secretary-accountant of the hall, a man who had to be literate and versed in accounting,[14] began reading names from a carefully prepared list (which had been reviewed beforehand by a committee of elders presiding over the pork division). Elders were summoned by sequence of age (oldest first) and collected their shares, at which point they left the hall or lingered to watch the proceedings. No one thanked the accountant, nor the butcher. The pork was conferred by right of descent from Man Sai-go. A handful of older women entered the hall, picked up their husband's shares, and departed immediately; they did not speak, nor did anyone speak to them (it was assumed that their husbands were too ill to come themselves).

The number of shares granted to each elder varied according to his age: 60–69 years, one share; 70–79 years, two shares; 80–89 years, three shares; 90 years+, five shares.[15] Male *xin ding* (新丁 infants, literally "new male") born during the year were granted three shares each.[16] Elders who held formal offices, such as *jiazhang* (家長 lineage-branch master, oldest surviving male of the branch) received additional shares. Among the Man, elderhood was a formally recognized status that began at sixty *sui* (歲 years of age) as calculated by the lunar calendar (everyone becomes one year older on the first day of the first lunar month).[17]

Women were not recognized as elders, nor were they celebrated in the same manner as their husbands when they aged. They did, however, have their own set of rites that also involved pigs—but these celebrations were unrelated to the ancestral cult, and village men were barely aware that such rites existed.[18]

What, one might well ask, happened to the shares of raw pork that male elders collected at the ancestral halls?[19] In most cases the meat was taken home and turned over to wives or daughters-in-law, who proceeded to cook it as the main course of a special family meal. First servings went to the elder who received the meat, second servings were reserved for sons in order of seniority, third servings to daughters and daughters-in-law, and fourth (and final) serving to the senior woman of the household. Women thus consumed the ancestral pork *after* it had been *transformed* (by cooking) into a family-focused meal in individual homes.

ANNUAL RITES AT THE TOMB

Seven months after the pork division rites in San Tin's four ancestral halls, Man elders made the twenty-mile trek to the tomb of their founder (Man Sai-go) for the annual *chongyang jie* (重陽節 autumn rites)[20]—held on the sixteenth day of the ninth lunar month (October 15, 1970). This was the first of seven Man tombs visited that day and on successive days by Man lineage elders who walked into the nearby hills to commemorate important ancestors in order of seniority. Until the 1925 completion of Castle Peak Road in the northern New Territories, Man elders had to take a ferry from San Tin's pier on the nearby Shenzhen River to Castle Peak Bay along the southwest coast of the New Territories; from there they walked into the nearby (steep and treacherous) hills to reach Man Sai-go's tomb. Accordingly, only a small party of able-bodied elders, accompanied by young men of the San Tin *xun ding* (巡丁 village patrol), made the journey each year. Since the late 1950s, the annual pilgrimage has been facilitated by lorries, school buses, and private autos—making it possible for larger groups of elders to visit the tomb.

In 1970, approximately two hundred Man elders, together with the San Tin School Band and many younger (male) supporters, visited Man Sai-go's tomb (see figure 13.3). The rites began with elders lining up in generation sets—with the *zuzhang* leading the rites. Many elders wore the long gray gown that was the signature of elderhood in the 1960s and 1970s. Eight roast pigs were arrayed on the stone platform in front of the tomb. The lineage master began the proceedings by slicing the nose of each pig with a sharp chopper-knife, a process referred to as *"da kai"* (打開 opening) the meal for the ancestor, allowing him to *shi* (食 eat, *sihk* in Cantonese) the meat. The master also made a series of presentations to the ancestor with trays containing food and incense.[21] Sets of elders in generation sequence then assembled in front of the tomb and bowed in unison (one generation at a time), performing the *kau-tau*.[22] A long string of firecrackers was then ignited, creating a deafening noise that ended this phase of the rites.

Next the local schoolmaster read (in a loud voice) a prepared script,[23] written in formal Chinese, which few in the assemblage—other than the speaker himself—could understand (see figure 13.3).[24] The text, which I photographed and (with the help of colleagues) translated, was essentially a report to the ancestor outlining the success of his various financial investments and landowning enterprises. The San Tin School Band then played loudly and energetically for several minutes, ending with the deafening noise from another string of firecrackers. This signaled the end of the rites, and the assemblage began the long trek downhill to a parking lot where they boarded buses and a handful of automobiles for the long drive back to San Tin. The pig carcasses were carried downhill (by a team of hired workers) to a lorry and later divided in San Tin.

The tomb rites held in 1977 followed essentially the same format, except for the innovation of a group *kau-tau* (standing together and bowing the head

FIGURE 13.3
Man elders and San Tin villagers at Man Sai-go's tomb. Photo by James L. Watson.

three times in unison) by emigrants who worked and/or lived permanently in Europe and Canada. Many young men joined this collective performance and loudly applauded when it concluded. This innovation in performative structure was clearly a preview of the ritual transformations I observed in 2009 (outlined in the opening section of this essay). By the mid-1970s, San Tin had already begun its transformation from a tight-knit farming village to an emigrant community supported by a transnational diaspora. One innovation, however, was not yet obvious: no women participated in the 1977 rites.[25] This, too, would change dramatically by the early 2000s.

TOMB RITES AND THE CELEBRATION OF MALE SUPREMACY

In the days following the rites at Man Sai-go's tomb, six other key ancestors were visited at their hillside tombs and presented with (varying numbers of) whole pigs, most of which were carried back to San Tin[26] where they were divided into shares and distributed to participating elders—as a reward for making the arduous journey into the hills. Like the shares distributed at the ancestral halls, this pork was also taken home and divided among the elder's male descendants. Great efforts were made to ensure that every married male of reproductive age ate some of this meat (even if they lived in Kowloon or other parts of Hong Kong). Women consumed portions of the meat *after* males of the family had eaten (or were provided with) their portions.

The rites at the ancestors' tombs were treated, therefore, as the supreme celebration of patrilineal ideology—an exclusive domain of men, in the imagined absence of women.[27] Pork offered at the tomb was perceived as the symbolic bridge that linked living males with their deceased ancestor. How was this bridge conceptualized? The main benefit of eating ritual pork (according to Man and Teng lineage males interviewed between 1969 and 1978), was that it conveyed health, good luck, and—most importantly—*male offspring*.

The physical structure of the tomb, an armchair-shaped enclosure[28] made of bricks and granite, was thought to capture and store the good *feng shui* of the site.[29] Once each year, on the appointed day and in the presence of the ancestor's assembled descendants, the geomantic "energy" or "*charge*"[30] was transferred *from* the ancestral bones *into* the pork—transforming it into a special meal. One could not, in other words, approach the tomb at other times and expect to interact (or communicate) with the ancestor.[31]

The logic of this exchange implied that the ancestor was responsible for the success and efficacy of the lineage—and *not* the women who carried the *xin ding* to birth. Thus, the pork presented at an ancestor's tomb was not an ordinary meal; it was thought to transmit the essential element necessary for the perpetuation of the ancestral line.[32] It is also important to note that the ancestor *paid for his own rites*—through an estate established in his name by his sons and grandsons.[33]

HOMECOMING RITES: DIASPORA FORMATION AND THE ROLE OF WOMEN

Half a century has passed since I first witnessed the elaborate rites of pork division and ancestor worship in San Tin. The village has been transformed from a tightly knit, closed community into a post-emigrant settlement. For many former residents who live abroad, it is now a site of nostalgia—the *laojia* (老家 old home, *lou ga* in Cantonese) back in the Chinese countryside.[34] Few Man diasporics have any personal memories of San Tin as a living (or *lived-in*) community. They may have visited the village during trips to Hong Kong but have not lingered more than two or three hours.

In the 1960s and 1970s, San Tin was physically transformed as returning emigrants replaced their old single-story village houses with multi-story dwellings suitable for two or three families.[35] By the mid-1980s, the majority of these housing units were unoccupied and their steel gates locked—awaiting the occasional visit of their overseas owners. During walks through San Tin's narrow lanes in the 1990s and early 2000s, I seldom encountered any residents at all (in contrast to the bustling activity of the 1960s and 1970s). Unlike other villages near the Hong Kong/China border, San Tin's empty houses were not rented to outsiders; they stood ready for reoccupation—although that prospect faded as the builders grew old and joined their ancestors.

Meanwhile, the annual rites at Man Sai-go's tomb have become a "destination" for diasporics and their children (both male and female) who live and work abroad. These rites now resemble the homecoming/reunion celebrations (complete with picnics) that many American extended families enjoy during the summer holiday season.[36] The Man ancestor's story and the complexities of lineage history are of secondary interest to this new generation of celebrants. What draws them to the events is the opportunity to meet far-flung cousins and establish new ties of mutual benefit.

These changes are by no means restricted to New Territories diasporics who live in Europe and Canada. Anne-Christine Trémon describes similar rites among Hakka immigrants who return from homes in Tahiti to their ancestral village in Shenzhen, China—five miles north of San Tin.[37] The annual "homecoming" rite among migrants who work in Chinese cities and return to their natal villages for lunar New Year celebrations is a common feature of contemporary Chinese society. What is notable about these rites is the leading role played by *women* (wives, daughters, and daughters-in-law) who serve as the primary organizers and the leading participants.[38] Yan Yunxiang relates these developments to the demise of patriarchal decision-making and the emergence of a *conjugal* family system—in which husbands and wives act as coequals.[39] Yan further argues that women now command equal, if not (in many cases) superior authority.[40] Transformations of this nature are also evident among Chinese diasporics throughout the Pearl River Delta region.

When I first began research in San Tin (1969–1970) most emigrants had left their wives and children in the village, where they lived with their in-laws in expanded households (supported by monthly remittances). The village was essentially devoid of working-age men and was managed by a cohort of active—perhaps even empowered—male elders. The village was awash with children; they were everywhere and ever-present (evidenced by my photographs of that era). By the late 1970s and early 1980s, the village scene had changed as whole families joined the emigrant workers in London, Manchester, Amsterdam, Dusseldorf, and Toronto (among other places). Daughters and wives began serving as greeters and cashiers in Man-owned restaurants.[41] Children were dispatched to local primary schools and—within a matter of months—they were speaking colloquial English, Dutch, or German. Family size among Man diasporics decreased to one or two children—paralleling developments in Hong Kong, which had, by the early 2000s, one of the lowest fertility rates in the world.[42] San Tin was also changing as the last generation of emigrant retirees returned to the village to spend their remaining years in (often) unhappy isolation.[43]

The radical transformations of Man Sai-go's tomb rites that I witnessed in 2009 should not have been a surprise given these social developments. Daughters began playing a role in Man Clan Association activities in Hong Kong and London in the 1990s.[44] Descendants (both male and female) of my old San Tin friends now hold advanced degrees from some of Britain's leading universities. Man daughters, daughters-in-law, and mothers obviously felt entitled to worship at the ancestor's tomb during the 2009 rites and consume a portion of his roast pork. It also became obvious during my interviews in the early 2000s that young male diasporics had no idea where they ranked in the generation sequence (counting down from Man Sai-go), nor did they care. The majority had little interest in the history of San Tin or the New Territories. In effect, the Man *lineage*—as a male-dominated, patrilineal, patrilocal institution—no longer exists. That centuries-old kinship organization has been replaced by an amorphous, transnational *clan*—defined by shared surname and voluntary participation.[45] In this regard, today's Man diasporics parallel developments among Chinese who settled in Southeast Asia, North America, and the Pacific islands during the late nineteenth and early twentieth centuries.[46]

The pigs are still front and center in the Man clan rites—but the reverence with which they were once held is gone. They are, today, poignant reminders of past glories, and only a handful of older celebrants have any memory of (or knowledge regarding) the complex rituals of pork division. During the 2009 rites, two of these men stood on the sideline with me, watching their grandchildren performing the rites. They had a glimmer of tears in their eyes. Were they thinking about the past? Or, was it the dust—swirling down from the nearby hills on a brilliant fall day?

ACKNOWLEDGMENTS

This essay is dedicated to my friend and Harvard colleague, K. C. Chang, with whom I shared a lifelong interest in ancestral rites. The author is indebted to many colleagues who kindly shared observations and comments on ancestor worship and pig sacrifice: Eugene Anderson, Dale Bratton, Chan Kwok-shing, Selina Chan, David Faure, Patrick Hase, David Johnson, Liu Tik-sang, Ellen Oxfeld, Jon Unger, Rubie Watson, and Yan Yunxiang. Stanley Wong Sheung-yan helped with the research in San Tin (1969–1970); thanks also to Teng Tim-sing and Jennifer Teng Ying-lan for their assistance in Ha Tsuen (1977–1978). And, finally, a special note of thanks to Man Tso-chuen and Teng Ying-lin, elders—and now ancestors—of two ancient and illustrious lineages.

NOTES

1. *Feng-shui* "stands for the power of the natural environment, the wind and the airs of the mountains and hills; the streams and the rain; and much more than that: the composite influence of the natural processes. . . . By placing oneself well in the environment *feng-shui* will bring good fortune." Stephan Feuchtwang, *An Anthropological Analysis of Chinese Geomancy* (Vientiane: Vithagna, 1974), 2.

2. There are, of course, many disagreements and debates among Sai-go's descendants regarding the size and depth of the patrilineage—complicated by the late-twentieth century dispersal of Man emigrants to Britain, Western Europe, and Canada; see James L. Watson, "Virtual Kinship, Real Estate, and Diaspora Formation: The Man Lineage Revisited," *Journal of Asian Studies* 63, no. 4 (2004): 893–891.

3. This was a largely symbolic position and, although it carried great respect, it did not reflect the implied legal and social authority. Until recent generations, the lineage master was almost invariably illiterate and innumerate—a consequence of wealth disparities within major lineages. Wealthy families tended to marry their sons early (ages fifteen to nineteen), thereby reproducing generations much more rapidly than poor families (who were lucky to marry sons by their mid-twenties or early thirties). The rich reproduced more rapidly, which means that the lineage master (the descendant closest in generation to the founder) was—until recent generations—invariably a poor man.

4. The notion that an inlaw (an *affine*) would attend, let alone participate in, a lineage rite would have been dismissed as preposterous by my male confidants during the 1960s and 1970s. Affines were, by definition, members of other—often competing—lineages. See Rubie S. Watson, "Class Differences and Affinal Relations in South China," *Man* 16 (1981): 593–615.

5. In 1970, and again in 1978, several girls, ages eight to eleven, were bused to the tomb (together with twenty brothers and male cousins) as members of the San Tin School Band. Although most of these schoolchildren were descendants of Man Sai-go, they did not engage in any worship activities or consume pork at the tomb.

6. Report on the Census of the Colony for 1911, *Hong Kong Sessional Papers*, 1911/1917.

7. See James L. Watson, "Chattel Slavery in Chinese Peasant Society: A Comparative Analysis." *Ethnology* 15 (1976): 361–375.

8. Maurice Freedman, *Chinese Lineage and Society: Fukien and Kwangtung* (London: Athlone, 1966), 20–21; James L. Watson, "Chinese Kinship Reconsidered: Anthropological Perspectives on Historical Research," *China Quarterly* 92 (1982): 608–609.

9. These rites were abolished by Communist activists in the 1950s and 1960s; see Anita Chan, Richard Madsen, and Jonathan Unger, *Chen Village: Revolution to Globalization*, 3rd edition (Berkeley: University of California Press, 2009), 363–366; and Jing Jun, *The Temple of Memories: History, Power, and Morality in a Chinese Village* (Stanford,

CA: Stanford University Press, 1996), 79–85. San Tin's reengagement with Man lineage communities in Chinese territory is discussed in Watson, "Virtual Kinship." For a study of a Man lineage village on the Chinese side of the border, see Guo Man and Carsten Herrmann-Pillath, "Lineage, Food and Ritual in a Chinese Metropolis," *Anthropos* 114 (2019): 195–207.

10. See James L. Watson and Rubie S. Watson, "Fieldwork in the Hong Kong New Territories, 1969–1997," in *Village Life in Hong Kong: Politics, Gender, and Ritual in the New Territories* (Hong Kong: Chinese University Press, 2004) for a detailed account of research procedures underlying this and other projects carried out in Ha Tsuen and San Tin Districts. In both villages we went through a lengthy vetting procedure before we were allowed to rent houses.

11. Tablets are carved wooden slats (approximately one foot high and three inches wide) mounted on small stands that are arranged by generation sequence on large altars at the back wall of ancestral halls. There were (in 1969) 279 tablets in Man Sai-go's hall. A discussion of tablets and their complex arrangements can be found in Hugh Baker's study of Cantonese lineage organization. Hugh D. R. Baker, *A Chinese Lineage Village: Sheung Shui* (Stanford, CA: Stanford University Press, 1968), 54–61.

12. This was a major organizational and logistical feat, orchestrated by San Tin's elder council. The halls held their pork distributions during a single day—beginning in the early morning and progressing until the conclusion of the rites at Man Sai-go's hall in the late afternoon.

13. The character *fen* has a double meaning in this context; it is both an active verb ("to divide" or "to cut up") and a noun ("share" or "portion").

14. In the 1960s and 1970s, most elders could read at a minimal level but could not keep ledgers or write letters.

15. Note that the number of shares skips from three (at age 80–89) to five (at age 90+). The number four (Cantonese *sei*) is a homophone for death (Cantonese *sei*) and is therefore avoided in ritual contexts. Lineage branches with only a handful of older men sometimes recognized males aged 50–59 as elders, but this was rare.

16. Baker (*A Chinese Lineage Village*, 52) notes that, at age ninety, elders in nearby Sheung Shui village received a pig's head and "a whole wooden tub of pork." An elite lineage in Jiangxi Province distributed shares of meat (presumably pork) on a complex scale depending on age and attainment of imperial degrees; the tiered age-scale for shares increased every five years; Hu Hsien-chin, *The Common Descent Group in China and Its Functions*, no. 10 (New York: Viking Fund Publications in Anthropology, 1948), 126). Another elite lineage in Jiangxi distributed meat-cakes (rather than sections of uncooked pork) to the resident elders; see Fred R. Brown, "Clan Customs in Kiangsi Province," *Chinese Recorder and Missionary Journal* (August 1922), 519.

17. Solar calendar birthdays, marking individual dates of birth, did not become a part of village culture until the mid-1980s when nearby restaurants, including McDonald's,

began to offer birthday celebrations for local children; see James L. Watson, "McDonald's in Hong Kong," in *Golden Arches East: McDonald's in East Asia* (Stanford, CA: Stanford University Press, 1997), 103–105.

18. Rubie S. Watson, "Girls' Houses and Working Women: Expressive Culture in the Pearl River Delta, 1900–1941," in *Women and Chinese Patriarchy: Submission, Servitude and Escape*, eds. Maria Jaschok and Suzanne Miers (London: Zed Press. 1994).

19. San Tin's three branch ancestral halls (founded in honor of Man Sai-go's third-, fifth-, and tenth-generation descendants) also distributed pork on February 3, 1970, making it a busy day for local elders. My sixty-five-year-old neighbor received ten catties (1 *jin*,= 1.3 English pounds) of raw sliced pork that day. Most of his allotment was distributed to family members who did not attend the rites.

20. Baker (*A Chinese Lineage Village*, 66–69) provides an excellent description of these annual rites practiced by the Liu lineage in the New Territories village of Sheung Shui.

21. Items included: one cup of wine, one bowl of steamed pork intestines, one bowl of fruit, one set of sugar candies, one cup of tea, and one bundle of incense sticks.

22. Stand to attention, bow from the waist, go down on knees, lean forward, and touch head to ground; repeat three times.

23. Referred to as *jiwen*, a formal document handwritten on red paper that was later burned on-site, indicating that it was perceived as a form of prayer (regardless of its apparent mundane content).

24. The San Tin Schoolmaster of the late 1960s and 1970s was a highly educated man who was renowned for his mastery of formal written Chinese (*wenyin*) and calligraphy. He was consulted widely by villagers who needed help in preparing stone inscriptions that appear in local temples and ancestral halls.

25. A handful of adult women attended as observers (from the sidelines), and several Man daughters (ages eight to twelve) performed with the San Tin School Band.

26. Minor estates sometimes held hillside banquets for elders at their ancestor's tomb. A single whole pig was butchered on site, cooked, and eaten. Only a handful of elders and younger men were involved. Elders called this banquet *chi shantou* (eating on the mountain top) and explained that "in the past" all tomb rites culminated in a tomb-side banquet; for a description of this style of banqueting, see Baker, *A Chinese Lineage Village*, 68.

27. Accordingly, among the Man and Teng, the bond of affinity was dismissed as a troublesome burden; see Rubie S. Watson, "Class Differences and Affinal Relations in South China," *Man (new series)* 16, no. 4 (1981): 593–615. In contrast, the kinship systems that emerged in Taiwan and parts of north China were less exclusive and more open to affines; see, for example, Emily Ahern, *The Cult of the Dead in a Chinese Village* (Stanford, CA: Stanford University Press, 1973); Myron Cohen, "Lineage Organization in North China," *Journal of Asian Studies* 49, no. 3 (1990): 509–534; Burton Pasternak, *Kinship and Community in Two Chinese Villages* (Stanford, CA: Stanford University

Press, 1972); and Yan Yunxiang, *The Flow of Gifts: Reciprocity and Social Networks in a Chinese Village* (Stanford, CA: Stanford University Press, 1996).

28. Known in some regions of south China as *yizi fen* (chair tombs); see Chen Ningning, "Governmentality and Translation: Rethinking the Cultural Politics of Lineage Landscapes in Rural China," *China Quarterly* 247 (2021): 703–723.

29. This is one reason the physical sites of ancestral tombs are carefully guarded and watched by the *xun ding* (village patrolmen) of larger lineages. The first sign that a lineage was losing its dominance was the intrusion of rogue burials in its traditional hillside territory.

30. Several elders described the process as one of transferring spiritual *qi* (Cantonese: *hei* steam or force) into the meat. One elder observing of the 1970 rites used the metaphor of electric batteries to explain the exchange. For a full description of this exchange, see Rubie S. Watson, "Remembering the Dead: Graves and Politics in Southeastern China," in *Death Ritual in Late Imperial and Modern China*, ed. James L. Watson and Evelyn Rawski (Berkeley: University of California Press, 1988).

31. Communication with Cantonese ancestors thus differed from other systems of ancestral interchange, such as those prevailing in parts of Africa; see, for example, John Middleton, *Lugbara Religion: Ritual and Authority among an East African People* (Oxford: Oxford University Press, 1960).

32. There are, of course, many tempting parallels in other religious traditions, the most obvious being the Christian Eucharist. In my earlier years, under the influence of an essay presented by Edmund Leach at a London University Intercollegiate Seminar, I pursued this line of inquiry. With age, however, I have calmed down and prefer to leave such speculations to my younger colleagues.

33. Such estates are known by the ancestor's posthumous name (*hao*) and are recorded under that name in Hong Kong government land records. For a full discussion of ancestral estates, see Rubie S. Watson, *Inequality among Brothers: Class and Kinship in South China* (Cambridge: Cambridge University Press, 1985), 68–72.

34. See, for example, Maggi W. H. Leung, "Notions of Home among Diaspora Chinese in Germany," in *The Chinese Diaspora: Space, Place, Mobility, and Identity*, eds. Laurence Ma and Carolyn Cartier (New York: Rowman and Littlefield, 2003); Khun-eng Kuah-Pearce, *Rebuilding the Ancestral Village: Singaporeans in China* (Hong Kong: Hong Kong University Press), 102–109; and Anne-Christine Trémon, *Diaspora Space-Time: Transformations of a Chinese Emigrant Community* (Ithaca, NY: Cornell University Press, 2022).

35. These compounds (which Man emigrants called "sterling houses") had barred windows and steel gates; most were left unoccupied for months, even years, by their owners; see James L. Watson, *Emigration and the Chinese Lineage: The Mans in Hong Kong and London* (Berkeley: University of California Press, 1975) 155ff.

36. "Homecoming" or "reunion" rites are popular summer events for many American kindreds and emerged during the late nineteenth century along with a growing

interest in genealogical reconstruction; see, for example, François Weil, *Family Ties: A History of Genealogy in America* (Cambridge, MA: Harvard University Press, 2013), 180–202. Graveyards and tombs are often the focus of these events; see Jack Glazier, *Been Coming through Some Hard Times: Race, History, and Memory in Western Kentucky* (Knoxville: University of Tennessee Press, 2012), 181–187.

37. Trémon, *Diaspora Space-Time*, 151–160.

38. Ellen Oxfeld, *Drink Water, but Remember the Source: Moral Discourse in a Chinese Village* (Berkeley: University of California Press, 2010), 158–166. Oxfeld also notes (email of December 5, 2020) that "married-out daughters of the lineage seem to be playing a more prominent and enthusiastic role in the ancestral rites." Jonathan Unger (email of November 30, 2020) describes a similar development in "Chen Village," a rural community located in the Shenzhen Special Economic Zone, twenty miles north of San Tin (see footnote 9 above). See also Gonçalo Santos, *Chinese Village Life Today: Building Families in an Age of Transition* (Seattle: University of Washington Press, 2021), 53ff.

39. Yan Yunxiang, "The Triumph of Conjugality: Structural Transformation of Family Relations in a Chinese Village," *Ethnology* 36 (1997): 191–212.

40. Yan notes that villagers describe this as "girl power"; see Yan Yunxiang, "Young Women and the Waning of Patriarchy in Rural North China," *Ethnology* 45 (2006): 105–123. These developments are also described in *Chinese Families Upside Down*, ed. Yan Yunxiang (Leiden: Brill, 2021). See also Michelle King's essay in this volume (chapter 7) for an account of women's increasing command of respect in the Chinese publishing industry.

41. Watson, *Emigration and the Chinese Lineage*, 119–131.

42. Hong Kong's fertility decline paralleled simultaneous developments in Britain and Western Europe—the primary destinations of Man emigrants; see Rubie S. Watson, "The Anatomy of Fertility Decline in Hong Kong," *Harvard China Review* 6, no. 1 (2010): 34–36.

43. In 1997, while I was walking through San Tin, a young woman ran out of a large renovated house and threw a pan of water into a nearby drain. Instinctively I said: "*Josan, sihk-jou fan mei- ya?*" (Cantonese for "Good morning, have you eaten yet?"), the colloquial way of asking "How are you today?" She responded in English, "I'm sorry; I don't speak Chinese." She was a Thai healthcare worker who had been hired to take care of the aging grandmother of an emigrant family in London. The retiree, who also spoke English, said she did not want to end her years in a "foreign place like England."

44. The International Man Clan Association held its 1997 annual meeting in Kowloon, with women appearing for the first time onstage with the assembled management committee. There is also a Man Clan Association Headquarters in London's Soho District, 29 Romilly Street. For a discussion of this organization, see Gregor Benton and Edmund Gomez, *The Chinese in Britain, 1800-Present: Economy, Transnationalism, Identity* (London: Palgrave Macmillan, 2008), 161–162.

45. For a discussion of "lineage" versus "clan" in the Chinese historical context, see Watson, "Chinese Kinship Reconsidered."

46. See, for example, Steven B. Miles, *Chinese Diasporas: A Social History of Global Migration* (Cambridge: Cambridge University Press, 2020); G. William Skinner, *Chinese Society in Thailand: An Analytic History* (Ithaca, NY: New York, 1957); and Anne-Christine Trémon, "Flexible Kinship: Shaping Transnational Families among the Chinese of Tahiti," *Journal of the Royal Anthropological Institute* 23 (2017): 42–60.

AFTERWORD: CHINESE FOOD FUTURES

JIA-CHEN FU, MICHELLE T. KING, AND JAKOB A. KLEIN

Chinese foodways are modern, and Chinese culinary modernity can offer lessons for understanding culinary modernity elsewhere around the world. This volume has endeavored to demonstrate these two points by bringing together scholarly work that expands our temporal sense of the modern and complicates the idea that modern Chinese foodways should be understood simply or primarily as responses to techno-scientific, economic, and cultural developments in the North Atlantic "core" of an emerging capitalist global food system since the nineteenth century. Instead, a serious exploration of Chinese culinary modernity offers insights on dynamic processes that were by no means unique to China yet nonetheless subject to and reflective of the particularities of Chinese historical experiences. Not only do aspects of what we now understand as modern Chinese food predate the First Opium War (1839–1842), but as seen in the chapters of this volume, Chinese foodways—be it the development of soy sauce into a proto-industrial mass food (chapter 2), the medically pluralist discourses on dietary health that preceded, intertwined, and reinterpreted vitamin knowledge of the early twentieth century (chapter 4), cookbooks written by and for female domestic cooks (chapter 7), or the importance of diasporic Chinese restaurant cuisines in the formation of Peruvian national identity (chapter 12)—are themselves constitutive of global culinary modernity.

While the hyperbole (whether triumphant or ominous) of an "Asian century" should be treated with a healthy dose of skepticism, it is hard to dispute the increasingly dominant roles played by China as both producer and consumer in the global food system. Chinese food futures and global food futures are inescapably intertwined. This fact has often been recognized in relation to the impact of China's "nutrition transition" (in particular, the shift away from grain-based diets and the growing demand for animal-derived foods) on food production and distribution systems, food supplies, sustainability, climate change, animal welfare, food safety, and zoonotic diseases.[1] However, the implications of "rising China" for foodways—from everyday dietary cultures to elite cuisines—have received much less attention. In this afterword, we attempt to sketch out some of these implications. Rather than offer firm predictions, we put forward a series of themes that are likely to be relevant in the coming years and through which possible food futures may be imagined, discussed—and researched.[2]

TECHNOSCIENCE AND NEO-TRADITIONALISM

China is pioneering technoscientific research and implementation in food production, distribution, and preparation. This includes seeking technoscientific solutions to address the challenges connected to food system intensification itself, including but not limited to the introduction of vertical pig farming, investments in seed technology, or utilizing unique identification systems for livestock.[3] Some of these measures, especially those in areas such as food safety and quality, may serve to allay consumers' anxieties while heightening their sense of control and agency. Blockchain technologies, for example, are being deployed to allow high-end consumers in the People's Republic of China (PRC) to scan a QR code attached to a slaughtered chicken that has been ordered and delivered to their door, so that they can learn about its life on the farm.[4] And, at a more grassroots level (at least prior to the COVID-19 pandemic), in cities like Shenzhen, WeChat was being used to facilitate the direct sale of farm goods to networks of urban consumers in a practice known as "*shequ tuangou*" (社区团购 community group buying).[5] Other innovations promise greater "convenience" to the consumer through online and drone food delivery services, where complex food supply chains are streamlined with the help of Information and Communication Technologies (ICTs) (chapter 6). While these transformations have been underway for some time, many have undoubtedly been accelerated by the COVID-19 pandemic, worsening geopolitical relations across the globe, and corresponding government responses by China, Hong Kong, and Taiwan. Coupled with state-led initiatives to ensure Chinese food security are business-led innovations in digital food that are impacting practices around the world, where Chinese apps such as Eleme (饿了么 literally "Are you hungry yet?")—and apps modeled on these such HungryPanda (熊猫外卖), which was launched in Manchester in 2016—are used for ordering food among Chinese communities overseas and increasingly by others, too.[6]

Some of the social and cultural implications of the revolution in digital food are already becoming apparent and will continue to be an important area for researchers.[7] As with industrial food in nineteenth-century Europe, digital food is contributing to the transformation of class relations, gendered relations, and to reconfigurations of domestic and public "cooking" (chapter 6).[8] For instance, like networks supplying wealthy Chinese households with imported baby formula, blockchain technologies may increase social inequalities by enabling elites to protect themselves by bypassing ordinary food supply chains.[9] Further, the digital food revolution may contribute to a radical transformation of the catering sector and the practices and social significance of China's diverse and often vibrant public food spaces, be they restaurants, teahouses, hawkers' stalls, or fresh food markets ("wet markets"). Even before the COVID-19 pandemic began in 2020, many of these public food spaces were being transformed by urban reconstruction and policies of "spatial cleansing."[10] Will the revolution in digital food contribute to the demise of Chinese urban public food cultures, or will it provide tools for its (post-COVID) regeneration?

Parallel to the growing significance of science and technology for Chinese foodways, we are also seeing an increase in the demand for artisanal and local foods, often couched in languages of Chinese "tradition" and "heritage" (chapter 1), as well as the proliferation (at least until recently) in Taiwan, Hong Kong, and the PRC of "ethical" movements advocating, for instance, animal protection, vegetarianism, and consumer-producer connections.[11] This demand, as in the case of some digital technologies, is partly driven by anxieties surrounding the changing food system, in relation to food safety, taste, and dietary health. Arguably, the interest in "heritage" and "ethical" foods is partly a reaction against the role of technoscience in food production and processing.

At the same time, the move toward "heritage" Chinese foods and the emergence of "alternative" food networks often draw on new technologies, whether it be in production, processing, distribution, or retail. One example is the development of "cultured meat" and other alternatives to animal flesh, such as Hong Kong's OmniPork; research in this area will need to take into account the thousand-year history of Chinese Buddhist vegetarian cuisine and its "mock meats." Media—both traditional media and new media—have been important drivers of the domestic and global interest in Chinese local and "heritage" foods, as in the case of the demand stimulated by the *Bite of China* (2012, 2014, and 2018) television series and facilitated through online retail platforms. More recently, Chinese content creators (e.g., Li Ziqi, Dianxi Xiaoge, Wayne Shen of "foodiechina888") on social media platforms, such as YouTube, TikTok, and Instagram, who are popular in the global Sinosphere and beyond, celebrate all manner of "Chinese" food, be it local, traditional, or trending foods (chapter 10). As Zhiyi Yang has observed, food media activists do not only draw on the language of rural foods; they also celebrate historical culinary connections to Chinese elite literati culture, a form of what she calls "Sinophone classicism."[12] These media productions tie in with wider trends toward culinary nationalism and regionalism, to which we now turn.[13]

CULINARY NATIONALISMS AND REGIONALISMS

In China, Hong Kong, and Taiwan, we are likely to see the continued proliferation of new or transformed local and regional culinary identities and practices. In part, this will be driven by the ongoing demand, mentioned earlier, for local, artisanal, and heritage foods. Food producers and tourism providers, in some cases supported or guided by local officials, will seek to attract visitors and customers by branding, certifying, and sometimes inventing distinctive local and ethnic foods, drinks, and consumption styles (chapters 1 and 9).

We will also see the creation of new or reinvented local and regional foodways in response to climate change, changing food production and supply chains, and migrations. For example, Jakob Klein has discussed the reinvention of potato-based delicacies in Yunnan and Inner Mongolia, in response to government attempts to improve national food self-sufficiency through the promotion of potato production.[14] Yunshan Li has documented the creation by migrant mine workers from southwest China of "Sichuan-style noodles" in Zoucheng, Shandong Province, in northern China, and their development over time into an emblem of local identity in Zoucheng—a story that in some ways resonates with the creation of "Taiwan beef noodles" by mainlander immigrants in Taiwan in the mid-twentieth century.[15]

In some cities, distinctive culinary identities will evolve, not (or not only) through the development of distinctive local or regional foods but through the popularity of cuisines from outside and the development of "diversity-receptive" culinary habits and predispositions.[16] The rise of Japanese cuisine in Shanghai and Tianjin has come to define and shape Chinese urban food culture (chapter 11). Similarly, Chenjia Xu documents the growth of a confident "culinary cosmopolitanism" in Beijing, one dimension of the capital's emergence in the 2010s as a global city.[17] Cosmopolitan food cultures in places like Beijing, Shanghai, Hong Kong, and Taipei also draw our attention to questions about the relationship between taste and status in Chinese-speaking societies and to the unequal geographies of taste and their articulation with place-based identities and stereotypes.

Alongside the divergences of local foodways, we are also likely to see a growing convergence at national scales of culinary practices and identities. These are in part driven by the same processes as the diversities. Across towns and cities in twenty-first-century China, as in post-independence India as famously discussed by Arjun Appadurai,[18] labor migrations, class mobilities, tourism, and media productions are creating the conditions for people to taste each other's foods and to imagine and articulate culinary similarities and differences. Food supply networks and retail and restaurant chains operating at a national scale may also increase experiences of culinary convergence and create new opportunities for trying out foods from other regions. Through the influence of social media, new culinary trends in one place are quickly picked up elsewhere, as Wu notes in chapter 9 in relation to the Hubei "Tujia" custom of "smashing bowls" liquor.

But we should not forget the importance of nation- and state-building projects. These include an emphasis in the PRC on increasing self-sufficiency in food and on domestic, consumption-led growth; barriers to international trade, migration, and cultural exchange in the context of disputes over trade balances, territories, and human rights; and protective measures to counter viral epidemics and pandemics, from African swine fever to avian flu to COVID-19. In other words, far from the ever-increasing global flows and trade liberalizations imagined by many in the 1990s and 2000s, the reinforcement of national boundaries is likely to persist for some time and coincide awkwardly with global interdependencies in food and other areas—and this is likely to have a significant effect on Chinese foodways.[19]

Relatedly, we should pay close attention to the culinary implications of discourses on nationhood in the PRC, Hong Kong, and Taiwan (chapter 8), be they popular, state-led, or both.[20] These debates on culinary belonging are played out not only domestically but also globally in spheres such as UNESCO's list of intangible cultural heritage, in media productions such as *A Bite of China*, and in social media across the Sinosphere.[21] We would do well to investigate what kinds of culinary diversities will fit into discourses and policies on Chinese or Taiwanese nation-building projects, and which will not. For example, while some distinctive ethnic minority foodways are celebrated as part of national culture, others (or the same ones, in other settings) may be losing ground to Han-associated practices due to a combination of ecosystem destruction, settler colonialism, and political suppression.[22] And what of Hong Kong's and Macau's unique culinary styles and eating establishments, some of which are profoundly shaped by the experience and structures of British and Portuguese colonialisms? What role will they play following the demise of Western colonialism and the rise of an increasingly vocal Chinese cultural nationalism?

CONVERGENCES AND DIVERGENCES IN CHINESE DIASPORIC FOODWAYS

As with foodways in China, Hong Kong, and Taiwan—and in ways that will often be connected to developments there—we are likely to witness both convergences and divergences in Chinese diasporic foodways. An obvious place to begin is with changing patterns of migration and settlement. While Chinese international migration from the mid-nineteenth to mid-twentieth centuries was complex and diverse, nevertheless, Chinese migrants to places such as Southeast Asia (which, of course, already had long-standing Chinese trade networks and settlements), the Americas, Australia, South Africa, and Europe were overwhelmingly rural and working class and had origins in China's southeastern provinces. In several countries of settlement, many Chinese (not least for reasons of race-based exclusion) came to occupy service industries and often the food and restaurant trades, in many cases catering to non-Chinese.[23]

Beginning in the 1960s and 1970s in countries such as the United States and Canada, and more broadly since the 1990s with the rise of China and other East Asian economies, rural and working-class Chinese migrants have been joined by students, professionals, and tourists with origins from across China and the Chinese-speaking world. Many of these new migrants are middle-class, and some are quite affluent. Increasingly, Chinese restaurants in Europe, Japan, and North America cater to the tastes of Chinese migrants and settlers. Migrant-run, "mom-and-pop" restaurants and food and drink retailers are now joined by a growing number of highly capitalized chains, often with headquarters in China, Taiwan, and elsewhere in East Asia—but in some cases also with bases in countries of settlement. Advertising "authenticity" (in the sense of being grounded in regional traditions in China or Taiwan) and orienting themselves toward the latest gastronomic trends in global cities like Shanghai, Taipei, and Singapore,[24] these establishments also attract many non-Chinese "culinary adventurers"[25] and "cultural omnivores,"[26] and have the potential for a revolutionary impact on the experiences, perceptions, and statuses of "Chinese cuisines" outside Chinese communities.

This Chinese culinary transnationalism is likely to be affected by the growing barriers to mobility alluded to in the previous section and which of course also includes new barriers to migration established by governments in Europe, North America, and elsewhere. While more research is needed on new Chinese migration and cuisines in the West, we know even less about the culinary dimensions of increasing Chinese migration and investment in Latin America, Africa, the Middle East, and Central Asia. At the same time, China's ever-more muscular nationalism is also giving rise to new waves of migrants and political refugees from places such as Tibet, Xinjiang, and Hong Kong. These people are already having an impact on public foodways in their countries of settlement. For example, Hong Kong migrants arriving in Britain since 2021 on British National (Overseas) visas have quickly been establishing Hong Kong-style *chahchaanteng* (茶餐廳 cafés).[27]

While new waves of migrants are rapidly changing Chinese foodways overseas, and in some respects bringing them into closer alignment with contemporary Chinese foodways in East Asia, many countries—also outside Asia—already have well-established and diverse Chinese communities, in some cases dating back several generations. In countries such as Canada and the United Kingdom, where Chinese settlement has been geographically dispersed, the nature of Chinese communities and cuisines continues to vary greatly between, for example, rural areas and major cities. Despite the challenges of doing so, especially outside of metropolitan centers with large numbers of East Asian residents and businesses, many people of Chinese descent maintain and renew Chinese culinary practices, often over several generations. As Watson suggests in this volume (chapter 13), return visits to ancestral homelands, for example on important ritual occasions, may connect diasporic communities to Chinese geographies of taste and strengthen bonds of kinship and identity, even as ritual practices are radically transformed.

Yet the culinary practices of long-standing Chinese communities in the diaspora are not static, nor are ancestral homelands peoples' only culinary reference points. Localized forms of Chinese cuisines in the diaspora, be they exo-cuisines (foods served to outsiders) and endo-cuisines (foods for cultural insiders), or some combination of both, are in many cases resilient. Some are or may become the focus for revival and reinvention. For example, chef Lucas Sin of the fast-casual chains Nice Day and Junzi Kitchen restaurants in the greater New York City area has discussed the challenges facing the North American Chinese takeout restaurants, despite their ongoing popularity. He and other culinary colleagues see opportunities in initiatives to revitalize and "modernize" Chinese-American takeout food.[28]

In some cases, long-standing Chinese immigrant cuisines have become embedded in constructions of "creolized" regional and national cuisines, as in Malaysia and Singapore,[29] Peru (chapter 12), and Hawai'i.[30] In countries such as the United States and Britain, Chinese chefs have long been pigeonholed as producers of "ethnic cuisine" and therefore have struggled to be recognized as legitimate producers of high cuisines.[31] Recently, Chinese restaurants run by diasporic Chinese are beginning to break into national "fine dining" spheres in some Anglophone countries—and even in France.[32] What is their relationship to the high-end Chinese restaurants, chefs, and restaurateurs from East Asia, some of whom cater internationally but predominantly to East Asian elites? To what extent will Chinese chefs, be they diasporic or based in China, cooking Chinese cuisines, come to join (or even displace!) French and Japanese chefs and set the standards in the "transnational culinary field"?[33]

As early as 1997, James L. Watson noted in his edited volume on McDonald's in East Asia that Hong Kong "has itself become a major center for the *production* of transnational culture, not just a sinkhole for its *consumption*."[34] Today, the same may be said for cities, countries, and regions across China and East Asia. If it once was possible to claim that the North Atlantic was the "core" of an emerging "world food system" in the nineteenth century,[35] in the twenty-first century, China and East Asia is a core region in a multipolar, global food system.

This has important implications for understanding the future of global culinary modernity. Increasingly, Chinese experiences and solutions will serve as harbingers of and perhaps models for change elsewhere. For example, the proactive role played by provincial Chinese governments (along with producers and restaurant industries) in underdeveloped, inland areas such as Shanxi Province in promoting regional foodways on both national and international stages may offer models for state-backed, food-led rural and regional development initiatives elsewhere.[36] At the same time, China's ongoing struggles to adequately address its food safety problems may serve as a warning of another dimension of state-led market economies, namely the tendency towards an over-centralization of regulatory regimes and policy formation.[37] Indeed, these examples reflect how food and foodways continue to be implicated in a long-standing tension in modern Chinese state practice, between drives toward centralization and standardization on the one hand and the encouragement of regional differentiation and flexibility on the other.

Yet China and Chinese foodways are crucial to global culinary modernity not only for the models or warnings that China may offer to others but also because it is increasingly inseparable from culinary modernity writ large. Of course, Chinese culinary modernities have long been intertwined with other culinary modernities, such as in the profound role Chinese immigrant businesses have had in shaping the modern foodways of countries such as the United States[38] and Peru (chapter 12). Increasingly, especially since the country's accession to the World Trade Organization in 2001, China's food exports and food imports play a central role in global food production and consumption. As suggested earlier in this afterword, this is important not only when addressing issues related to global food supplies.

With China and East Asia becoming ever-more dominant within the "global hierarchy of value,"[39] we must also pay close attention to the role that Chinese food consumers, food business, chefs, and gastronomes are likely to have in shaping transnational culinary tastes and trends. In short, interdisciplinary research into diverse Chinese culinary experiences, practices, values, movements, and innovations is now central to understanding the unfolding of global culinary modernity.

NOTES

1. See, for example, Lester Brown, *Who Will Feed China?: Wake-Up Call for a Small Planet* (Norton 1995); Emiko Fukase and Will Martin, "Who Will Feed China in the 21st Century? Income Growth and Food Demand and Supply in China," *The World Bank Policy Research Working Paper 6926* (June 2014): 1–52; Vaclav Smil, *China's Past, China's Future: Energy, Food, and Environment* (New York: Routledge, 2004); Hongzhou Zhang, *Securing the "Rice Bowl": China and Global Food Security* (Singapore: Palgrave Macmillan, 2019).

2. Some of these themes were addressed by artist and programmer Xiaowei Wang, chef Lucas Sin, and food writer Fuchsia Dunlop in an online roundtable panel on "Chinese Food Futures," which we held in April 2022. The roundtable was hosted at Emory University and chaired by Jia-Chen Fu. https://www.youtube.com/watch?v=qLVYxTKpg7Q&t=3s.

3. Daisuke Wakabayashi and Claire Fu, "China's Bid to Improve Food Production? Giant Towers of Pigs," *New York Times*, February 8, 2023, https://www.nytimes.com/2023/02/08/business/china-pork-farms.html; GT Staff reporters, "Seeds Self-Reliance Tops Govt Agenda," *Global Times*, April 11, 2022, https://www.globaltimes.cn/page/202204/1259047.shtml; Nurul Ain Razali, "Food Security Innovations Revealed: How Tech Advances Are Helping China Strengthen Its Supply Chains," *Food Navigator Asia*, March 23, 2022, https://www.foodnavigator-asia.com/Article/2022/03/23/innovations-in-food-security-china-utilises-ai-robotics.

4. Xiaowei Wang, *Blockchain Chicken Farm and Other Stories of Tech in China's Countryside* (New York: Macmillan, 2020).

5. Jakob Klein email communication with Wanlin Lu, SOAS University of London, January 20, 2021.

6. Zhongzhi He and Yuk Wah Chan, "Platform Economies and Home Virtuality: The Digitalized Life of International Chinese Students in the UK," paper presented at the 30th Anniversary Conference of the founding of the International Society for the Study of Chinese Overseas (ISSCO), San Francisco, November 11–12, 2022.

7. Tanja Schneider, Karin Eli, Catherine Dolan, Stanley Ulijaszek, eds., *Digital Food Activism*, (London: Taylor and Francis, 2018).

8. Sidney W. Mintz, *Sweetness and Power: The Place of Sugar in Modern History* (New York: Viking, 1985).

9. Amy Hanser and Jialin Camille Li, "Opting Out? Gated Consumption, Infant Formula, and China's Affluent Urban Consumers," *The China Journal* 74 (2015): 110–128.

10. Qian Forest Zhang and Zi Pan, "The Transformation of Urban Vegetable Retail in China: West Markets, Supermarkets, and Informal Markets in Shanghai," *Journal of Contemporary Asia* 43, no. 3 (2013): 497–518; Michael Herzfeld, "Spatial Cleansing: Monumental Vacuity and the Idea of the West," *Journal of Material Culture* 11 (1–2) (2006): 127–149.

11. See, for example, Jakob A. Klein, "Buddhist Vegetarian Restaurants and the Changing Meanings of Meat in Urban China," *Ethnos* 82, no. 2 (2017): 252–276; Yi-Chieh Lin, "Sustainable Food, Ethical Consumption and Responsible Innovation: Insights from the Slow Food and 'Low Carbon Food' Movements in Taiwan," *Food, Culture & Society* 23, no. 2 (2020): 155–172; and Hao-tzu Ho, "Cosmopolitan Locavorism: Global Local-Food Movements in Postcolonial Hong Kong," *Food, Culture & Society* 23, no. 2 (2020): 137–154.

12. Zhiyi Yang, "Sinophone Classicism: Chineseness as Temporal and Mnemonic Experience in the Digital Era," *The Journal of Asian Studies* (2022), 1–15. https://doi.org/10.1017/S0021911822000596.

13. Michelle T. King, ed., *Culinary Nationalism in Asia* (London: Bloomsbury Academic, 2019); Michelle T. King, "What Is 'Chinese' Food? Historicizing the Concept of Culinary Regionalism," in the special issue on Chinese culinary regionalism, *Global Food History* 6.2 (Summer 2020): 89–109.

14. Jakob A. Klein, "Ambivalent Regionalism and the Promotion of a New National Staple Food: Reimagining Potatoes in Inner Mongolia and Yunnan," *Global Food History* 6, no. 2 (2020): 143–163.

15. Yunshan Li, "'Sichuan-Style Noodles Are Not in Sichuan, but in Zoucheng': A Story of the Struggle for Positions in a Northern Chinese City from the 1970s to 2010s," unpublished MA thesis, (SOAS University of London, 2022).

16. C. Claire Hinrichs, "The Practice and Politics of Food System Localization," *Journal of Rural Studies* 19 (2003): 33–45.

17. Chenjia Xu, "From Culinary Modernism to Culinary Cosmopolitanism: The Changing Topography of Beijing's Transnational Foodscape," *Food, Culture, & Society* 26, no. 3 (2023): 775-792.

18. Arjun Appadurai, "How to Make a National Cuisine: Cookbooks in Contemporary India," *Comparative Studies in Society and History* 30, no. 1 (1988): 3–24.

19. James L. Watson, "Update: McDonald's as Political Target: Globalization and Anti-globalization in the Twenty-First Century," in *Golden Arches East: McDonald's in East Asia*, 2nd ed., ed. James L. Watson (Stanford, CA: Stanford University Press, 2006), 183–197; Alan Smart and Josephine Smart, "Time-Space Punctuation: Hong Kong's Border Regime and Limits on Mobility," *Pacific Affairs* 81, no. 2 (2008): 175–193.

20. Selina Ching Chan, "Tea Cafes and the Hong Kong Identity: Food Culture and Hybridity," *China Information* 33, no. 3 (2018): 311–328.

21. Philipp Demgenski, "Culinary Tensions: Chinese Cuisine's Rocky Road toward International Intangible Cultural Heritage Status," *Asian Ethnology* 79, no. 1 (2020): 115–135; Fan Yang, "*A Bite of China*: Food, Media, and Televisual Negotiation of National Difference," *Quarterly Review of Film and Video* 32, no. 5 (2015): 409–425. https://doi.org/10.1080/10509208.2015.999223

22. Brendan Galipeau, "Tibetan Wine Production, Taste of Place, and Regional Niche Identities in Shangri-La, China," in *Trans-Himalayan Borderlands: Livelihoods, Territorialities, Modernities*, ed. Dan Smyer Yü and Jean Michaud (Amsterdam: Amsterdam University Press, 2017), 207–228; Emily Yeh, *Taming Tibet: Landscape Transformation and the Gift of Chinese Development* (Ithaca: Cornell University Press, 2013).

23. Heather Ruth Lee, "Entrepreneurs in the Age of Chinese Exclusion: Transnational Capital, Migrant Labor, and Chinese Restaurants in New York City, 1850–1943" (Ph.D. diss., Brown University, 2014).

24. Haiming Liu, *From Canton Restaurant to Panda Express: A History of Chinese Food in the United States* (New Brunswick, NJ: Rutgers University Press, 2015).

25. Lisa Heldke, "Let's Cook Thai: Recipes for Colonialism," in *Food and Culture: A Reader*, 3rd ed., eds. Carole Counihan and Penny Van Esterik (London: Routledge, 2013), 394–408.

26. Josée Johnston and Shyon Baumann, *Foodies: Democracy and Distinction in the Gourmet Foodscape* (New York: Routledge, 2010).

27. Ho-Yee Mak, "Caacaanteng in the UK: Hong Kong Migrants, Food, Space and Identity," unpublished MA Thesis (SOAS University of London, 2022).

28. Cathy Erway, "More Than 'Just Takeout," *New York Times*, June 21, 2021. https://www.nytimes.com/2021/06/21/dining/american-chinese-food.html.

29. Tan Chee-Beng, "Nyonya Cuisine: Chinese, Non-Chinese and the Making of a Famous Cuisine in Southeast Asia," in *Food and Foodways in Asia: Resource, Tradition and Cooking*, eds. Sidney C.H. Cheung and Tan Chee Beng (London: Routledge, 2011), 172–182.

30. Samuel Hideo Yamashita, *Hawai'i Regional Cuisine: The Food Movement that Changed the Way Hawai'i Eats* (Honolulu: University of Hawai'i Press, 2019).

31. Krishnendu Ray, *The Ethnic Restaurateur* (London: Bloomsbury Academic, 2016).

32. Lise Gibet, "En rouge et noir: Les restaurants chinois à Paris, indicateurs des reconfigurations locales et transnationales," in *Mobilités et mobilisations Chinoises en France*, eds. Chuang Ya-Han and Trémon Ann-Christine (Marseille: Terra HN éditions, 2020).

33. James Farrer, "Shanghai's Western Restaurants as Culinary Contact Zones in a Transnational Culinary Field," in *The Globalization of Asian Cuisines*, ed. James Farrer (New York: Palgrave Macmillan, 2015), 103–124.

34. James L. Watson, "McDonald's in Hong Kong: Consumerism, Dietary Change, and the Rise of a Children's Culture," in *Golden Arches East: McDonald's in East Asia*, 2nd ed., ed. James L. Watson (Stanford, CA: Stanford University Press, 2006), 108.

35. Jack Goody, *Cooking, Cuisine and Class: A Study in Comparative Sociology* (Cambridge: Cambridge University Press, 1982).

36. David L. Wank, "Knife-Shaved Noodles Go Global: Provincial Culinary Politics and the Improbable Rise of a Minor Chinese Cuisine," in *The Globalization of Asian Cuisines*, ed. James Farrer (New York: Palgrave Macmillan, 2015), 187–208.

37. John K. Yasuda, *On Feeding the Masses: An Anatomy of Regulatory Failure in China* (Cambridge: Cambridge University Press, 2018).

38. Yong Chen, *Chop Suey, USA: The Story of Chinese Food in America* (New York: Columbia University Press, 2014).

39. Michael Herzfeld, *The Body Impolitic: Artisans and Artifice in the Global Hierarchy of Value* (Chicago: The University of Chicago Press, 2003).

RECOMMENDED WORKS

Complete citations for all works cited in this volume can be found in the corresponding endnotes for each chapter. Rather than compiling a bibliography of works cited here, we have provided readers with an up-to-date bibliography of Chinese food studies. The list below focuses primarily though not exclusively on English-language scholarship, including both foundational classics and very recent works.

Anderson, E. N. *The Food of China*. New Haven, CT: Yale University Press, 1988.

Anderson, E. N. *Food and Environment in Early and Medieval China*. Philadelphia: University of Pennsylvania Press, 2014.

Anderson, E. N., Teresa Wang, and Victor Mair. "Ni Zan, Cloud Forest Hall Collection of Rules for Drinking and Eating." In *Reader in Traditional Chinese Culture*, edited by Victor Mair, Nancy Steinhardt, and Paul R. Goldin, 444–455. Honolulu: University of Hawai'i Press, 2005.

Arnold, Bruce Makoto, Tanfer Emin Tunç, and Raymond Douglas Chong, eds. *Chop Suey and Sushi from Sea to Shining Sea: Chinese and Japanese Restaurants in the United States*. Fayetteville: The University of Arkansas Press, 2018.

Ash, Robert Fairbanks. "Population Change and Food Security in China." In *Critical Issues in Contemporary China*, edited by Czeslaw Tubilewicz, 143–166. New York: Routledge, 2006.

Ash, Robert Fairbanks. "Squeezing the Peasants: Grain Extraction, Food Consumption and Rural Living Standards in Mao's China." *The China Quarterly* 188 (2006): 959–998.

Bai, Junfei, Thomas I. Wahl, Bryan T. Lohmar, and Jikun Huang. "Food Away from Home in Beijing: Effects of Wealth, Time and 'Free' Meals." *China Economic Review* 21, no. 3 (2010): 432–441. https://doi.org/10.1016/j.chieco.2010.04.003.

Balbi, Mariella. *Los chifas en el Peru: Historia y recetas*. Lima: Universidad de San Martín Porras, 1999.

Banh, Jenny, and Haiming Liu, eds. *American Chinese Restaurants: Society, Culture and Consumption*. New York: Routledge, 2020.

Bello, David. *Across Forest, Steppe, and Mountain: Environment, Identity, and Empire in Qing China's Borderlands*. Cambridge: Cambridge University Press, 2016.

Benn, James A. *Tea in China: A Religious and Cultural History*. Honolulu: University of Hawai'i Press, 2015.

Billé, Franck. "Cooking the Mongols/Feeding the Han: Dietary and Ethnic Intersections in Inner Mongolia." *Inner Asia* 11, no. 2 (2009): 205–230.

Bray, Francesca. *Agriculture*, in *Science and Civilisation in China*, ed. Joseph Needham, vol. 6, *Biology and Biological Technology*, pt. 2. Cambridge: Cambridge University Press, 1984.

Bray, Francesca. *The Rice Economies: Technology and Development in Asian Societies*. Berkeley: University of California Press, 1994.

Bray, Francesca. *Technology, Gender, and History in Imperial China: Great Transformations Reconsidered*. Oxford: Routledge, 2013.

Bray, Francesca. "Agriculture." In *Cambridge History of China: Volume Two, The Six Dynasties 220–581*, edited by Albert E. Dien and Keith N. Knapp, 355–373. Cambridge: Cambridge University Press, 2019.

Bray, Francesca, Barbara Hahn, and John Bosco Lourdsamy. *Moving Crops and the Scales of History*. New Haven, CT: Yale University Press, 2023.

Bray, Francesca, Peter A. Coclanis, Edda L. Fields-Black, and Dagmar Schäfer, eds. *Rice: Global Networks and New Histories*. Cambridge: Cambridge University Press, 2015.

Brown, Lester. *Who Will Feed China?: Wake-Up Call for a Small Planet*. New York: W. W. Norton & Company, 1995.

Brown, Miranda. "Mr. Song's Cheeses: Southern China, 1368–1644." *Gastronomica: The Journal of Critical Food Studies* 19, no. 2 (2019): 29–42.

Buck, John Lossing. *Land Utilization in China: A Study of 16,786 Farms in 168 Localities, and 38,256 Farm Families in Twenty-Two Provinces in China, 1929–1933*. Nanking: University of Nanking, 1937.

Buell, Paul D., and E. N. Anderson. *A Soup for the Qan*. Leiden: Brill, 2010.

Buell, Paul D., E. N. Anderson, Montserrat de Pablo, and Moldir Oskenbay. *Crossroads of Cuisine: The Eurasian Heartland, the Silk Roads, and Foods*. Leiden: Brill, 2020.

Cesàro, M. Cristina. "Polo, Läghmän, So Säy: Situating Uyghur Food Between China and Central Asia." In *Situating the Uyghurs Between China and Central Asia*, edited by Ildikó Bellér-Hann, M. Cristina Cesàro, and Joanne Smith Finley, 185–202. London: Routledge, 2007.

Chan, Selina Ching. "Food, Memories, and Identities in Hong Kong." *Identities: Global Studies in Culture and Power* 17, no. 2–3 (2010): 204–227.

Chan, Selina Ching. "Tea Cafes and the Hong Kong Identity: Food Culture and Hybridity." *China Information* 33, no. 3 (2018): 311–328.

Chan, Yuk Wah, and James Farrer, eds. "Special Issue: Asian Food and Culinary Politics: Food Governance, Constructed Heritage and Contested Boundaries." *Asian Anthropology* 20, no. 1 (2020).

Chang, K. C., ed. *Food in Chinese Culture: Anthropological and Historical Perspectives*. New Haven, CT: Yale University Press, 1977.

Chau, Adam Yuet. "Hosting as a Cultural Form." *L'Homme* 231–232, no. 3–4 (2019): 41–66.

Chen, Nancy N. *Food, Medicine, and the Quest for Good Health*. New York: Columbia University Press, 2009.

Chen, Yong. *Chop Suey, USA: The Story of Chinese Food in America*. New York: Columbia University Press, 2014.

Chen, Yujen. "Ethnic Politics in the Framing of National Cuisine: State Banquets and the Proliferation of Ethnic Cuisine in Taiwan." *Food, Culture and Society* 14, no. 3 (2011): 315–333. https://doi.org/10.2752/175174411x12961586033483.

Chen, Yujen. *"Taiwancai" de wenhuashi: Shiwu xiaofei zhong de guojia tixian* [Cultural history of "Taiwanese cuisine": The embodiment of nationhood in food consumption]. Taipei: Linking, 2020.

Cheng, Sea Ling. "Back to the Future: Herbal Tea Shops in Hong Kong." In *Hong Kong: The Anthropology of a Chinese Metropolis*, edited by Grant Evans and Maria Tam, 51–76. Richmond: Curzon Press, 1997.

Cheung, Gordon C. K., and Chak Yan Chang. "Cultural Identities of Chinese Business: Networks of the Shark-Fin Business in Hong Kong." *Asia Pacific Business Review* 17, no. 3 (2011): 343–359.

Cheung, Gordon C. K., and Edmund Terence Gomez. "Hong Kong's Diaspora, Networks, and Family Business in the UK: A History of the Chinese 'Food Chain' and the Case of the W. Wing Yip Group." *The China Review* 12, no. 1 (2012): 45–72.

Cheung, Sidney C. H. "Consuming 'Low' Cuisine after Hong Kong's Handover: Village Banquets and Private Kitchens." *Asian Studies Review* 29, no. 3 (2005): 259–273.

Cheung, Sidney C. H., ed. *Rethinking Asian Food Heritage*. Taipei: Foundation of Chinese Dietary Culture, 2014.

Cheung, Sidney C. H. *Hong Kong Foodways*. Hong Kong: Hong Kong University Press, 2022.

Cheung, Sidney C. H., ed. *Berkshire Encyclopedia of Chinese Cuisines, 5 Volume Set*. Great Barrington, MA: Berkshire Publishing Group, forthcoming.

Cheung, Sidney C. H., and Tan Chee-Beng, eds. *Food and Foodways in Asia: Resource, Tradition and Cooking*. London: Routledge, 2007.

Ching, May-bo. "Chopsticks or Cutlery? How Canton Hong Merchants Entertained Foreign Guests in the Eighteenth and Nineteenth Centuries." In *Narratives of Free Trade: The Commercial Cultures of Early US-China Relations*, 99–114. Hong Kong: Hong Kong University Press, 2011.

Ching, May-bo. "The Flow of Turtle Soup from the Caribbean via Europe to Canton and Its Modern American Fate." *Gastronomica* 16, no. 1 (2016): 79–89.

Cho, Lily. *Eating Chinese: Culture on the Menu in Small Town Canada*. Toronto: University of Toronto Press, 2010.

Christmas, Sakura. "Japanese Imperialism and Environmental Disease on a Soy Frontier, 1890–1940." *Journal of Asian Studies* 78, no. 4 (2019): 809–836. https://doi.org/10.1017/s0021911819000597.

Clements, Jonathan. *The Emperor's Feast: A History of China in Twelve Meals*. London: Hodder & Stoughton, 2022.

Cody, Sacha. "Contending the Rural: Food Commodities and Regimes of Value in China." *Gastronomica: The Journal of Critical Food Studies* 18, no. 3 (2018): 42–53.

Coe, Andrew. *Chop Suey: A Cultural History of Chinese Food in the United States*. Oxford: Oxford University Press, 2009.

Cooper, Eugene. "Chinese Table Manners: You Are How You Eat." *Human Organization* 45, no. 2 (1986): 179–184.

Croll, Elisabeth. *The Family Rice Bowl: Food and the Domestic Economy in China*. London: Zed Press, 1983.

Croll, Elisabeth. *China's New Consumers: Social Development and Domestic Demand*. London: Routledge, 2006.

Crook, Steven, and Katy Hui-Wen Hung. *A Culinary History of Taipei: Beyond Pork and Ponlai*. Lanham, MD: Rowman & Littlefield, 2018.

Cui, Zhenling, Hongyan Zhang, Xinping Chen, Chaochun Zhang, Wenqi Ma, Chengdong Huang, Weifeng Zhang, et al. "Pursuing Sustainable Productivity with Millions of Smallholder Farmers." *Nature* 555, no. 7696 (2018): 363–366. https://doi.org/10.1038/nature25785.

Das, Mukta. "Making It in China: Informality, Belonging and South Asian Food in Guangzhou, Macau and Hong Kong." PhD diss., SOAS University of London, 2019.

Davis, Mike. *The Monster at Our Door: The Global Threat of Avian Flu*. New York: New Press, 2005.

Demgenski, Philipp. "Culinary Tensions: Chinese Cuisine's Rocky Road toward International Intangible Cultural Heritage Status." *Asian Ethnology* 79, no. 1 (2020): 115–135.

Deng, Ting. "Chinese Immigrant Entrepreneurship in Italy's Coffee Bars: Demographic Transformation and Historical Contingency." *International Migration* 58, no. 3 (2020): 87–100.

Dikötter, Frank. *Mao's Great Famine: The History of China's Most Devastating Catastrophe, 1958–1962*. London: Bloomsbury, 2010.

Ding, Mei. "Cultural Intimacy in Ethnicity: Understanding Qingzhen Food from Chinese Muslims' Views." *Journal of Contemporary China* 29, no. 121 (2020): 17–30. https://doi.org/10.1080/10670564.2019.1621527.

Dott, Brian R. *The Chile Pepper in China: A Cultural Biography*. New York: Columbia University Press, 2020.

Du Bois, Christine M., Chee-Beng Tan, and Sidney Mintz, eds. *The World of Soy*. Urbana: University of Illinois Press, 2008.

DuBois, Thomas David. "China's Dairy Century: Making, Drinking and Dreaming of Milk." In *Animals and Human Society in Asia*, edited by Rotem Kowner, Guy Bar-Oz, Michal Biran, Meir Shahar, and Gideon Shelach-Lavi, 179–211. Cham: Palgrave Macmillan, 2019.

DuBois, Thomas David. "Many Roads from Pasture to Plate: A Commodity Chain Approach to China's Beef Trade, 1732–1931." *Journal of Global History* 14, no. 1 (2019): 22–43. https://doi.org/10.1017/S1740022818000335.

DuBois, Thomas David. "China's Old Brands: Commercial Heritage and Creative Nostalgia." *International Journal of Asian Studies* 18, no. 1 (2021): 45–59. https://doi.org/10.1017/S1479591420000455.

DuBois, Thomas David. "Fast Food for Thought: Finding Global History in a Beijing McDonald's." *World History Connected* 18, no. 2 (2021): 1–13.

DuBois, Thomas David. "There's a Body in the Kitchen! A Cook's-Eye View of Sichuan Cuisine." *KNOW: A Journal on the Formation of Knowledge* 5, no. 1 (2021): 1–26. https://doi.org/10.1086/712998.

DuBois, Thomas David. *China in Seven Banquets: A Flavourful History*. London: Reaktion Books, 2024.

Dunlop, Fuchsia. *Invitation to a Banquet: The Story of Chinese Food*. London: Particular Books, 2023.

Edgerton-Tarpley, Kathryn. "From 'Nourish the People' to 'Sacrifice the Nation': Changing Responses to Disaster in Late Imperial and Modern China." *The Journal of Asian Studies* 73, no. 2 (May 2014): 447–469. https://doi.org/10.1017/s0021911813002374.

Edgerton-Tarpley, Kathryn. *Tears from Iron: Cultural Responses to Famine in 19th Century China*. Berkeley: University of California Press, 2008.

Elvin, Mark. *The Retreat of the Elephants: An Environmental History of China*. New Haven, CT: Yale University Press, 2004.

Fan, Ka Wai. "Feeding on Fancies with Recipe Books during the Period of China's Great Famine (1959–1961)." *Food, Culture & Society* 27, no. 1 (2024): 135–151. https://doi.org/10.1080/15528014.2022.2054503.

Farquhar, Judith. *Appetites: Food and Sex in Postsocialist China*. Durham, NC: Duke University Press, 2002.

Farquhar, Judith, and Qicheng Zhang. *Ten Thousand Things: Nurturing Life in Contemporary Beijing*. New York: Zone Books, 2012.

Farrer, James. "Domesticating the Japanese Culinary Field in Shanghai." In *Feeding Japan*, edited by Andreas Niehaus and Tine Walravens, 287–312. Cham: Palgrave Macmillan, 2017.

Farrer, James. "The Decline of the Neighborhood Chinese Restaurant in Urban Japan." *Jahrbuch für Kulinaristik—The German Journal of Food Studies and Hospitality* 2 (2018): 197–222.

Farrer, James, ed. *Globalization, Food and Social Identities in the Asia-Pacific Region.* Tokyo: Sophia University Institute of Comparative Culture, 2010.

Farrer, James, ed. *The Globalization of Asian Cuisines: Transnational Networks and Culinary Contact Zones.* London: Palgrave McMillan, 2015.

Farrer, James, and David Wank, eds. *The Global Japanese Restaurant: Mobilities, Imaginaries, and Politics.* Honolulu: University of Hawai'i Press, 2023.

Fei, Hsiao-t'ung. *Peasant Life in China: A Field Study of Country Life in the Yangtze Valley.* London: Kagan Paul, 1939.

Feng, Jin. "The Female Chef and the Nation: Zeng Yi's 'Zhongkui lu' (Records from the kitchen)." *Modern Chinese Literature and Culture* 28, no. 1 (2016): 1–37.

Feng, Jin. *Tasting Paradise on Earth: Jiangnan Foodways.* Seattle: University of Washington Press, 2019.

Fiskesjö, Magnus. "Participant Intoxication and Self-Other Dynamics in the Wa Context." *The Asia Pacific Journal of Anthropology* 11, no. 2 (2010): 111–127. https://doi.org/10.1080/14442211003720588.

Fiskesjö, Magnus. "On the 'Raw' and the 'Cooked' Barbarians of Imperial China." *Inner Asia* 1, no. 2 (1999): 139–168.

Fu, Jia-Chen. "Scientizing Relief: Nutritional Activism from Shanghai to the Southwest, 1937–1945." *European Journal of East Asian Studies* 11, no. 2 (2012): 259–282.

Fu, Jia-Chen. *The Other Milk: Reinventing Soy in Republican China.* Seattle: University of Washington Press, 2018.

Fu, Jia-Chen. "The Tyranny of the Bottle: Vitasoy and the Cultural Politics of Packaging." *Worldwide Waste Journal of Interdisciplinary Studies* 1, no. 1 (2018): 1–11. https://doi.org/10.5334/wwwj.11.

Fu, Jia-Chen. "Would Mr. Science Eat the Chinese Diet?" *East Asian Science, Technology, and Society* 16, no. 3 (2022). https://doi.org/10.1080/18752160.2022.2096815.

Galipeau, Brendan. "Tibetan Wine Production, Taste of Place, and Regional Niche Identities in Shangri-la, China." In *Trans-Himalayan Borderlands: Livelihoods, Territorialities, Modernities*, edited by Dan Smyer Yü and Jean Michaud, 207–228. Amsterdam: Amsterdam University Press, 2017.

Gibet, Lise. "En rouge et noir: Les restaurants chinois à Paris, indicateurs des reconfigurations locales and transnationales." In *Mobilités et mobilisations Chinoises en France*, edited by Ya-Han Chuang and Ann-Christine Trémon. Marseille: Terra HN éditions, 2020.

Gillette, Maris Boyd. *Between Mecca and Beijing: Modernization and Consumption Among Urban Chinese Muslims.* Stanford, CA: Stanford University Press, 2000.

Goossaert, Vincent. *L'interdit du boeuf en Chine: Agriculture, éthique et sacrifice.* Paris: Institut des Hautes Etudes Chinoises, 2005.

Greenspan, Anna. "Movable Feasts: Reflections on Shanghai's Street Food." *Food, Culture & Society* 21, no. 1 (2018): 75–88.

Guan, Jing, Jun Gao, and Chaozhi Zhang. "Food Heritagization and Sustainable Rural Tourism Destination: The Case of China's Yuanjia Village." *Sustainability* 11, no. 10 (2019): 2858. https://doi.org/10.3390/su11102858.

Guo, Man, and Carsten Herrmann-Pillath. "Lineage, Food and Ritual in a Chinese Metropolis." *Anthropos* 114, no. 1 (2019): 195–207. http://dx.doi.org/10.5771/0257-9774-2019-1-195.

Hansen, Anders Sybrandt, and Mikkel Bunkenborg, eds. "Special Issue: Moral Economies of Food in Contemporary China." *Journal of Current Chinese Affairs* 48, no. 3 (2019): 243–399.

Hanser, Amy, and Jialin Camille Li. "Opting Out? Gated Consumption, Infant Formula and China's Affluent Urban Consumers." *The China Journal* 74, no. 1 (2015): 110–128.

Hathaway, Michael J. *What a Mushroom Lives For: Matsutake and the Worlds They Make.* Princeton, NJ: Princeton University Press, 2022.

Hayford, Charles W. "Open Recipes Openly Arrived At: Mrs. Chao's 'How to Cook and Eat in Chinese' (1945) and the Translation of Chinese Food." *Journal of Oriental Studies* 45, no. 1–2 (December 2012): 67–87.

Hayford, Charles W. "The Several Worlds of Lin Yutang's Gastronomy." In *The Cross-Cultural Legacy of Lin Yutang: Critical Perspectives*, edited by Suoqiao Qian, 232–252. Berkeley: The Institute of East Asian Studies, 2015.

Hayward, Jane. "Beyond the Ownership Question: Who Will Till the Land? The New Debate on China's Agricultural Production." *Critical Asian Studies* 49, no. 4 (October 2017): 523–545. https://doi.org/10.1080/14672715.2017.1362957.

He, Hongzhong, Joseph Lawson, Martin Bell, and Fuping Hui. "Millet, Wheat, and Society in North China over the Very Long Term." *Environment and History* 27, no. 1 (2021): 127–154. https://doi.org/info:doi/10.3197/096734019X15463432086937.

Ho, Elaine Lynn-Ee. "African Student Migrants in China: Negotiating the Global Geographies of Power through Gastronomic Practices and Culture." *Food, Culture & Society* 21, no. 1 (2018): 9–24.

Ho, Hao-tzu. "Cosmopolitan Locavorism: Global Local-Food Movements in Postcolonial Hong Kong." *Food, Culture & Society* 23, no. 2 (2020): 137–154.

Hoefer, Jacob A., and Patricia Jones Tsuchitani. *Animal Agriculture in China: A Report of the Visit of the CSCPRC Animal Sciences Delegation.* Washington, DC: National Academy Press, 1980.

Höllmann, Thomas. *The Land of the Five Flavors: A Cultural History of Chinese Cuisine.* New York: Columbia University Press, 2010.

Hu Shiu-Ying. *Food Plants of China.* Hong Kong: Chinese University of Hong Kong Press, 2005.

Huang, H. T. *Fermentations and Food Science*, Science and Civilisation in China, ed. Joseph Needham, vol. 6, *Biology and Biological Technology*, pt. 5. Cambridge: Cambridge University Press, 2000.

Huang, Philip C. C. *The Peasant Economy and Social Change in North China.* Stanford, CA: Stanford University Press, 1985.

Huang, Philip C. C. "China's New-Age Small Farms and Their Vertical Integration: Agribusiness or Co-ops?" *Modern China* 37, no. 2 (2011): 107–134.

Hubbert, Jennifer. "Revolution Is a Dinner Party: Cultural Revolution Restaurants in Contemporary China." *China Review* 5, no. 2 (2005): 125–150.

Humphrey, Caroline, and David Sneath. *The End of Nomadism? Society, State and the Environment in Inner Asia.* Cambridge: The White Horse Press, 1999.

Ikeya, Kazunobu, ed. "The Spread of Food Cultures in Asia." *Senri Ethnological Studies* 100. Osaka: National Museum of Ethnology, 2019.

Imbruce, Valerie. *From Farm to Canal Street: Chinatown's Alternative Food Network in the Global Marketplace.* Ithaca, NY: Cornell University Press, 2015.

Isett, Christopher. *State, Peasant, and Merchant in Qing Manchuria, 1644–1862.* Stanford, CA: Stanford University Press, 2007.

Janku, Andrea. "Drought and Famine in Northwest China: A Late Victorian Tragedy?" *Journal of Chinese History* 2, no. 2 (2018): 373–391. https://doi.org/10.1017/jch.2018.4.

Jiao, Wenjun, and Qingwen Min. "Reviewing the Progress in the Identification, Conservation and Management of China-Nationally Important Agricultural Heritage Systems (China-NIAHS)." *Sustainability* 9, no. 10 (2017): 1–14.

Jing, Jun, ed. *Feeding China's Little Emperors: Food, Children, and Social Change.* Stanford, CA: Stanford University Press, 2000.

Jung, Yuson, Jakob A. Klein, and Melissa L. Caldwell, eds. *Ethical Eating in the Postsocialist and Socialist World.* Berkeley: University of California Press, 2014.

Kaufmann, Lena. *Rural-Urban Migration and Agro-Technological Change in Post-Reform China.* Amsterdam: Amsterdam University Press, 2021.

King, Michelle T. "The Julia Child of Chinese Cooking, or the Fu Pei-mei of French Food?: Comparative Contexts of Female Culinary Celebrity." *Gastronomica: The Journal of Critical Food Studies* 18, no. 1 (February 2018): 15–26. https://doi.org/10.1525/gfc.2018.18.1.15.

King, Michelle T. "Say No to Bat Fried Rice: Changing the Narrative of Coronavirus and Chinese Food." *Food and Foodways* 28, no. 3 (Fall 2020): 237–249. https://doi.org/10.1080/07409710.2020.1794182.

King, Michelle T. *Chop, Fry, Watch, Learn: Fu Pei-mei and the Making of Modern Chinese Food.* New York: W.W. Norton & Company, 2024.

King, Michelle T., ed. *Culinary Nationalism in Asia.* London: Bloomsbury Academic, 2019.

King, Michelle T., ed. "Special Issue on Chinese Culinary Regionalism." *Global Food History* 6, no. 2 (2020).

King, Michelle T., Jia-Chen Fu, Miranda Brown, and Donny Santacaterina, eds. "Rumor, Chinese Diets, and COVID-19: Questions and Answers about Chinese Food and Eating Habits." *Gastronomica: The Journal of Critical Food Studies* 21, no. 1 (Spring 2021): 77–82. https://doi.org/10.1525/gfc.2021.21.1.77.

Klein, Jakob A. "Redefining Cantonese Cuisine in Post-Mao Guangzhou." *Bulletin of the School of Oriental and African Studies* 70, no. 3 (2007): 511–537.

Klein, Jakob A. "Creating Ethical Food Consumers? Promoting Organic Foods in Urban Southwest China." *Social Anthropology/Anthropologie Sociale* 17, no. 1 (2009): 74–89.

Klein, Jakob A. "'For Eating, It's Guangzhou': Regional Culinary Traditions and Chinese Socialism." In *Enduring Socialism: Explorations of Revolution and Transformation, Restoration and Continuation*, edited by Harry G. West and Parvathi Raman, 44–76. New York: Berghahn Books, 2009.

Klein, Jakob A. "'There Is No Such Thing as Dian Cuisine!' Food and Local Identity in Urban Southwest China." *Food and History* 11, no. 1 (2013): 203–225.

Klein, Jakob A. "Connecting with the Countryside? 'Alternative' Food Movements with Chinese Characteristics." In *Ethical Eating in the Postsocialist and Socialist World*, edited by Yuson Jung, Jakob A. Klein, and Melissa L. Caldwell, 116–143. Berkeley: University of California Press, 2014.

Klein, Jakob A. "Changing Tastes in Guangzhou: Restaurant Writings in the Late 1990s." In *Consuming China: Approaches to Cultural Change in Contemporary China*, edited by Kevin Latham, Stuart Thompson, and Jakob Klein, 104–120. London: Routledge, 2006.

Klein, Jakob A. "Buddhist Vegetarian Restaurants and the Changing Meanings of Meat in Urban China." *Ethnos* 82, no. 2 (2017): 252–276.

Klein, Jakob A. "Heritagizing Local Cheese in China: Opportunities, Challenges, and Inequalities." *Food and Foodways* 26, no. 1 (2018): 63–83. https://doi.org/10.1080/07409710.2017.1420354.

Klein, Jakob A. "Transformations of Chinese Cuisines." In *Routledge Handbook of Chinese Culture and Society*, edited by Kevin Latham, 376–394. London: Routledge, 2020.

Knechtges, David R. "A Literary Feast: Food in Early Chinese Literature." *Journal of the American Oriental Society* 106, no. 1 (1986): 49–63.

Knechtges, David R. "Gradually Entering the Realm of Delight: Food and Drink in Early Medieval China." *Journal of the American Oriental Society* 117, no. 2 (1997): 229–239.

Ku, Robert Ji-Song, Martin F. Manalansan, and Anita Manur, eds. *Eating Asian America: A Food Studies Reader*. New York: New York University Press, 2013.

Kunze, Rui. "Stakes of Authentic Culinary Experience: Food Writing of Tang Lusun and Wang Zengqi." *Ex-position* 43 (June 2020): 109–133.

Kuo, Chung-Hao. *Pigs, Pork, and Ham: From Farm to Table in Early Modern China*. Honolulu: University of Hawai'i Press, forthcoming.

Kwan, Cheuk. *Have You Eaten Yet: Stories from Chinese Restaurants around the World*. New York: Pegasus Books, 2023.

Lai, Lili, and Judith Farquhar. "Nationality Medicines in China: Institutional Rationality and Healing Charisma." *Comparative Studies in Society and History* 57, no. 2 (2015): 381–406.

Lander, Brian, Mindi Schneider, and Katherine Brunson. "A History of Pigs in China: From Curious Omnivores to Industrial Pork." *The Journal of Asian Studies* 79, no. 4 (2020): 865–889.

Laudan, Rachel. *Cuisine and Empire: Cooking in World History*. Berkeley: University of California Press, 2015.

Lavelle, Peter B. *The Profits of Nature: Colonial Development and the Quest for Resources in Nineteenth-Century China*. New York: Columbia University Press, 2020.

Lee, Heather Ruth. *Gastrodiplomacy: Chinese Exclusion and the Ascent of New York's Chinese Restaurants, 1870–1949*. Chicago: University of Chicago Press, forthcoming.

Lee, Jennifer. *The Fortune Cookie Chronicles: Adventures in the World of Chinese Food*. New York: Twelve Press, 2008.

Lee, Seung-Joon. "Taste in Numbers: Science and the Food Problem in Republican Guangzhou, 1927–1937." *Twentieth-Century China* 35, no. 2 (April 2010): 81–103.

Lee, Seung-Joon. *Gourmets in the Land of Famine: The Culture and Politics of Rice in Modern Canton*. Stanford, CA: Stanford University Press, 2011.

Lee, Seung-Joon. "The Patriot's Scientific Diet: Nutrition Science and Dietary Reform Campaigns in China, 1910s–1950s." *Modern Asian Studies* 49, no. 6 (November 2015): 1808–1839.

Lee, Seung-Joon. "Expertise Marginalized: Quality Inspection and the Grain Market in Republican China." *Frontiers of History in China* 10, no. 4 (December 2015): 668–694.

Lee, Seung-Joon. "Airborne Prawn and Decayed Rice: Food Politics in Wartime Chongqing." *Journal of Modern Chinese History* 13, no. 1 (September 2019): 124–147.

Lee, Seung-Joon. "Canteens and the Politics of Working-Class Diets in Industrial China, 1920–1937." *Modern Asian Studies* 54, no. 1 (January 2020): 1–29.

Leong-Salobir, Cecilia, ed. *Routledge Handbook of Food in Asia*. London: Routledge, 2019.

Leppman, Elizabeth J. *Changing Rice Bowl: Economic Development and Diet in China*. Hong Kong: Hong Kong University Press, 2005.

Leung, Angela Ki Che, and Hallam Stevens, eds. *Crafting Everyday Food: Technology, Tradition, and Transformation in Modern East Asia*. Honolulu: University of Hawai'i, 2025.

Leung, Angela Ki Che, and Melissa L. Caldwell, eds. *Moral Foods: The Construction of Nutrition and Health in Modern Asia*. Honolulu: University of Hawai'i Press, 2019.

Li, Lillian M. *Fighting Famine in North China: State, Market, and Environmental Decline, 1690s–1990s*. Stanford, CA: Stanford University Press, 2007.

Li, Lillian M., and Alison Dray-Novey. "Guarding Beijing's Food Security in the Qing Dynasty: State, Market, and Police." *The Journal of Asian Studies* 58, no. 4 (1999): 992–1032.

Li, Shang-Jen. "Eating Well in China: Diet and Hygiene in Nineteenth-Century Treaty Ports." In *Health and Hygiene in Chinese East Asia: Policies and Publics in the*

Long Twentieth Century, edited by Angela Ki Che Leung and Charlotte Furth, 109–131. Durham, NC: Duke University Press, 2010.

Li, Yifei, and Judith Shapiro. *China Goes Green: Coercive Environmentalism for a Troubled Planet*. Cambridge, MA: Polity, 2020.

Liang, Limin. "Consuming the Pastoral Desire: Li Ziqi, Food Vlogging and the Structure of Feeling in the Era of Microcelebrity." *Global Storytelling: Journal of Digital and Moving Images* 1, no. 2 (2022). https://doi.org/10.3998/gs.1020.

Liang, Yan. "Reflections on a Braised Pig's Head: Food and Vernacular Storytelling in Jin Ping Mei." *Journal of the American Oriental Society* 134, no. 1 (2014): 51–68. https://doi.org/10.7817/jameroriesoci.134.1.0051.

Lin, Hsiang Ju. *Slippery Noodles: A Culinary History of China*. London: Prospect Books, 2015.

Lin, Yi-Chieh. "Sustainable Food, Ethical Consumption and Responsible Innovation: Insights from the Slow Food and 'Low Carbon Food' Movements in Taiwan." *Food, Culture & Society* 23, no. 2 (2020): 155–172.

Liu, Haiming. *From Canton Restaurant to Panda Express: The History of Chinese Food in the United States*. New Brunswick, NJ: Rutgers University Press, 2015.

Liu, Haiming. "Chop Suey as Imagined Authentic Chinese Food: The Culinary Identity of Chinese Restaurants in the United States." *Journal of Transnational American Studies* 1, no. 1 (2009). https://doi.org/10.5070/T811006946.

Liu, Tik-sang. "Custom, Taste and Science: Raising Chickens in the Pearl River Delta Region, South China." *Anthropology & Medicine* 15, no. 1 (April 2008): 7–18.

Lo, Vivienne, and Penelope Barrett. "Cooking up Fine Remedies: On the Culinary Aesthetic in a Sixteenth-Century Chinese *Materia Medica*." *Medical History* 49, no. 4 (2005): 395–422. https://doi.org/10.1017/s0025727300009133.

Lu, Hanchao. "Out of the Ordinary: Implications of Material Culture and Daily Life in China." In *Everyday Modernity in China*, edited by Madeleine Yue Dong and Joshua Goldstein, 22–51. Seattle: University of Washington Press, 2006.

Lu, Hanchao. "The Tastes of Chairman Mao: The Quotidian as Statecraft in the Great Leap Forward and Its Aftermath." *Modern China* 41, no. 5 (2015): 539–572.

Lu, Kecai, ed. *Zhonghua minzu yinshi fengsu daguan* [Grand spectacle of the culinary habits of the Chinese nation/nationalities]. Beijing: Shijie zhishi chubanshe, 1992.

Luo, Qiangqiang, Joel Andreas, and Yao Li. "Grapes of Wrath: Twisting Arms to Get Villagers to Cooperate with Agribusiness in China." *The China Journal* 77 (January 2017): 27–50. https://doi.org/10.1086/688344.

Ma, Zhen. "Sensorial Place-Making in Ethnic Minority Areas: The Consumption of Forest Puer Tea in Contemporary China." *The Asia Pacific Journal of Anthropology* 19, no. 4 (2018): 316–332. https://doi.org/10.1080/14442213.2018.1486453.

Mak, Veronica S. W. "The Revival of Traditional Water Buffalo Cheese Consumption: Class, Heritage, and Modernity in Contemporary China." *Food and Foodways* 22, no. 4 (2014): 322–347.

Mak, Veronica S. W. *Milk Craze: Body, Science, and Hope in China*. Honolulu: University of Hawai'i Press, 2021.

Mallory, Walter. *China: Land of Famine*. New York: American Geographical Society, 1927.

Manning, Kimberly Ens, and Felix Wemheuer, eds. *Eating Bitterness: New Perspectives on China's Great Leap Forward and Famine*. Vancouver: University of British Columbia Press, 2011.

Marks, Robert B. *Tigers, Rice, Silk, and Silt: Environment and Economy in Late Imperial South China*. Cambridge: Cambridge University Press, 1998.

Mazumdar, Sucheta. *Sugar and Society in China: Peasants, Technology, and the World Market*. Cambridge, MA: Harvard University Asia Center, 1998.

Mendelson, Anne. *Chow Chop Suey: Food and the Chinese American Journey*. New York: Columbia University Press, 2016.

Nakayama, Tokiko, ed. *Zhongguo yinshi wenhua* [Chinese food culture]. Beijing: Zhongguo shehui kexue chubanshe, 1992.

Newman, Jacqueline M. *Food Culture in China*. Westport, CT: Greenwood Press, 2004.

Oakes, Tim. "Eating the Food of the Ancestors: Place, Tradition, and Tourism in a Chinese Frontier River Town." *Ecumene* 6, no. 2 (1999): 124–145.

Oxfeld, Ellen. *Bitter and Sweet: Food, Meaning, and Modernity in Rural China*. Berkeley: University of California Press, 2017.

Pierson, Stacey, ed. *Visual and Material Cultures of Food and Drink in China, 200 BCE–1900 CE*. Cambridge: Cambridge Scholars, 2022.

Pilcher, Jeffrey M. "'Tastes Like Horse Piss': Asian Encounters with European Beer." *Gastronomica* 16, no. 1 (2016): 28–40. https://doi.org/10.1525/gfc.2016.16.1.28.

Pilcher, Jeffrey M. *Food in World History*. 3rd ed. New York: Routledge, 2023.

Pomeranz, Kenneth. *The Great Divergence: China, Europe, and the Making of the Modern World Economy*. Princeton, NJ: Princeton University Press, 2000.

Roberts, J. A. G. *China to Chinatown: Chinese Food in the West*. London: Reaktion Books.

Sabban, Françoise. "'Follow the Seasons of the Heavens': Household Economy and the Management of Time in Sixth-Century China." *Food and Foodways* 6, no. 3–4 (1996): 329–349. https://doi.org/10.1080/07409710.1996.9962046.

Sabban, Françoise. "Art et culture contre science et technique. Les enjeux culturels et identitaires de la gastronomie chinoise face à l'Occident." *L'Homme* 36, no. 137 (1996): 163–194. https://doi.org/10.3406/hom.1996.370040.

Sabban, Françoise. "*A Scientific Controversy in China over the Origins of Noodles*." Carnets du Centre Chine, March 15, 2013. https://cecmc.hypotheses.org/7663.

Sabban, Françoise. "An Experience of Otherness: Conceptions of Chinese and European Diplomats about Foreign Food Practices (18th-19th Century)." In *The Spread of Food Cultures in Asia*, edited by Kazunobu Ikeya, 23-34. Osaka: National Museum of Ethnology, 2019.

Sabban, Françoise. "Les nouvelles figures du gastronome chinois." In *L'imaginaire de la gastronomie*, edited by Julia Csergo and Olivier Etcheverria, 129–146. Chartres: Menu Fretin, 2020.

Sabban, Françoise. *La Chine par le menu. Cuisine, culture culinaire et traditions alimentaires chinoises*. Paris: Belles Lettres, 2024.

Sankar, Amal. "Creation of Indian–Chinese Cuisine: Chinese Food in an Indian City." *Journal of Ethnic Foods* 4, no. 4 (2017): 268–273. https://doi.org/10.1016/j.jef.2017.10.002.

Schäfer, Dagmar, ed. *Cultures of Knowledge: Technology in Chinese History*. Leiden: Brill, 2012.

Schmalzer, Sigrid. *Red Revolution, Green Revolution: Scientific Farming in Socialist China*. Chicago: University of Chicago Press, 2016.

Schmalzer, Sigrid. "Layer upon Layer: Mao-Era History and the Construction of China's Agricultural Heritage." *East Asian Science, Technology and Society* 13, no. 3 (September 2019): 413–441. https://doi.org/10.1215/18752160-7498416.

Schmalzer, Sigrid. "Prometheus and the Fishpond: A Historical Account of Agricultural Systems and Eco-Political Power in the People's Republic of China." *Made in China Journal* 7, no. 2 (September 2022): 124–131.

Schneider, Helen. *Keeping the Nation's House: Domestic Management and the Making of Modern China*. Vancouver: UBC Press, 2011.

Schneider, Mindi. "Dragon Head Enterprises and the State of Agribusiness in China." *Journal of Agrarian Change* 17, no. 1 (2017): 3–21. https://doi.org/10.1111/joac.12151.

Schneider, Mindi. "What, Then, Is a Chinese Peasant? Nongmin Discourses and Agro-industrialization in Contemporary China." *Agriculture and Human Values* 32, no. 2 (2015): 331–346.

Serventi, Silvano, and Françoise Sabban. *Pasta: The Story of a Universal Food*. Translated by Antony Shugaar. New York: Columbia University Press, 2002.

Shao, Qin. "Tempest over Teapots: The Vilification of Teahouse Culture in Early Republican China." *The Journal of Asian Studies* 57, no. 4 (1998): 1009–1041.

Si, Zhenzhong, Theresa Schumilas, and Steffanie Scott. "Characterizing Alternative Food Networks in China." *Agriculture and Human Values* 32, no. 2 (2015): 299–313.

Simoons, Frederick J. *Food in China: A Cultural and Historical Inquiry*. Boca Raton, FL: CRC Press, 1991.

Siu, Lok. "Chino Latino Restaurants: Converging Communities, Identities, and Cultures." *Afro-Hispanic Review*. Special Issue: Afro-Asia. 27, no. 1 (2008): 161–172.

Siu, Lok. "21st Century Food Trucks: Mobility, Social Media, and Urban Hipness." In *Eating Asian America: A Food Studies Reader*, edited by Robert Ko, Martin Manalansan, and Anita Mannur. New York: NYU Press, 2013.

Skinner, G. William. "Marketing and Social Structure in Rural China, Part 1." *The Journal of Asian Studies* 24, no. 1 (1964): 3–43.

Smil, Vaclav. *China's Past, China's Future: Energy, Food, and Environment.* New York: Routledge, 2004.

Smith, Hilary A. "Beyond Indulgence: Diet-Induced Illnesses in Chinese Medicine." *Historia Scientiarum* 27, no. 2 (2018): 233–253.

Smith, Hilary A. "Skipping Breakfast to Save the Nation: A Different Kind of Dietary Determinism in Early Twentieth-Century China." *Global Food History* 4, no. 2 (2018): 152–167.

Smith, Hilary A. "Food, Health, and Nutrition in Chinese History." *History Compass* 20, no. 1 (2022). https://doi.org/10.1111/hic3.12704.

Sterckx, Roel, ed. *Of Tripod and Palate: Food, Politics, and Religion in Traditional China.* New York: Palgrave Macmillan US, 2005.

Sterckx, Roel, Martina Siebert, and Dagmar Schäfer, eds. *Animals through Chinese History: From the Earliest Times to 1911.* Cambridge: Cambridge University Press, 2018.

Sun, Ping. "Your Order, Their Labor: An Exploration of Algorithms and Laboring on Food Delivery Platforms in China." *Chinese Journal of Communication* 12, no. 3 (September 2019): 308–323. https://doi.org/10.1080/17544750.2019.1583676.

Sutton, Donald S. "Consuming Counterrevolution: The Ritual and Culture of Cannibalism in Wuxuan, Guangxi, China, May to July 1968." *Comparative Studies in Society and History* 37, no. 1 (1995): 167–192.

Swislocki, Mark. *Culinary Nostalgia: Regional Food Culture and the Urban Experience in Shanghai.* Stanford, CA: Stanford University Press, 2008.

Swislocki, Mark. "Nutritional Governmentality: Food and the Politics of Health in Late Imperial and Republican China." *Radical History Review* 2011, no. 110 (May 2011): 9–35. https://doi.org/10.1215/01636545-2010-024.

Tagliacozzo, Eric, and Wen-Chin Chang, eds. *Chinese Circulations: Capital, Commodities, and Networks in Southeast Asia.* Durham, NC: Duke University Press, 2011.

Tan, Chee-Beng. "Family Meals in Rural Fujian: Aspects of Yongchun Village Life." *Taiwan Journal of Anthropology* 1, no. 1 (2003): 179–195.

Tan, Chee-Beng, ed. *Chinese Food and Foodways in Southeast Asia and Beyond.* Singapore: NUS Press, 2011.

Thaxton, Ralph A., Jr. *Catastrophe and Contention in Rural China: Mao's Great Leap Forward Famine and the Origins of Righteous Resistance in Da Fo Village.* Cambridge: Cambridge University Press, 2008.

Thompson, Stuart E. "Death, Food, and Fertility." In *Death Ritual in Late Imperial and Modern China*, edited by James L. Watson and Evelyn S. Rawski, 71–108. Berkeley: University of California Press.

Tilt, Bryan. "Smallholders and the 'Household Responsibility System': Adapting to Institutional Change in Chinese Agriculture." *Human Ecology* 36, no. 2 (2008): 189–199.

Tracy, Megan. "Pasteurizing China's Grasslands and Sealing in *Terroir*." *American Anthropologist* 115, no. 3 (2013): 437–451. https://doi.org/10.1111/aman.12027.

Trappel, René. "From Peasant to Elite: Reshaping Agriculture in Gansu Province." *China Perspectives* 2021/2 (2021): 9–18.

Tseng, Pintsang. *Cong tianqi dao canzhuo: Qingdai Taiwan hanren de nongye shengchan yu shiwu xiaofei* [From farm to table: The agricultural production and food consumption of the Taiwanese Han people in Qing Dynasty]. PhD diss., National Taiwan University, 2006.

Tseng, Pintsang. "Banzhuo: Qingdai Taiwan de yanhui yu hanren shehui" [Banzhuo: Banquets and Han Society in Qing Taiwan]. *New History* 21, no. 4 (2010): 1–55.

Tseng, Pintsang. "Cong huating dao jiulou qingmozhi rizhi chuqi Taiwan gonggong kongjian de xingcheng yu kuozhan (1895–1911)" [From private dining halls to drinking parlors: The formation and expansion of public spaces in modern Taiwan (1895–1911)]. *Chinese Dietary Culture* 7, no. 1 (2011): 89–142.

Tseng, Pintsang. "Shengzhu maoyi de xingcheng: 19 shiji moqi Taiwan beibu shangpin jinji de fazhan (1881–1900)" [Emergence of pig trade: Development of commercial economy in North Taiwan, 1881–1900]. *Taiwan Historical Research* 21, no. 2 (2014): 33–68.

Tseng, Pintsang, and Yujen Chen, "Making 'Chinese Cuisine': The Grand Hotel and Chuan-Yang Cuisine in Postwar Taiwan," *Global Food History* 6, no. 2 (March 2020): 110–127.

Tsing, Anna Lowenhaupt. *The Mushroom at the End of the World: On the Possibility of Life in Capitalist Ruins*. Princeton, NJ: Princeton University Press, 2015.

Tuxun, Shajidanmu. "The 'Ethnic' Restaurant: Migration, Ethnicity, and Food Authenticity in Shanghai." *China Perspectives* 2022/04 (2022): 59–68.

Veeck, Ann. "The Revitalization of the Marketplace: Food Markets of Nanjing." In *The Consumer Revolution in Urban China*, edited by Deborah S. Davis, 107–123. Berkeley: University of California Press, 2000.

Veeck, Ann, and Hongyan Yu. "The Transformation of Pig Feasts in Rural Northeast China." *Gastronomica* 17, no. 3 (2017): 58–67.

Veeck, Ann, Hongyan Yu, and Alvin C. Burns. "Consumer Risks and New Food Systems in Urban China." *Journal of Macromarketing* 30, no. 3 (2010): 222–237.

Waley-Cohen, Joanna. "The Quest for Perfect Balance: Taste and Gastronomy in Imperial China." In *Food: The History of Taste*, edited by Paul Freedman, 99–133. London: Thames & Hudson, 2007.

Wang, Di. *The Teahouse: Small Business, Everyday Culture, and Public Politics in Chengdu, 1900–1950*. Stanford, CA: Stanford University Press, 2008.

Wang, Di. *The Teahouse under Socialism: The Decline and Renewal of Public Life in Chengdu, 1950–2000*. Ithaca, NY: Cornell University Press, 2018.

Wang, Q. Edward. *Chopsticks: A Cultural and Culinary History*. Cambridge: Cambridge University Press, 2015.

Wang, Xiaowei. *Blockchain Chicken Farm and Other Stories of Tech in China's Countryside*. New York: Macmillan, 2020.

Wang, Zuoyue, and Peter Neushul. "Between the Devil and the Deep Sea: C. K. Tseng, Mariculture, and the Politics of Science in Modern China." *Isis* 91, no. 1 (2000): 59–88. https://doi.org/10.1086/384626.

Watson, James L. "From the Common Pot: Feasting with Equals in Chinese Society." *Anthropos* 82, no. 4 (1987): 389–401.

Watson, James L. "Feeding the Revolution: Public Mess Halls and Coercive Commensality in Maoist China." In *The Handbook of Food and Anthropology*, edited by Jakob A. Klein and James L. Watson, 308–348. London: Bloomsbury Academic, 2016.

Watson, James L., ed. *Golden Arches East: McDonald's in East Asia*. 2nd ed. Stanford, CA: Stanford University Press, 2016.

White, Tom. *China's Camel Country: Livestock and Nation-Building at a Pastoral Frontier*. Seattle: University of Washington Press, 2024.

White, Tom. "Pastoralism and the State in China's Inner Mongolia." *Current History* 120, no. 827 (2021): 227–232.

Whyke, William Thomas, Zhen Troy Chen, and Joaquin Lopez-Mugica. "An Analysis of Cultural Dissemination and National Image Construction in Chinese Influencer Li Ziqi's Vlogs and Its Impact on International Viewer Perceptions on YouTube." *The Journal of Chinese Sociology* 9, no. 14 (2022). https://doi.org/10.1186/s40711-022-00173-2.

Wilkinson, Endymion. "Chinese Culinary History." *China Review International* 8, no. 2 (2001): 285–304. https://doi.org/10.1353/cri.2001.0110.

Will, Pierre-Étienne. *Bureaucracy and Famine in Eighteenth-Century China*. Translated by Elborg Forster. Stanford, CA: Stanford University Press, 1990.

Will, Pierre-Étienne, and R. Bin Wong with James Lee. *Nourish the People: The State Civilian Granary System in China, 1650–1850*. Ann Arbor: University of Michigan Center for Chinese Studies, 1991.

Wu, David Y. H., and Sidney C. H. Cheung, eds. *The Globalization of Chinese Food*. London: Routledge Curzon, 2002.

Wu, David Y. H., and Tan Chee-beng, eds. *Changing Chinese Foodways in Asia*. Hong Kong: The Chinese University Press, 2001.

Wu, Xu. *Farming, Cooking, and Eating Practices in the Central China Highlands: How Hezha Foods Function to Establish Ethnic Identity*. Lewiston, NY: The Edwin Mellen Press, 2011.

Wu, Xu. "The Farmhouse Joy (*nongjiale*) Movement in China's Ethnic Minority Villages." *The Asia Pacific Journal of Anthropology* 15, no. 2 (2014): 158–177. https://doi.org/10.1080/14442213.2014.894556.

Wu, Xu. "Ethnic Foods as Unprepared Materials and as Cuisines in a Culture-Based Development Project in Southwest China." *Asian Ethnology* 75, no. 2 (2016): 419–439.

Xiao, Kunbing. "The Taste of Tea: Material, Embodied Knowledge and Environmental History in Northern Fujian, China." *Journal of Material Culture* 22, no. 1 (2017): 3–18. https://doi.org/10.1177/1359183516633901.

Xu, Chenjia. "From Culinary Modernism to Culinary Cosmopolitanism: The Changing Topography of Beijing's Transnational Foodscape." *Food, Culture & Society* 26, no. 3 (2023): 775–792. https://doi.org/10.1080/15528014.2022.2046990.

Yamashita, Samuel Hideo. *Hawai'i Regional Cuisine: The Food Movement that Changed the Way Hawai'i Eats.* Honolulu: University of Hawai'i Press, 2019.

Yan, Hairong, Chen Yiyuan, and Ku Hok Bun. "China's Soybean Crisis: The Logic of Modernization and Its Discontents." *The Journal of Peasant Studies* 43, no. 2 (2016): 373–395.

Yan, Yunxiang. "Of Hamburgers and Social Space: Consuming McDonald's in Beijing." In *The Consumer Revolution in Urban China*, edited by Deborah S. Davis, 201–225. Berkeley: University of California Press, 2000.

Yan, Yunxiang. "Food Safety and Social Risk in Contemporary China." *The Journal of Asian Studies* 71, no. 3 (2012): 705–729.

Yang, Fan. "*A Bite of China*: Food, Media, and Televisual Negotiation of National Difference." *Quarterly Review of Film and Video* 32, no. 5 (2015): 409–425. https://doi.org/10.1080/10509208.2015.999223.

Yang, Jisheng. *Tombstone: The Untold Story of Mao's Great Famine*, edited by Edward Friedman, Guo Jian, and Stacy Mosher. Translated by Stacy Mosher and Guo Jian. London: Allen Lane, 2012.

Yang, Xiaomin. *La fonction sociale des restaurants en Chine.* Paris: L'Harmattan, 2006.

Yasuda, John K. *On Feeding the Masses: An Anatomy of Regulatory Failure in China.* Cambridge: Cambridge University Press, 2018.

Yeh, Emily. *Taming Tibet: Landscape Transformation and the Gift of Chinese Development.* Ithaca, NY: Cornell University Press, 2013.

Yu, Shuenn-Der, ed. *Food Cultures and Technologies.* Taipei: Institute of Ethnology, Academia Sinica, 2022.

Yu, Shuenn-Der, and Jinghong Zhang, eds. "Special Issue: Reinventing a Tradition: East Asian Tea Cultures in the Contemporary World." *Asian Journal of Social Science* 50, no. 3 (2022): 167–236.

Yuan, Mei. *Recipes from the Garden of Contentment: Yuan Mei's Manual of Gastronomy.* Translated by Sean J. S. Chen. Great Barrington, MA: Berkshire Publishing Group, 2017.

Yue, Gang. *The Mouth That Begs: Hunger, Cannibalism, and the Politics of Eating in Modern China.* Durham, NC: Duke University Press, 1999.

Yue, Isaac, and Siufu Tang, eds. *Scribes of Gastronomy: Representations of Food and Drink in Imperial Chinese Literature.* Hong Kong: Hong Kong University Press, 2013.

Zhang, Hongzhou. *Securing the "Rice Bowl": China and Global Food Security.* Singapore: Palgrave Macmillan, 2019.

Zhang, Jinghong. *Puer Tea: Ancient Caravans and Urban Chic.* Seattle: University of Washington Press, 2014.

Zhang, Jinghong. "'A Sense of Life': The Abstruse Language of Taste in Chinese Culture." *Food, Culture & Society* 26, no. 1 (2023): 47–62.

Zhang, Lawrence. "A Foreign Infusion: The Forgotten Legacy of Japanese *Chadō* on Modern Chinese Tea Arts." *Gastronomica: The Journal of Critical Food Studies* 16, no. 1 (2016): 53–62.

Zhang, Ling, and Mindi Schneider, eds. "Introduction: Feeding, Eating, Worrying: Chinese Food Politics across Time." *Global Food History* 8, no. 3 (2022): 153–156.

Zhang, Qian Forrest, and John A. Donaldson. "The Rise of Agrarian Capitalism with Chinese Characteristics: Agricultural Modernization, Agribusiness and Collective Land Rights." *The China Journal* 60 (2008): 25–47.

Zhang, Qian Forest, and Zi Pan. "The Transformation of Urban Vegetable Retail in China: West Markets, Supermarkets, and Informal Markets in Shanghai." *Journal of Contemporary Asia* 43, no. 3 (2013): 497–518.

Zhao, Rongguang. *Man-Han quanxi yuanliu kaoshu* [Investigative account of the origins and development of the complete Manchu-Han banquet]. Beijing: Kunlun chubanshe, 2003.

Zhao, Rongguang, ed. *Zhongguo yinshi wenhua shi* [History of Chinese food and drink culture]. Ten volumes. Beijing: Zhongguo qinggongye chubanshe, 2013.

Zhao, Rongguang. *Zhongguo yinshi wenhua shi* [History of Chinese food and drink culture]. Shanghai: Shanghai renmin chubanshe, 2014.

Zhong, Shuru, Mike Crang, and Guojun Zeng. "Constructing Freshness: The Vitality of Wet Markets in Urban China." *Agriculture and Human Values* 37 (2020): 175–185.

Zhou, Shuwen. "Formalisation of Fresh Food Markets in China. The Story of Hangzhou." In *Integrating Food into Urban Planning*, edited by Yves Cabannes and Cecilia Marocchino, 247–263. London: UCL Press, 2018.

Zhou, Xun. *Great Famine in China, 1958–1962: A Documentary History*. New Haven, CT: Yale University Press, 2012.

Zhou, Zhang-Yue, Hongbo Liu, and Lijuan Cao. *Food Consumption in China: The Revolution Continues*. Cheltenham: Edward Elgar, 2014.

Zhu, Zizhen, and Shen Han. *Zhongguo cha-jiu wenhua shi* [History of China's tea and alcohol cultures]. Taipei: Wenjin chubanshe, 1995.

Zinda, John Aloysius, and Jun He. "Ecological Civilization in the Mountains: How Walnuts Boomed and Busted in Southwest China." *The Journal of Peasant Studies* 47, no. 5 (2020): 1052–1076. https://doi.org/10.1080/03066150.2019.1638368.

CONTRIBUTORS

E. N. Anderson, PhD, is an emeritus professor of anthropology at the University of California, Riverside. He received his PhD in anthropology from the University of California, Berkeley, in 1967. He has done research on ethnobiology, cultural ecology, political ecology, and medical anthropology, in several areas, especially Hong Kong, British Columbia, California, and the Yucatan Peninsula of Mexico. His books include *The Food of China* (1988), *Ecologies of the Heart* (1996), *Political Ecology of a Yucatec Maya Community* (2005), *The Pursuit of Ecotopia* (2010), *Caring for Place* (2014), *Everyone Eats* (2014), *Food and Environment in Early and Medieval China* (2014), and with Barbara A. Anderson, *Warning Signs of Genocide* (2013), *Halting Genocide in America* (2017) and *Complying with Genocide: The Wolf You Feed* (2020).

Francesca Bray (emerita professor of social anthropology, University of Edinburgh) is a historian of science, technology, and medicine in East Asia, specializing in gender and technology, the politics of historiography, and the history of agriculture. Recent publications on food history include *Rice: Global Networks and New Histories* (2015), "Feeding the Body National: Rice as Self in Malaysia and Japan" (in *Moral Foods*, eds. Leung and Caldwell, 2019), "Translating the Art of Tea" (in *Entangled Itineraries*, ed. Smith, 2019), *Moving Crops and the Scales of History* (2023), and "Food in Medieval China" (in *A Cultural History of Technology in the Medieval Period*, ed. Magnusson, 2024).

Yujen Chen is a professor in the Department of Taiwan Culture, Languages and Literature at the National Taiwan Normal University. She is the author of 「臺灣菜」的文化史：食物消費中的國家體現 [*Cultural History of "Taiwanese Cuisine": the*

Embodiment of Nationhood in Food Consumption] (2020). Her research interests cover food history in Taiwan and East Asia, food anthropology, and culinary literature. She has published papers in the journals *Global Food History, Food, Culture and Society*, and *Journal of Family History*.

Thomas David DuBois is a historian of modern China and a professor in the Beijing Normal University School of Chinese Language and Culture. DuBois combines the textual study of history with on-the-ground observation, an approach he first used in his work on rural religion life. His recent work on China's food chains has taken him from cattle markets in Tongliao to cooking school in Chengdu. His newest book is *China in Seven Banquets: A Flavourful History* (Reaktion Books 2024).

James Farrer is a professor of sociology and director of the Graduate Program in Global Studies at Sophia University in Tokyo. His research has focused on the contact zones of global cities, including ethnographic studies of foodways, nightlife, sexuality, and expatriate cultures. His recent publications include *The Global Japanese Restaurant: Mobilities, Imaginaries and Politics* (with David Wank) and *Globalization and Asian Cuisines: Transnational Networks and Contact Zones*. He now leads a public ethnography project on neighborhood Tokyo foodways (www.nishiogiology.org) and is member of the editorial collective producing *Gastronomica: The Journal of Food Studies*.

Jia-Chen Fu is an associate research fellow in the Institute of Modern History at Academia Sinica. She is the author of *The Other Milk: Reinventing Soy in Republican China* (University of Washington Press, 2018). She has written on the intertwining histories of Chinese food and science in such publications as *East Asian Science, Technology, & Society, Worldwide Waste* and *Moral Foods: The Construction of Nutrition and Health in Modern Asia*, edited by Angela Ki Che Leung and Melissa L. Caldwell. She is also on the editorial board for *Re:Past—Studies in the History of Nutrition* series with John Hopkins University Press.

Michelle T. King is an associate professor of history at the University of North Carolina at Chapel Hill. She is the editor of *Culinary Nationalism in Asia* (Bloomsbury Academic, 2019), which won the 2021 Best Edited Volume book award from the Association for the Study of Food and Society. She edited a 2020 special issue of *Global Food History* on Chinese culinary regionalism. She is the author of, *Chop Fry Watch Learn: Fu Pei-mei and the*

Making of Modern Chinese Food (W. W. Norton, 2024), for which she received a National Endowment for the Humanities Public Scholar Grant. Her articles on Chinese foodways have appeared in *Gastronomica, Global Food History* and *Food and Foodways*.

Jakob A. Klein is a senior lecturer (associate professor) in social anthropology at SOAS University of London. He has coedited five volumes on anthropological approaches to the study of China and/or food, including *Consuming China: Approaches to Cultural Change in Contemporary China* (Routledge, 2006) and *Ethical Eating in the Socialist and Postsocialist World* (University of California Press, 2014). He has authored journal articles and book chapters in publications such as *Food and Foodways, Ethnos, The China Quarterly, Food and History*, and *Social Anthropology*. Klein was chair of the SOAS Food Studies Centre (2016–2022) and an editorial board member of *Gastronomica* (2014–2018). He currently serves on the executive committee of *The China Quarterly*'s editorial board.

Angela Ki Che Leung is an emerita professor of the University of Hong Kong. She obtained her PhD at the École des Hautes Études en Sciences Sociales, Paris. She was a research fellow at the Academia Sinica, Taipei, from 1982 to 2008, chair professor of history at the Chinese University of Hong Kong from 2008–2010, and chair professor of history and Joseph Needham-Philip Mao Professor of Chinese Science and Civilization at the University of Hong Kong from 2011 to 2023. She has published on the history of Chinese charitable organizations and the history of medicine in China. She is currently writing a monograph on the history of Chinese soy sauce.

Françoise Sabban is professor emerita at the Centre for Modern and Contemporary China Studies, École des Hautes Études en Sciences Sociales, Paris. She has published widely on the history and anthropology of food. Her solo and co-authored publications include *Le temps de manger. Alimentation, emploi du temps et rythmes sociaux* (with M. Aymard & C. Grignon, 1993); *The Medieval Kitchen. Recipes from France and Italy* (with O. Redon & S. Serventi, 2000); *Pasta: The Story of a Universal Food* (with S. Serventi, 2002); *Atlante dell'alimentazione e della gastronomia, Risorse, Scambi, Consumi* (with M. Montanari, 2004); *Cucine, pasti convivialità* (with M. Montanari, 2004); *Un aliment sain dans un corps sain—Perspectives historiques*, Collection "À boire et à manger," no. 1 (with. F. Audouin-Rouzeau, 2007); and *La Chine par le menu. Cuisine, culture culinaire et traditions alimentaires* (2024).

Sigrid Schmalzer is a professor of history at the University of Massachusetts Amherst, where she teaches Chinese history and the history of science. Her publications include *The People's Peking Man: Popular Science and Human Identity in Twentieth-Century China* (Chicago, 2008), *Red Revolution, Green Revolution: Scientific Farming in Socialist China* (Chicago, 2016), and a children's picture book, *Moth and Wasp, Soil and Ocean: Remembering Chinese Scientist Pu Zhelong's Work for Sustainable Farming* (Tilbury, 2018).

Lok Siu is a professor of ethnic studies and Associate Vice Chancellor for Research at UC Berkeley. She is a cultural anthropologist with expertise in diaspora, transnational migration, belonging and citizenship, food, ethnography, and hemispheric Asian American studies. Her books include *Memories of a Future Home: Diasporic Citizenship of Chinese in Panama* (2005), *Asian Diasporas: New Formations, New Conceptions* (2007), *Gendered Citizenships: Transnational Perspectives on Knowledge Production, Political Activism, and Culture* (2009); *Chinese Diaspora: Its Development in Global Perspective (2021)*, and *Worlding LatinAsian: Cultural Intimacies in Food, Art, and Politics* (forthcoming).

Hilary A. Smith is an associate professor of history at the University of Denver. She works on the history of health, medicine, and nutrition in China. In addition to her book *Forgotten Disease: Illnesses Transformed in Chinese Medicine* (Stanford, 2017), she has published on the history of Chinese dietetics in the journals *Global Food History*, *Historia Scientiarum*, and *History Compass*, and in *Moral Foods: The Construction of Nutrition and Health in Modern Asia*, edited by Angela Ki Che Leung and Melissa L. Caldwell.

Chuanfei Wang received her PhD in global studies at Sophia University, Tokyo, in 2017. She is currently an assistant professor with the Faculty of Sustainability Studies at Hosei University in Tokyo, teaching sociology, global studies, sustainability studies, Japanese culinary globalization, and sustainable wine tourism. She is the author of numerous articles and book chapters on the Japanese wine industry, the globalization of Japanese culinary culture, and Japanese wine tourism.

James L. Watson is Fairbank Professor of Chinese Society and emeritus professor of anthropology at Harvard University. Watson's research has focused on Chinese emigration, ancestor worship, popular religion, family life, village organization, food systems, and the emergence of a post-socialist culture in China. He has worked with graduate students in Harvard's Department of Anthropology to investigate foodways in China, Russia, Eastern Europe, South Asia, and North America. Among other publications Prof. Watson is editor of *Golden Arches East: McDonald's in East Asia* (Stanford University Press), coeditor (with Melissa Caldwell) of *The Cultural Politics of Food and Eating* (Blackwell), and coeditor (with Jakob Klein) of *Handbook of Food and Anthropology* (Bloomsbury).

Xu Wu is a professor of anthropology at East China Normal University in Shanghai. He has a PhD in anthropology from University of Alberta in Canada and previously taught anthropology as an assistant professor at University of Saskatchewan (2006–2008). His research interests include food, ethnobiology, and placemaking in China's ethnic regions. He has published food study papers in journals like *Food & Foodways*, *Journal of Ethnobiology*, *The Asia Pacific Journal of Anthropology*, and *Modern China*. He was consultant for the final evaluation of UN Millennium Development Goals Fund "China Culture & Development Partnership Framework Project" (2011).

Fan Yang is an associate professor in the Department of Media and Communication Studies and a faculty affiliate in Asian Studies, Global Studies, and the PhD program in language, literacy, and culture at the University of Maryland, Baltimore County (UMBC). She is the author of *Faked in China: Nation Branding, Counterfeit Culture, and Globalization* (Indiana University Press, 2016). Yang's work intersecting cultural studies, global/transnational media studies, urban communication, postcolonial studies, and contemporary China has appeared in numerous journals. Her new book, *Disorienting Politics: Chimerican Media and Transpacific Entanglements* (University of Michigan, 2024).

INDEX

Page numbers in *italics* refer to illustrations.

Acurio Jaramillo, Gastón, 253, 264, 267
Ailuo Brain Tonic, 98
alcohol, 199–201
alfalfa, 31
algae chlorella, 123
Alibaba, 133
Amur, Russia, 74
ancestor worship, 272
Anderson, Marja, xi
Andrews, Bridie, 102
animal welfare, 221, 292, 293
anime, 239
Annales school, xii
Anqing, Anhui Province, 54
Appadurai, Arjun, 158, 295
apricots, 31
Aspergillus oryzae, 61
Australia, 81
Austronesians, 14, 169
avian flu, 5, 295

Ba people, 202–203
Baidu Deliveries, 133
bai-sahn (worshipping the spirit with burning incense), 272
Balbi, Mariella, 256

Bamanzi (Zhou leader), 202–203
bamboo shoots, 175
baogu baba (maize cake wrapped in *tong* leaves), 197
Barrett, Penelope, 51
bean paste, 31, 175
beans, 31, 33, 39, 50, 55–57, 62, 175, 264–265
beef, 10–11, 69–86, 237–238
beer, 31
Beijing, 26, 54, 71, 73, 234, 295
Bencao gangmu (*Compendium of Materia Medica*; Li Shizhen), xiii, 50, 115, 120, 121
beriberi, 91, 92, 96, 97
Bi Huai, 60
biji (jottings), 210
bingzi (cake), 197
Bite of China (television series), 294
black beans, 39, 175
Bloch, Marc, xii
boboji (Bobo chicken from Chengdu), 198
bodegas, 262, 266
Book of Changes, 154
Books of Documents and Rites of Zhou, 99

Braudel, Fernand, xii
broomcorn millet (*Panicum miliaceum*), 27, *29*
Bryan, Lettice Pierce, 155
bubble tea, 169
Buck, John Lossing, 33, 35
buckwheat, 33
Buddhists, 13, 99
buffalo, 37
Burroughs Wellcome, 99, 102

cabbage moths, 115
cai pu dan (fried eggs with dried radish), 181
caiquan (finger-guessing games), 204
California, 254–255
Cang'er (Siberian cocklebur; *Xanthium strumarium*), 123
canning, 76–77
Carolina Housewife, The (Rutledge), 155
Carter, Susannah, 155
cassava, 123
cattle, 70–72, 76
cenas (restaurants), 261, 262, 266
chafu (tea bran), 118
chahchaanteng (tea cafés), 297
chancho con tamarindo (*cerdo en salsa tamarindo* (pork with tamarind sauce), 266
Chang, K. C., xi, 3–4, 14
Chang Kyehyang, 154
Chang-Rodriguez, Eugenio, 268n1
Chaoshan beef hotpot, 81–82
cha-siu (roasted pig), 261–262
chaufa (fried rice), 261
Che Qianzi, 216, 217
cheese, 31
Chen, Botao, 104
Chen, Helen, 162
Chen, Joyce, 151, 159, 162–163, 164
Chen, Julie Yujie, 137–138, 141
Chen Shaoqing, 164
Chen Shui-bian, 183–184

Chen Taosheng, 61
Cheng, F. T., 153
Cheng, Libin T. (Zheng Ji), 95
Chengdu, 136, 222
chestnuts, 31, 36, 38
chi (eating), 213
Chi, chidexiao (Let's eat and laugh about eating; Shi Qiao), 221
Chiang Ching-Kuo, 183
Chiang Kai-shek, 13, 183
chicharrones de prensa (deep-fried pressed pork with bread), 262
chifa (Chinese food in Peru), 253–269, 291
Child, Julia, 163
China Commercial Publishing House, 213–214
China Eastern Railway, 74
Chinese Civil War (1945–1949), 12, 14
Chinese Gastronomy (Lin Yutang), 160
Chinese medicine, 102–105, 115, 119
Chongqing, 193
choudoufu (stinky tofu), 198
Chow, Eric (Zhou Shoukuan), 220
Chuan cai (Sichuan cuisine), 180
Chun jiu qiu song (Spring chive, autumn cabbage; Dai Aiqin), 216–217, 218
chunshu (*chouchun*; tree of heaven; *Ailanthus altissima*), 123
ciba (lunar New Year sticky rice cake), 197–198
Cixi, Empress Dowager, 73
climate change, 292, 294
Cochran, Sherman, 98, 102
cod liver oil, 99, *101*, 102
Compendium of Materia Medica (Li Shizhen), xiii, 50, 115, 120, 121
Compleat Housewife, The (Smith), 155
concentrated animal feeding operations (CAFOs), 8
congee, 25
cookbooks (*shipu*), 151–168, *172*, 291
cooking oil, 175

cooking robots, 129, *130*, 131, 140, 143
Cooking with the Chinese Flavor (Lin Yutang), 160
"coolies," 255, 259–260
corn borers, 115
cotton, 35, 259
COVID-19 pandemic, 5, 131–134, 139, 143, 235, 245, 292, 293, 295
cows, 175
Cuba, 255
Cuixiu Tang (Cuixiu Hall), 57
cujiu (vinegar alcohol), 200
Cultural Revolution, 197, 211, 212, 220, 234
cutworms, 115
cyanide, 123

dabao (taking food home), 177
Da County, Sichuan, 114, 117, 119, 121
Dai Aiqun, 216–219
dai shipin (food substitutes), 123
danhuang rou (steamed pork with salted egg yolk), 181
Daoists, 99
Dashengkui (livestock firm), 71
dates, 31, 38
Dawson, Thomas, 155
De Da (restaurant), 234
Deng Xiaoping, 4, 15–16, 76
Dianxi Xiaoge, 223–224, 294
dianxin (snacks), 175
Diao zui (To be picky about food; Wang Lang), 219
digital food, 293
digua xifan (sweet potato porridge), 181
Dingfeng, 54
domestic servants, 260
Domino's, 133
Donghuifang (restaurant), 171
donkeys, 36, 37–38, 44
  meat of, 77
dragon-head enterprises, 39, 41
drinking, 199–201

drought, 26
Drummond, Jack Cecil, 92
Dulong people, 200

Eleme (food delivery platform), 133, 135, 138, 141, 293
English Huswife, The (Markham), 155
Enjoy Manor (tea company), 2
Enshi, 192–196, *199*, 200–202, 205
Er Fan Zhu (restaurant), 241, *242*

famine, 9, 97–98, 109, 111, 124
Fan, Jinmin, 58
fancaiguan (Western restaurants), 232–233
Farmers' Seed Network China, 26
Farquhar, Judith, 119
fast food, 79, 232
fazhi stewed beef, 70–71
Fei Xiaotong, 195
Febvre, Lucien, xii
fen zhurou (pork division), 17, 272, 274–278, *279*, 280–281, 283
Feng, Jin, 154
Feng Ruizhai, 117
Feng Wantong, 54
Fermentations and Food Science (Huang), xiii
fertilizer, 109, 113
fiber, 113
First Opium War (1839–1842), 291
First Sino-Japanese War (1894–1895), 245
fish, 72, 175, 261
flatbreads (*bing*), 33
fleas, 116
fondas (taverns), 260–261, 264–265, 266
food away from home (FAFH), 79–80, 81
food delivery (*waimai*), 132–143
Food in Chinese Culture (Chang), xi–xii, 3–4, 14

food safety, 5, 80–81, 141, 144, 245, 246, 292, 293, 299
food tourism, 78–79
foodways, defined, 6–7
food writing, 209–226
foxtail millet (*Setaria italica*), 27, *28*, 31, 34, 35, 40, 42
Frugal Housewife, The (Carter), 155
Fu Peimei, 151, 159, 163–165, 180
Fujian Province, 54, 58
  Fuzhou, 61
Fukushima nuclear disaster (2011), 246
Fukuzawa Yukichi, 74
Funk, Casimir, 91
Furong Garden ("ancient town"), 206n7
Fuzhou, Fujian Province, 61

ganba (dried beef), 72
Gansu Province, 35
Gao Hua, 123
Gao Lian, 51
garlic (*dasuan*; *Allium sativum*), xii, 78, 117, 120–122, 175
Geertz, Clifford, 243
General Tso's chicken, 179
Genso Sushi, 244
geographic indication (GI), 78
"ghost restaurants," 141
Gilroy, Paul, 254, 263
ginger, xii, 51, 175
Girls' City, 193–199, 204–206
globalization, 8
Globally Important Agricultural Heritage Systems (GIAHS), 26, 37
goats, 31
gongsuo (collective guilds), 53
Good Huswives Jewell (Dawson), 155
Gran Chifan (restaurant), 256
Grand Canal, 32
Great Leap Forward (1958–1961), 11, 76, 110–125
Guangdong Province, 54, 58, 256–258
  Shunde, 222

Guangzhou, 111, 136
Guanling Mountains, 81
guano, 259
guanxi (social relations), 206
Guangxi Province, 200
Guanyun (food brand), 77
Gudu de meishijia ("Solitary Gourmet"; manga and television program), 239–240
Guizhou Province, 193, 200
gyoza, 246
gyūnabe (beef hotpot), 73, 82

Ha Tsuen, Hong Kong, 273, 274, 283
Hakka cuisine, 183, 184
Han people, 13, 15, 193
Hani people, 200
Haochi (Che Qianzi), 216
Harbin, 74
he shuaiwan jiu (drinking smashing-bowl alcohol), 192–193
He, Xianlin, 39
Hebei Province, 193
  Wangjinzhuang, 25, 27, 35–44
Heiseiya (restaurant chain), 244
Henan Province, 33, 193
herbs, 109–128
  medicinal, 36, 38, 113
Heroes of the Water Margin, 70–71, 72
Hessler, Peter, 143
Hirohito, emperor of Japan, 171
hiyashi chūja (cold Chinese noodles), 246–247
honey, 39
Hong Guangzhu, 66n51
Hong Kong, xi, 17, 58, 158, 271–283, 295–298
horse meat, 77
Hosie, Alexander, 53, 56
hotpot, 73, 81, 82
Hou Hsiang-ch'uan, 95, 96–97
Hou Ki (mythical figure), 27
housewives, 155–157

Hu Peiqiang, 168n38
Hu Shiu-Ying, xii
Hu Yumei, 54
huaishu (scholar tree; *Sophora japonica*), 123
Huang, H. T., xiii, 6, 49–52
Huang Yuanshan, 151, 159–160, 164
Hubei Province, 15, 134–135, 191–206
Hui, Yuk, 137
Hulunbuir, 74, 81
Humane Elixir, 98
Hunan Province, 54, 193, 200
Hyakuyun, 234

Importance of Living, The (Lin Yutang), 160
Incas, 264
indentured servants, 255, 257, 259
India, 231, 295
indigenization, 183–184
information and communication technology (ICT), 11–12, 131, 141, 143, 292
Instagram, 165
intestinal worms, 116
Ireland, 81
irrigation, 76
Ishige, Naomichi, 49
Izakaya He Feng (restaurant), 241
izakayas (restaurants), 239–244, 247–248
Isett, Christopher, 56

Japan, 14, 61
  beef consumption in, 74, 76
  cuisine of, 229–251, 295 domesticity in, 156
  Hong Kong invaded by, 274 Taiwan ceded by, 178
Japanese cuisine, 229–251, 295
Jia, Sixie, 30
jian shimuyu (fried milkfish), 181
Jiangnan region, 52, 57

Jiangshan Lou (restaurant), 171, 173, *174*
Jiangsu Province, 57
  Yangzhou city, 13, 222
jiangyou (soy sauce), 50
Jiating shipu (Li Gong'er), 156
jiexinjiu (middle-of-the-street alcohol), 200
Jinding (food company), 79
jingjiuge (toasting songs), 204
Jingshi gaodeng shiye xuetang (Peking Higher College on Industrial Study), 61
Jining, Shandong Province, 54
jiuzhuo pengyou (drinking friends), 201
Joint Office for Scientific Research on Native Pesticides, 118, 121
jok-jeung (lineage head), 272
Joyce Chen Cookbook, The (Chen), 151, 162
Junzi Kitchen (restaurant chain), 298

Kádár, János, 74
Kamon (izakaya), 240
kang yangyu (baked small potato), 198
Kangxi Emperor, 36, 40, 41
Kanpai Classic (restaurant), 237, *238*, 244
kau-tau (bowing one's head), 272, 278–279
Kentucky Housewife, The (Bryan), 155
KFC, 79, 141, 234
Khrushchev, Nikita, 74–76
Kiessling (restaurant), 234
Koch, Robert, 98–99
Korea, 74, 81, 154
Kung, Lan-chen, 96
Kuong Tang (restaurant), 257
Kurogi (restaurant), 235, 247
Kuroki Jun, 235
Kyuhap ch'ongsŏ (Home encyclopedia for women in the inner chamber; Yo Pinghŏgak), 154

Ladurie, Emmanuel Le Roy, xii
Lai, Lili, 119
lanmenjiu (blocking-the-door alcohol), 200
Lanzou beef noodles (*Lanzhou niurou lamian*), 78–79, 82
laofan (scooped porridge), 35
Lean, Eugenia, 93
Lee Teng-hui, 183
Lei, Sean Hsiang-Lin, 102
Li Ciming, 59, 60
Li Gong'er, 156–157
Li Shizhen, xii–xiii, 50–51, 115, 120
Li, Yunshan, 294
Li Ziqi, 165, 222–224, 294
Liang, Shiqiu, 210, 217
Liao, Chengzhi, 234
lice, 116
Lin (retiree), 201–202
Lin Hsiang Ju, 159–161
Lin Tsuifeng, 159–161
Lin Yutang, 4–5, 160–161, 209, 209
Lippincott, J. B., 162
Lisu people, 200Liubiju, 54
Liubiju, 54
Lo, Vivienne, 51
lomo saltado (beef with onions and potatoes), 263–264
Longshengwang (trading company), 73
lotus root paste, 113
Luanzhou ("ancient town"), 206n7
Lu cai (Shandong cuisine), 180
Lu Wenfu, 211–213, 214, 219–220
Lu Xun, 209

M on the Bund (restaurant), 245
Ma Baozi, 78
Macao, 222, 296
Madam Tusán (restaurant chain), 253, 267
maize, 33, 35, 36, 40, 197, 200–201
malatang (spicy hot pot), 198
maltose, 113

Man lineage, 271, 274, *276*, 277
Man-Made Blood, 98
Man Sai-go, 271–283
manaijiu, 200
Manchu people, 13, 55
Manchuria, 33, 35
  soybean and soy sauce production in, 55–58
manga, 239
mantequerías, 261, 262, 266
Mao Zedong, 11, 13, 46n29, 75, 213
  food scarcity and, 111, 112, 123
mao'er yan ("kitten eyes"; herb), 117
Markham, Gervase, 155
Matthews, Glenna, 155
Mayoubu (food company), 79
McDonald's, 79, 80, 82, 141, 234, 247, 285–286n17, 298
medicinal herbs, 36, 38, 113
Meilin (food company), 77
Meishijia (The gourmet; Lu Wenfu), 211–213, 214
meishijia (gourmets), 16
Meituan (food delivery platform), 133, 134–135, 138, 141, 143
melons, 31
Men Yut (restaurant), 257
Mennell, Stephan, 155
Mexico, 182, 232
mian (noodles), 33, 113
Miao people, 200
mibaba (fermented rice cake), 197
Michelin Guides, 222, 237, *238*
mifen (rice noodles), 198
migration, 56, 179, 185–186, 231, 232, 254–256, 259–260
milk, 31
millet, 10, 25–44, 200
Millettia pachycarpa, 115
Ming dynasty, 13, 41, 51
Mintz, Sidney, 58, 62
Mongolia, Mongols, 13, 33, 76, 77, 200, 271, 294

M on the Bund (restaurant), 245
mosquitos, 117
moths, 115
Much Depends on Dinner (Visser), 69
mulberries, 31
mules, 37
mung beans, 31
Murakami House (izakaya chain), 244
Musings of a Chinese Gourmet (F. T. Cheng), 153
Muslims, 13, 78
My Country and My People (Lin Yutang), 160, 209

Narisawa (restaurant), 245
Nationally Important Agricultural Heritage Systems (NIAHS), 26, 37
Needham, Joseph, xiii, 6
neijuan (involution), 243
Ni Zan, 50
night blindness, 91, 92
niujiaojiu (drinking alcohol from bull horns), 200
Nice Day (restaurant chain), 298
Niuzhuang, 58
nongjia fan (peasant food), 25–26
nongjiale (farmhouse joy), 201
noodles (*mian*), 33, 113
Northern Song Dynasty, 30, 32
nutrition science, 90–98, 102–106
    as discipline and profession, 94–95

Okachimachi (restaurant), 241
OmniPork (meat substitute), 294
Oxfeld, Ellen, 288n38

Pan Guangdan, 195
Pan Peizhi, 168n38
Panama-Pacific Exposition (1915), 41
Panicum miliaceum (broomcorn millet), 27, 29
pān-toh (Taiwanese feast), 176–177

Paxson, Heather, 43
Peace, Navigation, and Friendship Treaty (1974), 259
Peimei shipu (Pei Mei's Chinese cook book), 151, 163, 180
pellagra, 92, 97
Peng Chang-kuei, 179
peppers, 31, 36, 38, 43, 51
persimmons, 38
Peru, 17, 253–269, 291, 299
pesticides, 12, 25, 109–125
Phillips, Carolyn, xii
pickle shops, 52–55, 59
pigs, 115, 262, *275*
Pinghu, Zhejiang Province, 54
Pinglai Ge (restaurant), 173
Pingyao, Shanxi Province, 72–73, 83n12, 84n14
Pingyao County Food Factory, 77
Pingyao pickled beef, 72–73, 82
Pizza Hut, 79
poisoning, 122–124
pork
    Chinese consumption of, 5
    Chinese Peruvian consumption of, 261–262
    ritual division of, 17, 272, 274–278, *279*, 280–281, 283
    in Taiwan, 175
potatoes, 264, 294
Potter, Jack, xi, xii
poultry, 77
Pratt, Mary Louise, 262
processed foods, 8, 9
prostitution, 233
Pu Yi, Emperor, 233

Qi Rushan, 159–160, 164
Qi Ying, 159
Qiang people, 200
Qimin yaoshu (Essential techniques for the common people; Jia Sixie), xiii, 30–31, 38, 42

Qingdao, Shandong Province, 74
Qing Dynasty, 12–13, 33, 49, 51–61, 170
qingkejiu (highland barley wine), 200
Qingming Shanghe Yuan ("ancient town"), 206n7
Qinling Mountains, 35
Qinzhou, Shanxi Province, 25, 27, 36–37, 39–41, 43–44
qinzhouhuang (Qinzhou yellow) millet, 40–42
qinzhou xiaocai (porridge and small dishes), 181
Qinzhou Yellow Millet Group (QYMG), 40–41, 44
quanjiu (pressuring others to drink), 204
quannong (promoting agriculture), 32
quinoa, 46–47n32

railroads, 74, 255, 259
rainfall, 30
ramen, 246
Randolph, Mary, 155
Ray, Krishnendu, 235
Red Cliff Rhapsody (Su Shi), 116
Red House (restaurant), 234
renao ("heat-noise"), 204
Republican China
 modernization in, 35, 49
 pickle shops in, 54, 55
 vitamins in, 89–108
resin, 113
restaurants, 16–17, 75, 201, 222
 beef in, 78, 81
 for Chinese diaspora, 297
 fast-food, 79, 232
 "ghost," 141
 in hotels, 234
 izakayas, 239–244, 247–248 Japanese, in China, 229–251, 295 market reforms linked to, 15

 in Peru, 253–269, 291
 in Taiwan, 171, 173, 174, 178–181 in U.S., 254–255
rice, 35, 129, 247, 261
 millet supplanted by, 30, 32–33
 in Taiwan, 175, 178
rice blast disease, 115, 117, 120
rice borers, 118
rickets, 91, 92, 96–97
robots, 129, *130*, 131, 140, 143, *144*
Rodriguez, Humberto, 268n1
Rokusan Gardens, 233
Rokusantei restaurant, 233
roujiabing (meat-filled bun), 198
Rozin, Elisabeth, xii
rubber, 113
Rural Rejuvenation Strategic Plan (*xiangcun zhenxing zhanlüe guihua*), 81
Russia, 74
Russo-Japanese War (1904–1905), 171
Rutledge, Sarah, 155

Sadamu (food company), 79
salt, 50, 51, 175
salting, 72
San De (drug company), 99, *100*
San Joy Lao (restaurant), 257
San Tin, Hong Kong, 271–283
sashimi, 240
scallions, 51
scarcity, 90–91, 111–114, 123, 124
Schafer, Ed, xi
Schneider, Mindi, 41
Science and Civilization in China, xiii, 6, 116
scientization, 10
scurvy, 92
Sennou Atsushi, 237
sesame oil, 51
Setaria italica (foxtail millet), 27, *28*, 31, *34*, 35, 40, 42

Shaanxi Province, 30, 33, 35
Shandong Province, 33, 35, 71
　Jining, 54
　Lu cai, 180
　Qingdao, 74
　Zoucheng, 294
Shanghai, 231, 295
　food delivery in, 136–137
　Japanese district in, 233
　pickle shops in, 54–55
　restaurants in, 17, 222, *230*, 232–233, 234–235, *236*, 237, 240, 244, 245, 295
　slaughtering in, 73
　as trading hub, 56, 57, 58
Shanxi Province, 25, 33, 48 n46, 72–73, 299
　Pingyao, 72–73, 83n12, 84n14
　　Qinzhou, 25, 27, 36–37, 39–41, 43–44
shaobing (cake baked over a charcoal fire), 198
shaokao (barbecue), 198
She County, 37
sheep, 31, 175
sheji (deities of soil and grain), 32
Shejian shang de Zhongguo (A bite of China), 221, 296
Shen Hongfei, 215–216, 217
Shen, Wayne, 294
Shen ye shitang (Late night diner; anime and television program), 239
Shennong, 116
Shenyang, Liaoning Province, 73
Shenzhen, 131, 136, 282, 292
Shenzhen River, 274
shequ tiangou (community group buying), 292
shi (fermented soybean paste), 31
Shi, Yiren, 104
shiba fan yao (eighteen conflicting medicines), 117, 119

Shicang, 59
Shikishima (restaurant), 233
Shimuyu wan tang (milkfish ball soup), 183–184
Shixian hongmi (Guide to the great mysteries of food), 52
Shu, Qiao, 221
Shufu no tomo (*The housewife's companion*), 156
Shui people, 200
shuijiu (rice beer), 200
Shunde, Guangdong Province, 222
Shunzhi, Emperor, 56
Sichuan Province, 13, 77, 119, 200
Sin, Lucas, 298
Singapore, 158, 297, 298
slaughtering, 73, *275*
Smith, Eliza, 155
Sokolov, Raymond, 163
Songhe Lou (restaurant), 220
Song Town ("ancient town"), 206n7
sorghum, 33, 35, 200
Southeast Asia, 16, 185–186
Soviet Union, 77, 112
soybeans, 31, 33, 50, 55–57, 62, 175
soy sauce, 10, 49–67, 175, 291
spring onions, 175
Stemona, 115
sterility, 91, 92
straw, 31
Su Shi, 116
sugar, 58, 61, 259
Suiyuan shidan (Recipes from the garden of contentment; Yuan Mei), 152, 209
Sun, Ping, 136, 137–138, 141
Sun Yat-sen, 5, 13
sushi, 229, 230, 235, 238–239, 245
Suzhou, 53, 57
sweet potatoes, 175, 181, 184
Szent-Györgyi, Albert, 92

tacu-tacu (rice, beans, and vegetables), 264–265
Tai cai (everyday Taiwanese food), 178, 180–181
Taierzhuang ("ancient town"), 206n7
Taihang Mountains, 25, 35, 36
Taiwan, 15, 158, 159, 169–189, 295, 296
Taiwan ryôri, 170–174, 185
Take Out (film), 133
tamarind, 266
Tang Lusun, 210
tannin, 113
Tao Xiaotao, 157
Tao Xingzhi, 157
Taobao (online platform), 36
Taomu pengrenfa (Grandma Tao's cooking methods), 157
taopian gao (walnut cake), 197–198
tapioca starch, 123
taro, 177, 184
taxation, 32
tea, 1–2, 8
Temple of the Soil and Grain, 33
Tencent, 133
tenshin don (Tianjin rice), 247
Tiadong ji (Flavoring the pot), 71–72
Tian Yiheng, 50
Tianjin, 17, 73, 231, 233–234, *236*, 240–241, 243, 248, 295
Tiao ding ji (The harmonious cauldron), 52
Tibet, 13, 200, 297
tofu, 113
Ton Po (restaurant), 257
tonggen (*Paulownia* roots), 115
Tonquin Sen (restaurant), 256
tourism, 78–79, 186, 191–197, 200, 233
Trémon, Anne-Christine, 282
tributes, 33
Tseng Pintsang, 173
tu (local or native science), 114–115, 118, 122
Tujia Girls' City, 193–199, 204–206

tunongyao zhi (records of native pesticides), 118–119
turnips, 31
Tusicheng (Chieftain City), 195
Typhoon Lekima (2019), 138

Ukraine, 75
Umsik timibang (Recipes of tasty foods; Chang Kyehyang), 154
UNESCO (United Nations Educational Scientific, and Cultural Organization), 1, 222, 296
urbanization, 10, 134, 143
Uyghurs, 13

Virginia Housewife, The (Randolph), 155
Visser, Margaret, 69
vitamins, 12, 89–108, 291
Vladivostok, 74

Wa people, 200
Wafu (restaurant), 234
wan guo (rice ball cakes), 184
Wang Hulin, 39
Wang Lang, 210, 218–219
Wang Peiyu, 72
Wang Shu, 75–76
Wang Xiaoyu, 152–153, 160
Wang Zengqi, 210, 217, 218
Wangjinzhuang, Hebei Province, 25, 27, 35–44
Watami (food grand), 245
Watson, Rubie, xii, 274
WeChat, 140–141, 245, 292
weiminsu (fakelore), 201
wenbing (warm-factor disorders), 115
Weng Tonghe, 59
wet markets, 293
wheat, 33, 35
World Trade Organization (WTO), 80, 299
Wu Chengluo, 67n55
Wu Hsien, 95, 97, 102
Wu Jiang-shan, 171, 173

Wu, Ku, 15
wu ren ji (humanless machines), 143
Wufengshan (village), 201
Wuhan, 133
Wusho zhongkui lu (*Madam Wu's kitchen records*), 153–154

Xi Jinping, 5, 134
Xi'an, 192
Xiang, Biao, 135, 139
Xiang cai (Hunan cuisine), 180
xiaochi (snacks), 176, 185
Xieshi zhuyi (Gatrographism), 215–216
Xingshenglei (trading company), 73
xingwei (character of taste and smell), 110, 116, 117, 119
Xinjiang people, 13, 297
Xu, Chenjia, 295
Xuanzong, Emperor, 132
Xueren tan chi (*Personalities' talk on food*), 214

yakitori, 239, 240
Yan Yunxiang, 247, 282
Yan'an, 35, 114
yang (foreign science), 114, 115, 118, 122
Yang, Empress, 132
Yang, Zhiyi, 294
Yangzhou city, Jiangsu Province, 13, 222
Yellow Emperor's Inner Canon, 103
Yellow River, 30, 71
Yi fan yi shijie (A world in a bowl of rice; Zhou Huacheng), 221–222
Yi Lu Yi Xian Seafood Izakaya (restaurant), 241
Yi Pinghŏgak, 154
Yijie chunqiu (Annals of the world of medicine), 104
Yingyang gailun (Wu Hsien), 95–96
yiwei (unusual foods), 199
yogurt, 31
Yongzheng period, 52

YouTube, 222
youxiang (deep-fried rice-soybean-meat dumpling), 198
Yu Yimeng, 135
Yuan Mei, 152–153, 158, 160, 161, 209, 210
Yuanmingyuan ("ancient town"), 206n7
Yuanshan dianxinpu (Yuanshan snacks and desserts cookbook), 159
Yuanshan shipu (cookbook), 151, 159
Yuanshan xicanpu (Yuanshan western cuisine cookbook), 159
Yue cai (Guangdong cuisine), 180
Yueshi (Epicure), 221
Yugu 1 millet, 40
Yunnan Province, 193, 200, 294
Yutang, 54

zajiu (sucking alcohol), 200
Zeck, Yvonne, 164
Zeng Guofan, 58–59
Zeng Yi, 154–155, 156
zha huazhi (deep-fried squid balls), 177
zha yutou (deep-fried taro), 177
Zhan, Yang, 139
Zhang, Carolyn, 142
Zhao Rongguang, 55
Zhao Xianzhang, 211, 212
Zhe cai (Zhejiang cuisine), 180
Zheng Fuguang, 60
Zheng Ji (Libin T. Cheng), 95, 105–106
Zheng Zhenwen, 95
Zhili Province, 33
Zhitang tan chi (Talk on food), 214
Zhong, Raymond, 142
Zhong Shuhe, 215, 216
Zhongguo pengren (Chinese cuisine), 214
Zhongkui lu (Kitchen records; Zeng Li), 154
Zhongyi nongye (Chinese medicine agriculture), 124–125
Zhou Dynasty, 32, 202–203

Zhou Huacheng, 221–222
Zhou Shoujuan (Eric Chow), 220
Zhou Zuoren, 209, 214, 215
Zhu Dahe, 137
Zhu Weiping, 124
zhutongjiu (special alcohol in bamboo tubes), 200
Zilicheng (trading company), 72
zoonotic diseases, 292
Zoucheng, Shandong Province, 294
Zunsheng bajian (Eight approaches to nourishing life; Gao Lian), 51